**Principles** of
*Fire Protection*

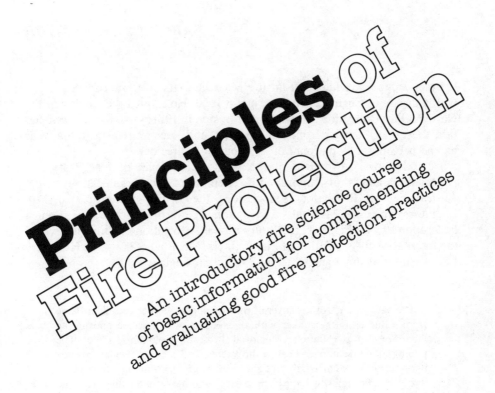

# Principles of Fire Protection

An introductory fire science course of basic information for comprehending and evaluating good fire protection practices

### Percy Bugbee

NATIONAL FIRE PROTECTION ASSOCIATION
Batterymarch Park, Quincy, MA 02269

## ABOUT THE AUTHOR

Percy Bugbee graduated from the Massachusetts Institute of Technology with a degree in Chemical Engineering in 1920. He entered the employ of the National Fire Protection Association in Boston in 1921, served as the Association's General Manager from 1939 to 1969, and since retirement in that year has been Honorary Chairman of the Board of Directors.

He is Honorary President of the World Conference of Fire Protection Associations and is an Honorary Life Member of such organizations as the Australian Fire Protection Association, the British Institution of Fire Engineers, the Fire Marshals Association of North America, the French Federation of Fire Fighters, the International Association of Fire Cheifs, and the International Association of Fire Fighters. Mr. Bugbee is also an Honorary Life Member of the Society of Fire Protection Engineers.

The background for the cover of this text shows a 9-inch turbulent flame from a Bunsen burner, taken with an exposure of 13 microseconds in the laboratory of the Battelle Memorial Institute, Columbus, Ohio. The photographic technique is called the "schlieren" method, which depends upon refraction of an edge of a light beam by gradients in the refractive index of gas through which the beam passes. Another term for this technique is "optical inhomogeneity." (Photo by Battelle Memorial Institute)

Cover design by Frank Lucas

Fifth Printing
November 1985

Copyright © 1978
National Fire Protection Association
All Rights Reserved

NFPA No. TXT-4
ISBN: 0-87765-084-5
Library of Congress No.: 76-41640
Printed in U.S.A.

# CONTENTS

Introduction     xi

CHAPTER 1     **Fire — The Destroyer**     1

FIRE — A TWO-SIDED GOD, 1; Fire in Prehistoric Times, 2; Fire Protection During the Roman Empire, 3; Early Fire Protection Regulations, 3; Major Fires in Early Times, 4; Fire Protection in Colonial America, 5; Establishment of the First Paid Fire Department, 5; Establishment of Mutual Fire Societies, 7; Formation of Salvage Corps, 7; The Growth of Paid Fire Departments, 7; The Formation of Fire Insurance Companies, 7; Progress of Fire Protection in the 19th Century, 8; Nineteenth Century Apparatus and Equipment, 11; Fire Hydrants, 11; Use of Fire Hose, 11; Advances in Fire Apparatus, 12; Major Hydraulic Studies, 12; Fire Alarm and Extinguishing Systems, 12; Fire Loss During the Earlier 20th Century, 12; Fire Loss During World War II, 14; Post World War II Conflagrations, 16; Evaluation of Modern Methods of Prevention and Control, 17; Fire Protection Defenses and Agencies, 17; Evolution and Scope of the Fire Department, 18; Fire Protection Agencies, 20; SUMMARY, 20; ACTIVITIES, 20; BIBLIOGRAPHY, 21.


CHAPTER 2     **Firesafety for People and Property**     22

ASSESSING LIFE SAFETY, 22; Life Loss and Injury from Fire, 23; Property Loss from Fire, 23; People Factors, 25; People Activities, 27; FUNDAMENTALS OF BUILDING DESIGN, 28; Concepts of Egress Design, 30; Flow Rates, 32; Numbers of Means of Egress, 33; The *Life Safety Code,* 34; THE HAZARDS OF OCCUPANCIES, 35; Firesafe Building Design, 37; Types of Occupancies, 37; Residential Occupancies, 38; Health Care Occupancies, 40; Educational and Assembly Occupancies, 41; Mercantile Occupancies, 42; Business Occupancies, 42; Industrial Occupancies, 43; Industrial Fire Brigades, 43; The Industrial Fire Risk Manager, 43; Industrial Fire Brigade Organization, 44; Traffic and Exit Drills, 45; External Traffic Control, 45; Internal Traffic Control, 45; PUBLIC EDUCATION AND COMMUNITY RELATIONS, 46; Fire Prevention Program, 46; Methods of Public Education, 47; Fire Department Activities in Community Relations, 48; SUMMARY, 48; ACTIVITIES, 49; BIBLIOGRAPHY, 49.

CHAPTER 3    **Characteristics and Behavior of Fire**    51

THE UNPREDICTABILITY OF FIRE, 51; Heat Measurement, 54; Heat Transfer, 55; Conduction, 56; Radiation, 56; Convection, 57; Sources of Ignition, 57; Chemical, 57; Electrical, 57; Mechanical, 59; Nuclear, 59; EXPLOSIONS, 60; Chemical Explosions, 60; Mechanical Explosions, 61; Atomic Explosions, 61; Thermal Explosions, 61; Deflagration, 61; Detonation, 62; PRODUCTS OF COMBUSTION, 62; Fire Gases, 62; Carbon Monoxide, 62; Carbon Dioxide, 63; Hydrogen Sulfide, 63; Sulfur Dioxide, 63; Ammonia, 63; Hydrogen Cyanide, 63; Hydrogen Chloride, 64; Nitrogen Dioxide, 64; Acrolein, 64; Phosgene, 64; Flame, 64; Heat, 64; Burns, 65; Smoke, 65; Oxygen Deficiency, 66; FIRE AND EXPLOSION CONTROL, 66; Use of Water, 66; Removal of Oxygen, 67; Fuel Removal, 68; Flame Inhibition, 68; SUMMARY, 68; ACTIVITIES, 69; BIBLIOGRAPHY, 70.

CHAPTER 4    **Fire Hazards of Materials**    71

COMBUSTIBLE SOLIDS, 71; Wood, 72; Physical Properties, 72; Moisture Content, 73; Ignition of Wood, 74; Flashover, 75; Flame Spread, 75; Fire Loading, 76; Smoke, 76; Storage, 76; Plastics, 76; Manufacture, 77; Fire Behavior, 77; Storage, 77; Textiles, 78; Natural Fiber Textiles, 78; Synthetic Textiles, 79; Flame Retardant Textiles, 80; Storage, 80; FLAMMABLE AND COMBUSTIBLE LIQUIDS, 80; Classification by Properties, 82; Flammable Liquids, 82; Combustible Liquids, 83; Characteristics, 83; Flash Point, 83; Vapor Pressure, 84; Boiling Point, 84; Specific Gravity, 84; Evaporation Rate, 84; Ignition Temperature, 85; Storage and Handling, 85; GASES, 85; Classification by Properties, 86; Flammable Gases, 86; Nonflammable Gases, 86; Toxic Gases, 86; Reactive Gases, 86; Physical Properties, 86; Compressed Gases, 87; Liquefied Gases, 87; Cryogenic Gases, 87; Classification by Usage, 88; Gas Laws, 89; Boyle's Law, 89; Charles's Law, 89; Gay-Lussac's Law, 89; Gas Fires, 90; BLEVEs, 90; Combustion Explosions, 90; Odorizing, 91; Gas Standards, 92; Comprehensive Reference, 92; HAZARDOUS MATERIALS, 92; Corrosive Chemicals, 92; Inorganic Acids, 92; Halogens, 93; Storage and Fire Protection for Corrosive Chemicals, 94; Corrosive Vapors, 94; Radioactive Materials, 95; Characteristics of Radioactive Materials, 95; Handling and Storage of Radioactive Materials, 95; Fire Protection for Radioactive Materials, 96; Transportation, 96; FIRE FIGHTING AND CONTROL, 98; Radioactive Emergencies, 98; Extinguishment of Fire, 100; Overhaul, 101; Chemical Emergencies, 101; Extinguishment of Fire, 102; Transportation Emergencies Involving Hazardous Materials, 103; Life Safety Hazards, 105; Action at Emergencies, 105; SUMMARY, 108; ACTIVITIES, 109; BIBLIOGRAPHY, 110.

CHAPTER 5    **Investigating the Fire Loss Problem**    112

THE NEED FOR INVESTIGATIONS, 112; The Purpose of Fire Investigations, 113; The Scope of the Investigation, 113; Fire Ignition Sequence, 113; Fire

Development, 114; Fire Casualties, 114; Conducting the Investigation, 115; Fire Scene Evidence, 115; The Preliminary Investigation, 115; Fire Scene Analysis, 116; RECORDING THE FIRE PROBLEM, 117; The Purpose of Fire Reports, 117; Types of Fire Reports, 118; Basic Incident Report, 118; Basic Casualty Report, 118; Fire Reporting Systems and Their Objectives, 119; Fact Finding, 121; Fact Processing, 122; Fact Use, 122; ANALYZING FIRE LOSSES, 123; Large-loss Fire Analysis, 123; Structural Defects, 123; Building Contents, 124; Fire Protection Weaknesses, 124; The Principal Causes of Fire, 125; Smoking and Related Fires, 126; Electrical Equipment Fires, 127; Open Flames and Sparks, 127; Flammable Liquids Fires, 128; Fireworks and Explosives, 128; Analysis of Conflagrations, 128; Prevention of Fire Losses, 131; Large-loss Fire Prevention, 131; Prevention of Fire Ignition, 131; Prevention of Life Loss, 132; "Group Fire" Prevention, 132; THE GROWING ARSON PROBLEM, 132; Investigation of the Arson Fire, 133; The Arsonists, 133; Determining Fire by Arson, 134; Responsibility for the Investigation, 135; Details of the Investigation, 135; Criminal Procedures for Arson, 136; The Model Arson Law, 136; Other Arson-related Laws, 137; Arson Prevention, 137; SUMMARY, 138; ACTIVITIES, 139; BIBLIOGRAPHY, 139.

CHAPTER 6    **Firesafe Building Design and Construction**    141

FUNDAMENTALS OF FIRESAFETY DESIGN, 141; Objectives of Firesafety Design, 142; Life Safety, 142; Property Protection, 142; Continuity of Operations, 143; Fire Hazards in Buildings, 143; Smoke and Gas, 143; Heat and Flames, 143; Building Elements and Contents, 143; Elements of Building Firesafety, 144; BUILDING AND SITE PLANNING FOR FIRESAFETY, 145; Firesafety Planning for Buildings, 145; Fire Fighting Accessibility to Building's Interior, 146; Ventilation, 148; Connections for Sprinklers and Standpipes, 148; Firesafety Planning for Sites, 148; Traffic and Transportation, 149; Fire Department Access to the Site, 149; Water Supply to the Site, 149; EXPOSURE PROTECTION, 149; INTERIOR FINISH, 150; Types of Interior Finish, 150; Wood, 152; Steel, 152; Concrete, 153; Glass, 153; Gypsum, 153; Masonry, 154; Plastics, 154; Test Procedures, 154; CONFINEMENT OF SMOKE AND FIRE, 156; Fire Loading, 157; The Standard Time-temperature Curve, 157; Fire Loads by Occupancy, 157; Fire Doors, 158; Types of Doors, 158; Smoke Control, 159; SUMMARY, 160; ACTIVITIES, 162; BIBLIOGRAPHY, 162.

CHAPTER 7    **Fire Protection Systems and Equipment**    164

WATER AS AN EXTINGUISHING AGENT, 164; The Physical Properties of Water, 164; The Extinguishing Properties of Water, 165; The Cooling Capacity of Water, 166; The Smothering Capabilities of Water, 167; Water Fog, 167; SPRINKLER SYSTEMS, 168; Development of Sprinkler Protection, 168; Value of Sprinkler Protection, 169; Life Safety, 169; Property Protection, 169; Sprinkler Performance, 170; Sprinkler System Installation, 170; The NFPA

Sprinkler System Standard, 172; Types of Sprinkler Systems, 172; Wet-pipe Systems, 173; Regular Dry-pipe Systems, 174; Preaction Systems, 174; Deluge Systems, 175; Combined Dry-pipe and Preaction Systems, 175; Limited Water Supply Systems, 176; Outside Sprinkler Systems, 176; Sprinkler Heads, 176; STANDPIPES, 177; Types of Standpipe Systems, 178; WATER SPRAY FIXED SYSTEMS, 179; Use of Water Spray Fixed Systems, 180; Extinguishment, 180; Controlled Burning, 180; Exposure Protection, 180; Prevention of Fire, 181; Application of Water Spray Fixed Systems, 181; FOAM EXTINGUISHING SYSTEMS, 181; Use of Foam Systems, 182; CARBON DIOXIDE SYSTEMS, 183; Extinguishing Properties of Carbon Dioxide, 184; Smothering, 184; Cooling, 184; HALOGENATED AGENTS AND SYSTEMS, 184; Halon 1211 (Bromochlorodifluoromethane), 186; Halon 1301 (Bromotrifluoromethane), 187; Extinguishing Characteristics of Halons, 187; DRY CHEMICAL EXTINGUISHING SYSTEMS, 188; Chemical and Physical Properties, 188; Stability, 188; Toxicity, 189; Particle Size, 189; Extinguishing Properties of Dry Chemicals, 189; Smothering Action, 190; Cooling Action, 190; Radiation Shielding, 190; Chain-breaking Reaction, 190; COMBUSTIBLE METAL EXTINGUISHING SYSTEMS, 191; PORTABLE FIRE EXTINGUISHERS, 192; Types of Portable Fire Extinguishers, 193; Vaporizing Liquids, 193; Liquefied Gases, 193; Carbon Dioxide, 193; Dry Chemicals, 194; Multipurpose Dry Chemical, 194; Dry Powder, 194; Application of Portable Fire Extinguishers, 194; FIRE APPARATUS EQUIPMENT, 196; Equipment Carried on Apparatus, 196; Forcible Entry Tools, 196; Communications Equipment, 197; Electric Lights and Generators, 197; Portable Pumps, 197; Protective Equipment for Fire Fighters, 198; Breathing Apparatus, 198; Resuscitators, 198; Smoke Ejectors, 198; Life Nets, 199; Life Guns, 199; Protective Clothing for Fire Fighters, 199; SUMMARY, 199; ACTIVITIES, 200; BIBLIOGRAPHY, 200.

CHAPTER 8    **Alarm and Detection Systems and Devices**    202

PUBLIC FIRE SERVICE COMMUNICATIONS, 202; Municipal Fire Alarm Systems, 202; Type A (Manual), 204; Type B (Automatic), 204; Fire Alarm Boxes, 204; Telegraph-type, 205; Telephone-type, 206; Radio-type, 207; The Communication Center, 208; Fire Department Radio, 209; AUTOMATIC AND MANUAL PROTECTIVE SIGNALING DEVICES, 210; Classification of Signaling Systems, 210; Central Station Systems, 211; Local Systems, 212; Auxiliary Systems, 213; Remote Station Systems, 213; Proprietary Systems, 214; Household Fire Warning Systems, 214; Automatic Fire Extinguishing Systems, 215; Power Supplies for Protective Signaling Systems, 215; FIRE DETECTION MECHANISMS AND DEVICES, 216; Heat Detectors, 216; Fixed-temperature Detectors, 217; Rate-of-rise Detectors, 218; Combined Rate-of-rise and Fixed-temperature Detectors, 219; Rate Compensation Devices, 219; Smoke Detectors, 219; Photoelectric Detectors, 220; Beam-type Detectors, 220; Ionization Detectors, 220; Resistance Bridge Detectors, 221; Sampling Detectors, 221; Flame Detectors, 221; Infrared, 221; Ultraviolet, 221; Photoelectric 221; Flame Flicker, 221; Gas Detectors, 222; Spacing and Location of Detection Equipment, 222; General Utilization of Detection Devices, 223; SUMMARY, 223; ACTIVITIES, 224; BIBLIOGRAPHY, 224.

CHAPTER 9                  **Municipal Fire Defenses**                  226

EVALUATION AND PLANNING OF PUBLIC FIRE PROTECTION, 226; Evaluation, 226; Public Fire Defenses, 227; Urban Fire Defenses, 227; Personnel, 228; Rural Fire Defenses, 228; Insurance Grading Schedules, 229; Water Supply, 230; Fire Department Strength, 231; Other Important Factors, 233; Communications, 234; Master Planning, 234; The ISO Grading Schedule and Master Planning, 235; Developing a Master Plan for Fire Protection, 235; WATER FOR FIRE PROTECTION, 236; Uses of Water Systems, 237; Rates of Consumption, 237; Fire Protection Requirements in Water Systems, 238; Evaluating System Capacity, 238; Pressure Characteristics of Systems, 238; WATER DISTRIBUTION SYSTEMS, 239; Sources of Supply, 239; Surface Supplies, 239; Ground Water Supply, 239; Selection of Supply, 239; Types of Systems, 240; Gravity System, 240; Pumping System, 240; Combination Systems, 240; SUMMARY, 240; ACTIVITIES, 241; BIBLIOGRAPHY, 242.

CHAPTER 10           **Fire Department Organization,**           243
                            **Administration, and Operation**

FIRE DEPARTMENT ORGANIZATION, 243; Fire Department Objectives, 244; Fire Department Structure, 245; Principles of Organization, 246; Line Functions, 247; Staff Functions, 248; Organizational Plans, 248; ADMINISTRATION AND MANAGEMENT, 248; Function of Management, 248; Personnel, 249; Recruitment, 249; Promotion Practices, 250; Staffing Practices, 250; Intergovernmental Relations, 251; Building Department, 251; Police, 252; Water Department, 252; Personnel Department, 252; Finance Department, 252; Purchasing, 252; Data Processing, 253; Personnel Utilization, 253; FIRE DEPARTMENT OPERATIONS, 253; Organizations for Fire Suppression, 254; Engine Company, 254; Ladder Company, 254; Rescue Company, 254; Tactical Control Units, 254; Task Forces, 254; Tactical Operations, 255; Size-up, 255; Rescue, 255; Exposures, 255; Confinement, 256; Extinguishment, 256; Ventilation, 256; Salvage, 256; Overhaul, 257; Nontactical Operations, 257; Prefire Planning, 257; Training, 257; Transportation Incidents, 258; Mutual Aid, 258; Fire Prevention, 259; Organizations for Fire Prevention, 260; Inspections, 261; Public Education, 262; Seasonal Activities, 263; Enforcement of Codes, 263; Investigation of Fires, 263; SUMMARY, 264; ACTIVITIES, 265; BIBLIOGRAPHY, 265.

CHAPTER 11                  **Codes and Standards**                   267

THE HISTORY AND DEVELOPMENT OF FIRE PROTECTION REGULATIONS, 267; Early Building and Fire Laws, 267; Development of Building and Fire Codes, 268; Relationships Between Building and Fire Codes, 270; Responsibility and Enforcement of Safety Provisions, 271; FORMATION OF CODES AND STANDARDS, 272; The Standards-making Process, 272; Publicizing Intent to Initiate a New Standard, 273; Formation of the Technical Committee, 274; Determining

Scope and Content of Standards, 274; Submission of Standard as a Draft, 274; Publication and Review of Report, 274; Adoption by the NFPA, 275; Enforcement of Codes and Standards, 275; Federal-level Authority, 276; State-level Authority, 276; Local-level Authority, 278; Adoption of Standards into Law, 279; TYPES OF CODES AND STANDARDS, 279; "Model" Fire Prevention Codes, 280; Model Building Codes, 280; *National Building Code,* 281; *Uniform Building Code,* 282; *Standard Building Code,* 283; *BOCA Basic Building Code,* 283; The Role of Standards in Building Codes, 283; FIRESAFETY STANDARDS-MAKING ORGANIZATIONS, 284; American National Standards Institute (ANSI), 284; American Society for Testing and Materials (ASTM), 285; National Fire Protection Association (NFPA), 286; SUMMARY, 288; ACTIVITIES, 288; BIBLIOGRAPHY, 289.

CHAPTER 12   **Fire Protection Organizations, Information**                                                291
                **Sources, and Career Opportunities**

THE SCOPE OF FIRE PROTECTION ORGANIZATIONS, 291; ORGANIZATIONS WITH FIRE PROTECTION INTERESTS, 292; The National Fire Protection Association, 292; Technical Standards, 292; Information Exchange, 293; Technical Advisory Services, 293; Public Education, 293; Research, 294; Services to Public Protection Agencies, 294; Fire Service Organizations, 294; Fire Marshals Association of North America (FMANA), 294; Fire Service Section (A Section of NFPA), 295; Society of Fire Protection Engineers (SFPE), 295; International Association of Fire Chiefs (IAFC), 295; International Association of Fire Fighters (IAFF), 296; International Association of Arson Investigators (IAAI), 296; International Association of Black Professional Fire Fighters (IABPFF), 296; International Fire Service Training Association (IFSTA), 296; International Municipal Signal Association (IMSA), 296; International Society of Fire Service Instructors (ISFSI), 296; Joint Council of National Fire Service Organizations (JCNFSO), 297; Fire Testing and Research Laboratories, 297; Private and Industrial Laboratories, 298; University and Federal Government Laboratories, 299; Insurance Organizations, 299; American Insurance Association (AIA), 300; American Mutual Insurance Alliance (AMIA), 300; Factory Mutual System (FM), 300; Industrial Risk Insurers (IRI), 301; Insurance Services Office (ISO), 301; Other Organizations Having Fire Protection Interests, 301; Federal Agencies Involved in Fire Protection, 301; Department of Commerce, 302; National Fire Prevention and Control Administration (NFPCA), 302; National Bureau of Standards (NBS), 302; Department of Labor, 302; Department of Agriculture, 303; U.S. Forest Service (USFS), 303; Farmers Home Administration, 303; Department of Housing and Urban Development, 303; Department of the Interior, 304; Department of Defense, 304; General Services Administration, 304; Department of Transportation, 304; Other Federal Agencies, 305; CAREER OPPORTUNITIES IN THE FIRE SCIENCES, 305; Fire Science Education, 307; Fire Protection Careers, 308; Careers in a Fire Department, 309; Fire Fighter, 309; Fire Apparatus Driver/Operator, 310; Paramedic, 311; SUMMARY, 312; ACTIVITIES, 312; BIBLIOGRAPHY, 313.

SUBJECT INDEX                                                                                                315

# INTRODUCTION

Fire is a fundamental force in nature. Without fire, life as we know it would not exist. Friendly fires heat our homes, cook our food, and help to generate our energy — in short, help make the world go 'round. But any force as powerful as fire also carries with it the potential for great harm; this destructive potential poses a threat to our lives, property, and resources.

Recently, we Americans have had to face the fact that the conservation of energy and natural resources is vital to the future well-being of our country. In America we have more television sets, more automobiles, and more dishwashers than any other country in the world; we also have *more fires* than any other country in the world.

In the United States alone, more than six thousand fires occur daily; they exact a toll of death, pain, and destruction that cannot fail to shock anyone acquainted with the facts. The majority of people, however, tend to think of a serious fire as they would a serious automobile accident or other tragedy — as something that happens to someone else. This is because the average person suffers a minor burn or experiences a small fire only once or twice in a lifetime, and thus the threat of a truly destructive fire seems improbable and remote. But these are also the people who find themselves unprepared, both physically and emotionally, when a large fire does occur. In too many cases this lack of preparation causes panic, death, and destruction that might have been avoided had the victims taken seriously the threat of fire and thus known how to prepare themselves accordingly.

Even though there are more fires today than ever before, it is also true that there are more people fighting them than ever before. People in many different fields and organizations, both private and public, are working to make fire protection technology and practices as sophisticated as possible. They include, among others, paid and volunteer fire fighters; federal, state, and local fire officials and fire investigators; fire insurance inspectors, raters, and agents; industrial safety and fire protection personnel; fire equipment manufacturers and salespersons; builders and building inspectors; electrical manufacturers, installers, and inspectors; and architects. New people enter these fields daily.

Therefore, it seems appropriate that there be a basic text on fire protection, a text whose objective it is to provide both fire science students and new fire service personnel with an overview of the fundamental methods of fire protection, prevention, and suppression — a text that could be used as a home study text, for basic in-service programs, and as an introductory college course in fire protection. This objective would be achieved through the inclusion of discussions of the various components of the fire protection system through presentation of subjects such as firesafety for people and property, the basic characteristics and behavior of fire, fire hazards of materials and buildings, fire protection equipment and systems, codes and standards for fire protection and prevention, fire fighting forces and how they operate, and the organizations, services, and careers in the field. Such a textbook should be a valuable resource guide in a world whose fire hazards increase proportionally with its complexity. The author sincerely hopes that *Principles of Fire Protection* will help to fulfill this objective.

*Percy Bugbee*

*Chapter One*

# Fire —
# The Destroyer

*Since prehistoric times, fire has been viewed as a force both beneficent and destructive. From the days of the Roman Empire to the present, many of the greatest losses of life and property can be attributed to fire. However, modern fire prevention techniques are helping to protect the human race from fire's destructive potential.*

## FIRE — A TWO-SIDED GOD

Natural phenomena such as fire, thunder, and lightning probably made early human beings wonder about the universe in which they lived. Frequently, early humans created their own explanations for what they imagined to be true. Having no reasonable scientific explanations, they imagined that natural phenomena were associated with the gods — gods that were often pictured in human terms. Because these early explanations for natural events sounded reasonable, they satisfied the curiosities of many people. And because there was no scientific knowledge to disprove the stories, they became accepted explanations about the universe. The stories grew with the telling and developed into the myths of the early heroes and gods.

Some of these early myths depicted fire as a "two-sided god" — one side having beneficial qualities, the other side having destructive qualities. Fire's two-sidedness was emphasized in the following excerpt from an address given in 1900 by Uberto C. Crosby, president of the National Fire Protection Association at that time:*

> A strange thing this fire from which we seek protection. No wonder the ancients worshipped it as a God, that primitive man guarded it jealously, keeping it constantly burning. A strange, inconsistent, two-sided God it has always

---

*From an address given by Uberto C. Crosby, then president of the National Fire Protection Association, at the NFPA's fourth Annual Meeting in New York City on June 26, 1900.

been to man; now giving comfort and blessing in manifold ways, and then, without warning, turning and destroying the objects of its benefaction. It warms and lights our homes, builds and runs our workshops and factories, furnishes the life and power of our modern civilization. It seems a friendly, beneficent factor, and yet at the very time it is showering blessings, with persistency and cunning, it seeks to destroy, and while we rest in financial security, breaks through the barriers with which we seek to surround it, and like a mighty avalanche sweeps away homes, blocks, towns, and cities in one common destruction. At the same time a friend and foe; in all ages and climes man has worked to obtain its blessings and, at the same time to prevent its ravages.

## *Fire in Prehistoric Times*

There is little doubt that early in the history of humanity fire laid waste to the crude shelters people built in order to protect themselves from the elements, that it threatened and sometimes took the lives of individuals, families, and small groups of people, and that it often destroyed the forested lands surrounding isolated tribes and early civilizations. We can speculate that most of the fires encountered early in the history of humanity were caused by lightning; thus, it is no wonder that fire became one of the first phenomena that early human beings came to worship as a god. This type of reverence is commented on in the following excerpt from Walter M. Haessler's book titled *The Extinguishment of Fire:*[1]

> The terrific energy of fire — the most important agent in civilization — the similarity of its effects with that of the sun, its intimate connection with light, its terrible and yet genial power, and the beauty of its changeful flame easily account for the reverence in which it was held in ancient times. At a period when cause and effect and form and essence were not distinctly separated, fire became an object of religious veneration, a distinguished element in mythology, an expressive symbol in poetry, and an important agent in the systems of cosmogony. It gained a place among the elements, and for a very long period was believed to be a constituent part in the composition of bodies. At the end of the 17th century, fire under the name of "phlogiston" was considered to be the source of all chemical action.

Haessler further describes fire as a fascinating phenomenon and an inexhaustible mystery, and defines it as follows:

> Fire is one of nature's most fascinating phenomena. It will ever be a subject of inexhaustible mystery in that the more it is studied, the more it leads to further discoveries and further questions. . . .
>
> . . . The most basic definition of a fire is: a combustion process sufficiently intense to emit heat and light. Note that our definition is quite broad and has not limited the chemical reaction to one involving only oxygen. . . .

## Fire Protection During the Roman Empire

Historically, the first recorded attempts to control the ravages of fire took place about 300 B.C. in Rome when fire fighting duties and night watch services were delegated to a band of slaves, the *Familia Publica,* supervised by committees of citizens. During the reign of Augustus Caesar (Gaius Julius Caesar Octavius) from 27 B.C. to 14 A.D., Rome developed what might be considered the first municipal-type fire department by organizing these slaves and citizens into a *Corps of Vigiles* (watch service). Decrees were issued stating the measures that all citizens should take to prevent and check fires. Although the *Corps of Vigiles* represents the first organized form of fire protection, night patrolling (performed by *Nocturnēs)* and night watch forces were its principal services. Thus, some of the *Vigiles* had duties more like those of police and soldiers than fire fighters.

The structure of the *Corps of Vigiles,* similar to the command structure of today's fire departments, consisted of seven divisions, each division being made up of one-hundred to one-thousand persons, with each person assigned a particular task. For example, some members *(Aquarii)* carried water to the fire scene in jars. Still later, aqueducts were built to carry water around the city, and hand pumps were developed to help get the water on the fire. Pump supervisors were titled *Siponarii,* and the earliest recorded fire chief was the *Praefectus Vigilum,* who was charged with the overall responsibility for the *Corps of Vigiles.* Roman law decreed that the *Quarstionarius,* the Roman equivalent of today's state fire marshal, determine the cause of all fires. During the time of the Roman Empire, leather hose came into use and large pillows were carried to the fire scene so that people trapped in taller buildings could jump onto them.

Marco Polo's account of the great Oriental civilization of the 13th century noted that a civil force of "watchmen" and "firemen," which had fire prevention duties, was maintained in Hang Chow (referred to by Marco Polo as "The Celestial City"), and could turn out "one to two thousand men" to deal with a fire. The force was divided into companies of ten men, five of whom were on watch by day and five by night.

## Early Fire Protection Regulations

Following the fall of the Roman Empire, there was an extended period of time when apparently little or no organized effort to prevent or control fires was made. About the only public regulation for fire protection was the "curfew" (from the French word meaning "cover fire"), a regulation requiring that fires be extinguished at a fixed hour in the evening.

Great Britain did not provide statutory authority to units of local government for night or other watchmen with fire apparatus until 1830. Blackstone's *A History of the British Fire Service*[2] mentions only two cases where the subject came up in England before the establishment of civil police forces. One was during the British Civil War in 1643 when, at Nottingham, a company of fifty women was organized to parade the town at night. The other was after the 1666 London fire when a fire watch from towers was proposed; however, it was not adopted.

One of the earliest fire protection regulations was adopted in Oxford, England, in the year 872, when a curfew was established requiring that fires be extinguished at a fixed hour in the evening. In the eleventh century a general curfew was established in England by William the Conqueror. It is believed that this law was passed more to help prevent uprisings and revolt than as a fire prevention measure.

In 1189, London's first Lord Mayor issued an ordinance requiring that new buildings have stone walls and slate or tile roofs, thus banning the previously widespread use of thatched roofs. In 1566, an ordinance requiring the safe storage of fuel for bakers' ovens was put into effect in Manchester, England, thus establishing what was probably the first enactment on a fire prevention subject not related to buildings themselves. In England, the first state action was the Parliamentary Act of 1583 forbidding candlemakers to melt tallow in dwelling houses. Still later, in 1647, wooden chimneys were banned, and after the London fire in 1666 a complete code of building regulations was adopted; commissioners to enforce the regulations were not appointed until 1774.

It was not until Edinburgh's Fire Brigade came into being in 1824 that public fire services began to develop more modern fire protection regulations and standards of operation. A surveyor named James Braidwood was appointed chief of the Edinburgh Fire Brigade. Braidwood wrote the first comprehensive handbook on fire department operation in 1830.[3] His handbook included some 396 standards and explained, for the first time, the kind of service a good fire department should perform.

On a world-wide basis, the early development of public fire protection regulations and types of public fire department organizations closely parallel those of Great Britain and the American colonies.

## *Major Fires in Early Times*

Because of the lack of adequate fire protection regulations, organizations, and equipment, early cities were fire prone. In 1752, Moscow suffered a major conflagration that destroyed 18,000 homes. Moscow was again demolished by fire in the War of 1812.

Constantinople (now Istanbul) was the greatest sufferer from conflagrations of any city on record, having suffered major fire disasters in 1729, 1745, 1750, 1756, 1782, 1791, 1798, 1816, and 1870. In more recent times, Constantinople suffered further major fire disasters in 1908, 1911, 1915, and 1918.

Many of the large cities in India, China, and Japan were wiped out by great fires, and while almost every child of school age has heard that "Nero fiddled while Rome burned," few remember that Rome burned again in 1764. Venice, Italy, was destroyed by fire in 1106 and again in 1577. London was ravaged by fire in 798, 982, 1212, and again in 1666 by the Great London Fire when some 436 acres were burned and 13,000 buildings valued at approximately $53,000,000 were destroyed.

## Fire Protection in Colonial America

In colonial America, night fire watches were instituted in the larger cities. In Boston in 1654, a "bellman" was put to work from 10 P.M. to 5 A.M., and in 1657 in New York, the city's fire wardens were supplemented by a company of eight night fire watch volunteers. These volunteers were called the "rattle watch" because of the large rattles they used to sound alarms. A night fire watch of four town criers was established in New York in 1687.

The night fire watch service was a community institution before there were municipal police forces.* The early night fire watch patrols were influenced by the necessity to control losses in insured properties at a time when there were no organized public fire forces. Such brigades helped get the new institution of fire insurance accepted by the public.

As a result of a disastrous fire in the town of Boston in 1631, the first fire ordinance in America was adopted. It prohibited thatched roofs and wooden chimneys, and was enforced by the "board of selectmen." In 1647, New Amsterdam appointed surveyors of buildings to control the worst fire hazards. In 1648, New Amsterdam appointed five municipal "fire wardens" who had general fire prevention responsibilities. Some people look upon this as the origin of the first fire department in North America.

Following a major fire on January 14, 1653, Boston selectmen were given authority to buy a fire engine. Instead, they contracted for the service of an engine to be brought to fires. There is no record as to the nature of this engine or any service performed. At the same time, additional fire protection regulations were adopted. The 1653 ordinance required all householders to keep a 12-ft swab for extinguishing roof fires. They were also required to maintain a ladder capable of reaching the ridge of the roof. At the same time, the town provided ladders, hooks, and chains for pulling down houses in the path of fires, and gunpowder was occasionally used for this purpose. It was decreed that the owners of the razed houses had no redress.

***Establishment of the First Paid Fire Department:*** Another large fire in Boston in 1679 led to the establishment of the first paid fire department in North America, if not in the world. As a result of a conflagration that

---

*Since the establishment of police forces, fire departments have still not been completely able to give up some night duties, mostly because police officers, in addition to their other duties, cannot be expected to detect conditions that could cause or contribute to fires. For example, district chiefs of fire departments in large cities frequently tour their districts at night.

*Fig. 1.1. As a result of a fire in 1653, Boston selectmen contracted for the use of a fire engine and adopted fire protection regulations requiring householders to keep a swab and a ladder tall enough to reach the ridge of the roof.* (From Library of Congress)

destroyed 155 principal buildings and a number of ships, building laws were adopted requiring stone or brick walls for buildings and slate or "tyle" roofs for houses. Boston imported a fire engine from England and employed twelve fire fighters and a fire chief named Thomas Atkins to operate it. Massachusetts adopted the practice of using paid municipal fire fighters on a call basis, instead of unpaid volunteer fire companies such as those organized later in the southern colonies. (See Fig. 1.1.)

In colonial American communities, each householder was required to keep two fire buckets on hand. When church bells rang to report a fire, people formed lines to pass water from wells or springs to the scene of the fire. When fire engines were obtained, companies were organized to staff the engines; however, citizens were still required to respond with buckets of water to fill the engines. In Boston in 1711, fire wardens were appointed to respond to fires with the members of their staffs and supervise the citizen bucket brigades. As late in history as 1810, Boston citizens were subject to a ten dollar fine for failure to respond to alarms with their buckets. (The laws in a number of states still impose penalties upon citizens who refuse to assist in fighting fires upon the orders of fire officers.) By 1715, Boston had six fire companies with engines of English manufacture. This was before either New York or Philadelphia had a single engine in service.

***Establishment of Mutual Fire Societies:*** Although fire fighting was handled by fire companies, the first of a number of mutual fire societies was formed in Boston in 1718. The members were the more affluent citizens who organized to assist each other in the salvage of goods exposed to fires in their homes or places of business. Their equipment was a screw driver, a bed key, and a bag in which to collect valuables. The bed key was a very important tool, for with it beds could be taken apart and brought outside a burning building. These fire societies became inactive about a century later when fire insurance became available to the more prosperous citizens.

***Formation of Salvage Corps:*** Following the organization of paid fire departments with steam engines in the principal cities of the United States in the latter part of the 19th century, insurance interests formed salvage corps to reduce water damage at fires. While the insurance salvage corps did much to develop the techniques of salvage work, improved fire fighting procedures plus the increased expense of operation led to their disbandment. Today, public fire departments handle most salvage operations at the fire scene.

***The Growth of Paid Fire Departments:*** Lack of discipline in the ranks of volunteer fire fighters, coupled with resistance to the introduction of steam pumping engines, led to the organization of paid fire departments in the larger cities. Following serious disorders at fires, a paid fire department with horsedrawn steam pumpers was placed in service in Cincinnati on April 1, 1853. In 1885 two steamers were delivered to New York City; however, the volunteer fire fighters refused to use them. Ten years later, the "Metropolitan Fire Department" (using steamers) replaced New York's volunteers. Boston reorganized on a steam engine basis in 1859, and employed paid engineers and drivers. A first-class steamer was rated at 900 gpm (gallons per minute) at 100 psi (pounds per square inch), and had a crew of about twelve. Most steamers were in the 500- to 600-gpm capacity range when "working at a fair speed."

## *The Formation of Fire Insurance Companies*

In England, the Great London Fire in 1666 stimulated at least some constructive action in that fire insurance companies were started primarily as a result of it. The fire insurance offices banned wooden chimneys, thatched roofs, and wooden roofs. In order to further protect their insured properties, the early fire insurance offices hired fire fighters and, in 1667, formed the first real fire brigades in England. These fire brigades were equipped with leather buckets, hooks, ladders, large syringes (the first fire extinguishers), and leather hose. Wooden water mains carried water around the city.

In *A History of the British Fire Service* (see Bibliography), G. V. Blackstone expresses the opinion that the history of English fire brigades should date from the formation of the insurance brigades in 1667. The insurance brigades formed by the insurance companies were without statutory authority or obligations, and the insurance company offices, not the government authorities, decided where the brigades would be located. The companies were located in the larger cities and in areas in which insured values were concentrated. The brigades of the various insurance offices were maintained in part for advertising value, and were, therefore, competitive. The bad features of this competition eventually led to consolidation of the fire companies in London into the London Fire Engine Establishment in 1833. The London Fire Engine Establishment was taken over by a Metropolitan Fire Brigade in 1865.

In 1736 Benjamin Franklin recommended the formation of a volunteer fire fighting force called the Union Fire Company, and served on it as America's first fire chief. Franklin also organized the first fire insurance company in the United States, the Philadelphia Contributionship. However, the actual job of fire fighting was performed either by fire companies operating under the authority of the municipality or by independent volunteer companies that owned their own stations and apparatus, although American insurance companies frequently contributed to the support of volunteer fire companies.

In 1835, the Manufacturers' Mutual Fire Insurance Company was established in Providence, Rhode Island. This company accepted only those mills utilizing good fire prevention and protection standards of the time.

## *Progress of Fire Protection in the 19th Century*

In general it can be said that slow progress was made in fire fighting and fire protection practices until the early part of the 19th century. In *A History of the British Fire Service* (see Bibliography), Blackstone notes that prior to the 19th century the only organized forces at work at the Great London Fire in 1666 were soldiers. Although London had had some crude fire engines since 1663, they were not properly maintained and there was no organized fire force. The 1666 London fire gave great impetus to the provision of fire engines and engine keepers in Great Britain. Edinburgh, Scotland, made a start in 1703 with the appointment of twelve firemasters. Beverly, in Yorkshire, established the first paid municipal fire brigade in England in 1726. In 1829, the first steam fire engine was created in England. However, not until the establishment of Edinburgh's Fire Brigade in 1824 did public fire services begin to develop modern standards of operation. James Braidwood, a surveyor, was appointed chief with eighty part-time aides. As fire fighters, Braidwood chose young men between the ages of seventeen and twenty-five, and required drills and night training. Braidwood, who wrote the first handbook on fire department operation in 1830 (see Bibliography), was chosen in 1833 by the insurance offices to head the new London Fire Engine Establishment. His death at the great Tooley Street fire led to the disbandment of the London insurance fire brigade and the formation of the public Metropolitan Fire Brigade.

In America, fire protection and fire prevention regulations still required major disasters before they were enacted and enforced, as can be evidenced by the Great Chicago Fire. On October 9, 1871, a sweeping conflagration destroyed most of Chicago. (Traditionally, "Mrs. O'Leary's cow" has been blamed for its start.) Annually, Fire Prevention Day, established in 1922 on the anniversary of this fire, is intended to serve as a reminder of the destructiveness of fire and the importance of its prevention.

Following the Great Chicago Fire, the Chicago City Council decreed that the city be rebuilt from brick and stone. Later, these regulations were relaxed to such an extent that the fire insurance companies became concerned and, in 1874, the National Board of Fire Underwriters, representing many insurance companies, demanded that the City Council enforce the building requirements. The National Board forced the issue by closing all insurance offices in Chicago for two months.

The beginnings of modern industrial fire protection and prevention were established in the textile mills of New England in the 1830s, and the first municipal fire alarm system was installed in Boston in 1852.* A paid fire department was organized in Cincinnati in 1853.

The automatic sprinkler system, one of the most important inventions for the control of fires, was conceived and put into use in the latter part of the 19th century. The great importance of the automatic sprinkler system was not fully realized at the time. In his book titled *Automatic Sprinkler & Standpipe Systems,* Dr. John L. Bryan presents the following operational definition for the basic concept of an automatic sprinkler system:[4]

> An automatic sprinkler system is a system of pipes, tubes, or conduits provided with heads or nozzles, that is automatically activated and (in some types) deactivated, utilizing the sensing of fire-induced stimuli consisting of light, heat, visible or invisible combustion products, and pressure generation, to distribute water and water-based extinguishing agents in the fire area.

According to Gorham Dana in *Automatic Sprinkler Protection,* the first recognized patent for a sprinkler system was issued in 1723 to a chemist named Ambrose Godfrey.[5] Godfrey's system consisted of a cask of fire extinguishing fluid, usually water, containing a pewter chamber of gunpowder. The chamber of gunpowder was connected with a system of fuses that were ignited by the flame of the fire, thus exploding the gunpowder and scattering the extinguishing liquid.

By the middle of the 19th century, additional improvements were reflected in the automatic sprinkler systems developed in England. The first automatic sprinkler system patented in the United States was developed by Philip W. Pratt in 1872 in Abington, Massachusetts. At about the same time, steam

---

*For the purposes of this text, the term *fire protection* shall refer to the detection and extinguishment of fire, the reduction of fire losses, the safeguarding of human life, and the preservation of property. *Fire prevention* refers primarily to measures directed towards avoiding the inception of fire, and is not synonymous with *fire protection.*

sprinkler systems were invented. From 1852 to 1885, perforated pipe systems were used extensively in textile mills throughout New England, and from 1874 to 1878 Henry S. Parmelee of New Haven, Connecticut, made continued design improvements on his invention, the first practical automatic sprinkler head. (See Fig. 1.2.) Both the design and the installation of the Parmelee automatic sprinkler system involved some basic design principles that are still being utilized in today's automatic sprinkler system installations.

Although primarily considered a 19th-century development, the perfection of the various types of automatic sprinkler systems and heads beyond the dry pipe automatic sprinkler system came about in the 20th century. The period from the 1920s to the 1950s marked the development and perfection of automatic sprinkler systems. In 1967, the multicycle automatic sprinkler system, designed so the water control valve can be automatically turned on and off by heat sensors, was developed. In 1972, the on-off sprinkler head was approved for manufacture and general use.

Three important organizations that would prove to exert a profound effect on the development of methods of fire protection and fire prevention were created during the 19th century. These three organizations were: (1) Underwriters Laboratories Inc., (2) Factory Mutual System, and (3) the National Fire Protection Association. The basic operations and functions of these three organizations, which are still in existence today, will be covered in detail in Chapter Twelve, "Fire Protection Organizations, Information Sources, and Career Opportunities," later in this book.

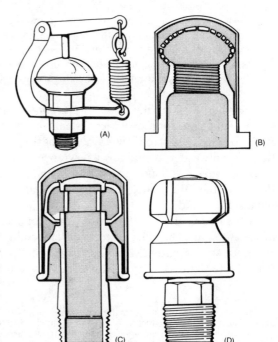

*Fig. 1.2. (A) The 1874 Parmelee automatic sprinkler head; (B) the 1875 Parmelee automatic sprinkler head; (C) and (D) the 1878 Parmelee automatic sprinkler heads. (From* Automatic Sprinkler Protection, *by Gorham Dana)*

## Nineteenth Century Apparatus and Equipment

Most of the basic fire fighting apparatus and equipment in use today were developed and put into service in the latter part of the 19th century, although some developments in the earlier part of the century reflected the later advances. For example, first mention of the use of an engine capable of taking suction was in New York City in 1806. The introduction of suction engines brought about objections from Boston fire fighters who felt that citizens would no longer respond to and help out at fires. The advent of suction engines also created a need for the construction of cisterns. These cisterns were located in much the same manner as present-day hydrants.

*Fire Hydrants:* From 1830 through the 1840s, the first fire hydrants were installed on public mains. Before that time some cities and towns depended on networks of wooden piping from which water could be obtained for fire fighting purposes. Because the primitive hydrants were unreliable and were supplied by three- or four-inch mains, cisterns long remained a principal source of water for fire engines. Boston Fire Chief John S. Damrell, in his testimony concerning Boston's 1872 conflagration, stated that the hydrants on four-inch mains were incapable of adequately supplying steamers, and that the cisterns were preferable and had been used to supply out-of-town engines. Damrell encouraged the installation of large mains which, following major fires in 1889 and 1893, were finally installed. Also installed in Boston were large pipes from the waterfront area to the conflagration district, thus enabling fireboats to supply pumpers. In the 20th century these mains were incorporated into the high-pressure fire main system.

*Use of Fire Hose:* In the early American colonies, the use of fire hose was a comparatively late development in fire fighting. In 1799, Boston imported a few short lengths of leather hose from England, thus allowing the nozzle to be advanced closer to the fire. (For more than a century, hose nozzles had been mounted directly atop the pumps.) Within a few years fire hose and fire hose reels became a necessary part of fire department equipment.

In Boston in 1871, 1½-in. fire hose was placed into service and woven-jacketed rubber-lined hose was introduced to replace leather hose. Expansion ring couplings with clear waterways replaced old-style couplings with restricted waterways, and increased interest in the standardization of hose threads came about following Chicago's fire of 1871 and Boston's fire of 1872. However, no significant progress was made until the NFPA was given the job of thread standardization following the Baltimore fire of 1904. The dimensions of the National (American) Standard for 2½-in. hose threads were selected because 70 percent of the existing threads could be recut to that standard. Water pressure relief valves were installed on all of the Boston Fire Department's fire engines, and shutoff nozzles were issued to fire companies so that for the first time "hosemen" could control water damage.

***Advances in Fire Apparatus:*** By the 1870s, self-propelled steam fire engines were in service in New York City, Boston, and Detroit, and in 1873 steam-propelled fireboats were introduced. In 1873, "Babcock" chemical engines were introduced to provide fast fire attack with ¾-in. hose, and the Babcock hose thread became the standard booster hose thread. Also put into service in 1873 were the first aerial ladders. While these forerunners of today's aerial ladders were capable of being manually extended to 97 ft, they were limited to 85-ft extensions as a result of some serious accidents. By 1882, water towers were in service for providing powerful streams of water to the upper floors of buildings, and in 1905 spring-assisted aerial ladders came into use. In the mid-1930s, power-operated 100-ft metal aerial ladders were introduced.

By 1910, the introduction of automobile fire apparatus was well underway; within two years, a 12-hour underwriters' test for automobile pumpers (first conducted in 1910) had become a routine procedure, and the first NFPA standard on automobile fire apparatus was adopted in 1914. Pumper pressure-volume ratings were standardized at 120 psi net pump pressure, which was the case for some forty years until the present 150 psi rating was established. The introduction of automobile fire apparatus gradually eliminated the separate chemical and hose wagons in many fire departments because each pumper could carry its own ancillary equipment.

***Major Hydraulic Studies:*** Much of the fundamental data currently employed in hydraulic work in fire protection was developed in a series of extensive investigations by John R. Freeman in 1888 and 1889, and by Boston city engineer William Jackson in 1893.[6] Freeman measured flows from nozzles, friction losses in fire hose, and suggested the standard 250-gpm fire stream. His suggestions led to the adoption of 3-in. fire hose by most fire departments. William Jackson conducted detailed tests of pumper performance, water tower operation, and practical fireground layouts with both 2½-in. and 3-in. hose with various sizes of nozzles, resulting in improved fire fighting procedures.

***Fire Alarm and Extinguishing Systems:*** A fire alarm telegraph system was installed in Boston in 1851, and fire alarm systems were in widespread use by the time the telephone was introduced in 1877. By the end of the 19th century, automatic detection and alarm systems and automatic sprinkler systems were becoming more common, thus adding to the efficiency of fire departments.

## Fire Loss During the Early 20th Century

At the beginning of the 20th century, the situation with respect to control of fires was vastly different from our present-day practices. Although in 1900 the larger American cities had paid fire departments, by today's standards these departments were poorly equipped and poorly trained. For the most part, fire fighters were mere water throwers. Building construction was fairly poor and fire protective devices in buildings were uncommon. Fire prevention procedures and practices were practically unknown. As a result of these deficiencies in firesafety, fires in the early 20th century resulted in tremendous losses of both life and property. The destructiveness of these fires, however, provoked constructive results in fire protection practices and improvements in building design. (See Table 1.1.)

Many of the conflagrations of the early 1900s were largely caused by a combination of wood-shingle roofs, strong winds, and a period of hot, dry weather. (See Fig. 1.3.) Table 1.2 shows a breakdown of wood-shingle conflagrations. As can be seen from this table, the southern regions were particularly suscep-

Table 1.1  Great Conflagrations, 1903-1942

| Date | Location | Property Destroyed | Life Loss | Constructive Results |
|---|---|---|---|---|
| Dec. 30, 1903 | Chicago, IL | Iroquois Theatre | 602 | Improvements in construction of, and fire protection for, theatres |
| Feb. 7, 1904 | Baltimore, MD | Eighty city blocks | 0 | Standardization of hose threads |
| April 18, 1906 | San Francisco, CA | 28,000 buildings | 674 | Reinforcement of windows, need for auxiliary water towers |
| Mar. 4, 1908 | Collinwood, OH | Lakeview Grammar School | 175 | Establishment of school fire drills |
| Mar. 25, 1911 | New York, NY | Triangle Shirtwaist Company | 145 | NFPA *Building Exits Code**  |
| Mar. 18, 1937 | New London, TX | Consolidated School | 294 | Need for state laws as safeguards for public buildings not subject to municipal ordinances and inspections |
| Sept. 3, 1939 | Off NJ coast | S.S. *Morro Castle* | 125 | International regulations on safety from fire at sea |
| April 23, 1940 | Natchez, MS | Rhythm Club | 207 | Banning of combustible decorations, improvement of exit facilities, and installation of emergency lighting equipment in nightclubs |
| Nov. 28, 1942 | Boston, MA | Cocoanut Grove | 492 | |

*The *Building Exits Code,* published by NFPA in 1927, provided a comprehensive guide to exits and related features of life safety from fire in all classes of occupancy. In 1966 the *Code* was changed to NFPA 101, *Code for Life Safety from Fires in Buildings and Structures,* and the contents were arranged in the same general order as contents of model building codes, as the *Code* is used primarily as a supplement to building codes.

Table 1.2  Wood-shingle Conflagrations

| Date | Location | Buildings |
|---|---|---|
| May 3, 1901 | Jacksonville, FL | 1,700 buildings |
| April 12, 1908 | Chelsea, MA | 3,500 buildings |
| June 25, 1914 | Salem, MA | 1,600 buildings |
| March 21, 1916 | Paris, TX | 1,440 buildings |
| March 22, 1916 | Nashville, TN | 648 buildings |
| March 22, 1916 | Augusta, GA | 682 buildings |
| May 21, 1917 | Atlanta, GA | 1,938 buildings |

tible to sweeping fires of this type. In spite of these and later fires, it took several decades to effectively reduce nationwide use of wood-shingle roofs.

*Fire Loss During World War II:* World War II provided the world with some of history's most graphic examples of the destructiveness of fire. Fires started by incendiary bombs in cities in England, Germany, and Japan caused

*Fig. 1.3. Wood-shingle roofs helped the spread of many fires, particularly in dry areas of the southern United States. Such a conflagration occurred on May 21, 1917, in Atlanta. In this fire a total of 1,938 buildings were destroyed or damaged.* (From Men Against Fire, by Percy Bugbee)

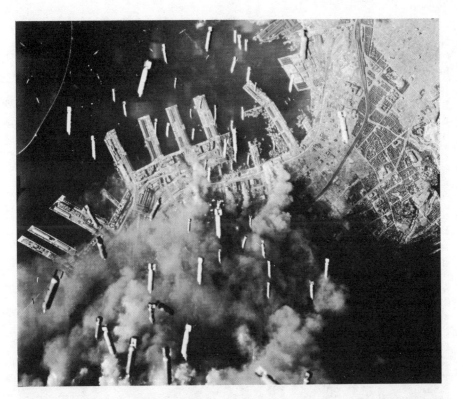

*Fig. 1.4. Aerial view of incendiary bombing of Kobe, Japan, in 1945. Many cities were destroyed by fire before the atomic bombs were dropped.* (From *Men Against Fire,* by Percy Bugbee)

almost unbelievable loss of life and property destruction. Approximately four-fifths of the wartime destruction in English cities was caused by fire damage. City after city in Germany was devastated by fire bombs. A series of fire attacks in the summer of 1943 destroyed 60 percent of Hamburg, an industrial city with a population of 1,760,000, and damaged an area of 30 square miles, burned out 12 square miles, made 750,000 people homeless, and killed from 60,000 to 100,000 persons. The many fires that started as a result of the fire attacks over a wide area of the city created a fire storm that no fire department in the world could handle.

In Japan, the effects of the fire bombings were obscured by the devastation caused by the atomic bomb. Two cities, Hiroshima and Nagasaki, were hit by atomic bombs, and sixty-five other Japanese cities were largely destroyed by incendiaries. In the atomic bomb attack on Nagasaki, some 35,000 people were killed. The attack on Hiroshima killed 70,000 to 80,000 people. In Tokyo on March 9, 1945, nearly 89,000 lives were lost from incendiary fires. Incendiary raids were a major cause of Japan's defeat in World War II. (See Fig. 1.4.)

## 16    Principles of Fire Protection

***Post World War II Conflagrations:*** Since World War II, the United States experienced a number of disastrous fires, in all types of buildings, that attracted at least temporary public attention. These fires occurred in ships, hotels, industries, hospitals, nursing homes, schools, and restaurants, and resulted in tragic life loss and staggering property loss. Accordingly, the tragedy of fire again provoked the need for firesafety regulations. After three major hotel fires in 1946 caused a total of 199 deaths, the American public became greatly aroused concerning firesafety in hotels. As a result, many states passed more stringent fire protection legislation for hotels.

In 1958, Chicago's Our Lady of the Angels Elementary School burned with a loss of ninety-five lives. (See Fig. 1.5.) The NFPA report of the fire stated: "It is the old story of open stairways, poor housekeeping practices, delayed discovery and delayed alarm, and lack of automatic fire protection in the building." Once again, this tragedy focused public attention on the fire threat to schools, and schools everywhere underwent extensive inspections; many defects in construction and fire protection were discovered and corrected.

*Fig. 1.5.   A view of the fire scene at Our Lady of the Angels grade school in Chicago. The December 1, 1958 fire took ninety-five lives.* (From *Men Against Fire,* by Percy Bugbee)

In November of 1963 in Fitchville, Ohio, a nursing home fire took sixty-three lives. The nursing home was located more than 7 miles from the nearest volunteer fire department, and there was no fire protection or alarm system in the home. This fire led to a congressional committee investigation of nursing home fires.

Table 1.3 identifies more recent major conflagrations in the United States and the factors contributing to their cause and spread. As can be noted from Table 1.3, conflagrations are seldom due solely to any one factor.

*Evaluation of Modern Methods of Prevention and Control:* Modern methods of fire prevention and control gradually evolved during the 20th century. From 1900 to the present, hundreds of devoted persons from federal, state, and local governments, from fire departments, from industry, and from insurance companies and bureaus have helped develop standards that cover the field of fire waste control. Today, the *National Fire Codes* of the NFPA are presented in sixteen volumes. As far back as 1903 a grading schedule to measure the water supply, fire fighting forces, and other factors for city and town fire protection was developed by the National Board of Fire Underwriters (now the American Insurance Association) and by the state insurance rating bureaus.

In the early 1900s only a handful of the larger cities maintained a fire prevention bureau in the fire department. Today, most paid fire departments devote personnel and time to fire prevention activities. The idea of a fire department inspection of dwellings, common today, began in 1915 in Portland, Oregon. The first state to create a state fire marshal's office was Ohio, about 1904. Most states now have such offices.

While much more needs to be done to reduce human suffering and the heavy economic costs of fire, over the last hundred years there has been a tremendous increase in our knowledge about fires, fire fighting, fire protection methods, and fire prevention measures.

## Fire Protection Defenses and Agencies

Destruction by fire and the lack of organization during fire fighting procedures in colonial America led to the formation of more organized methods of improving fire protection and prevention. These methods stemmed from the organization of fire departments to protect a municipality and its citizens from fires.

Today's fire department is a highly organized unit whose effectiveness depends upon the latest equipment and procedures for fighting the various types of fires caused by modern hazards. Chapter Nine, "Municipal Fire Defenses," will explore in greater detail the planning and evaluation of public fire protection and the water supply and distribution requirements for public fire protection. For purposes of this chapter, a brief overview of the evolution of fire department responsibilities and concerns is presented.

*18   Principles of Fire Protection*

***Evolution and Scope of the Fire Department:*** The growth of paid fire departments in the United States resulted largely from the lack of both discipline and

Table 1.3   Great Conflagrations, 1972-1977, and Causes Contributing to Their Spread

| Date | Location | Property Destroyed | Reported Loss |
|---|---|---|---|
| Feb. 7, 1972 | Wakefield, MA | 6 buildings | $1,500,000 |
| (Volatile chemicals, high winds, severe cold, congested and aging buildings) | | | |
| May 27, 1973 | Chicago, IL | 5 buildings involved | $25,000,000 |
| (Explosions damage sprinklers, broken water mains) | | | |
| June 12, 1973 | Philadelphia, PA | 4 buildings and 4 dwellings 2 fire fighters killed | Over $5,000,000 |
| (Explosions, flammable liquids) | | | |
| October 14, 1973 | Chelsea, MA | 300 buildings involved | $1,313,650 |
| (Poor water supply, dry weather, high winds, closely stored combustibles, narrow streets, no fire breaks, delayed reporting) | | | |
| November 5, 1973 | Indianapolis, IN | 5 buildings (2 high-rise) | $5,321,000 |
| (Delayed detection, tremendous radiant heat, sprinklers out of service) | | | |
| April 9, 1974 | Grand Junction, CO | 8 buildings | $2,594,000 |
| (Broken water main, high winds) | | | |
| April 9, 1974 | Cloudcroft, NM | 14,500 acres | $16,000,000 |
| (Towns of Sacramento and Weed and natural forest) | | | |
| May 22, 1974 | Chelsea, MA | 6 buildings | $3,592,327 |
| (Inadequate water supply, large closely located combustible buildings, flammable liquids helped spread, chemical explosions, hot and dry weather) | | | |
| September 8, 1974 | Burbank, CA | 7 structures involved | $5,826,832 |
| (Delayed detection, inadequate water, combustible movie sets and stages closely constructed) | | | |
| December 13, 1974 | Middleboro, MA | 4 buildings | $500,000 |
| (Wood-frame construction, delayed detection, inadequate water for sprinklers) | | | |
| May 21, 1976 | McKeesport, PA | 7 buildings | Over $5,500,000 |
| (Automatic sprinkler system control valves shut off in building where fire originated because of its being razed) | | | |
| November 26, 1976 | Belt, MT | 18 buildings, 24 derailed railroad cars, 19 automobiles | $2,340,000 |
| (Explosion of two 30,000-gallon propane tanks creating heat so intense it set fire to asphalt on the street) | | | |
| July 25, 1977 | Santa Barbara, CA | 140-200 structures; over 800 acres of land | $25,000,000 |
| (Wood-shingle roofs) | | | |

effective equipment at the fireground. In the latter half of the 19th century, fire departments developed improved techniques for reducing water damage and increasing salvage from fires. Technological advances in manufacturing methods led to such improvements as the development of more effective fire hose, automatically raised ladders, progress in the area of hydraulics, and increased use of mechanized apparatus. In turn, such improvements led to the need for specific organizational and training techniques within fire departments. In 1889 Boston established a drill school; in 1914, New York City established a "Fire College." In 1976 a National Fire Academy, under the aegis of the federal government, became a reality.

After World War I, developments such as adjustable spray nozzles and improved radio communications equipment increased the efficiency of fire departments and served to bring about more organized fire suppression capabilities for fire departments. The use of computers for record keeping, plotting, and assisting in dispatching, and pumpers that were designed to allow nozzle operators to control the rate of flow from the nozzle, have made fire departments more organized and efficient.

A fire department, like any other organization, is comprised of a group of people working together to achieve a common set of objectives. These objectives form the foundation of the organization and provide both purpose and direction to the organization. The NFPA's *Fire Protection Handbook* states that the traditional objectives commonly accepted by most fire departments are as follows:[7]

1. To prevent fires from starting.
2. To prevent loss of life and property when fire starts.
3. To confine fire to the place where it started.
4. To extinguish fires.

However, the scope of these objectives, as well as the scope of responsibilities within the fire department, has changed over the years because of the need for communal and organized fire fighting. Highly trained medical professionals in fire departments have broadened their responsibilities to include saving lives in disasters other than fires, such as airport crashes and automobile accidents. Today's fire chiefs and officers are no longer only fire scene leaders; they must also know how to train personnel, how to manage both physical and economic resources, and how to effectively manage an entire fire unit. All of these changes in roles and in fire department responsibility have evolved from years of increased technological development in the fire sciences.

NFPA 4, *Organization for Fire Services** recommends three areas as the scope of fire department responsibility: (1) control of combustibles and fire prevention work, (2) fire fighting and emergency services, and (3) govern-

---

*A revision of NFPA 4 came up for adoption at the 1977 NFPA Fall Meeting and was adopted. It has now been renumbered to NFPA 1201.

mental function. These three responsibilities will be discussed further in Chapter Nine, "Municipal Fire Defenses."

***Fire Protection Agencies:*** Fire's continuing threat to life and property has brought about the formation of fire protection agencies at federal, state, and private levels. These agencies have as their common objective the protection of life and property from fire. Chapter Twelve, "Fire Protection Organizations, Information Sources, and Career Opportunities," discusses several of these organizations in detail.

## *Summary*

The size and scope of fire waste, when viewed in total loss from the earliest times, is phenomenal. From the facts presented in this chapter, it is obvious that fire has always been and still is a destructive threat to all persons and property. However, it should not be concluded that the situation is hopeless and that no progress has been made to prevent and combat fire. Although there is much more that needs to be done to reduce the human suffering and heavy economic cost of fire, there has been a tremendous increase in knowledge about fire, fire fighting, fire science technology, and fire protection and prevention.

## *Activities*

1. Early fire protection regulations in England included such ordinances as curfews for fires, prohibition of thatched roofs and wooden chimneys, and safe storage of fuel for bakers' ovens. Describe some of the similarities between these regulations and the fire protection regulations instituted in Boston in the 17th century.
2. How was the structure of the Roman *Corps of Vigiles* similar in structure to today's municipal fire departments?
3. Identify the role of each of the following in improving fire protection and fire extinguishment in the 19th century.
    (a) Automatic sprinkler systems.
    (b) Fire hydrants.
    (c) Fire hose.
    (d) Fire alarm systems.
4. (a) How did fire loss in the early 20th century influence a greater awareness for life safety and firesafe building design?
    (b) What are some of the life and property firesafety methods currently employed by your community? Discuss how some other methods might be used advantageously by your community.
5. Discuss the relationship between mutual fire societies, salvage corps, and fire insurance companies. How did each of these groups contribute to fire protection regulations?
6. Identify some of the 19th century developments in fire fighting apparatus

and fire fighting equipment. In what ways did each of these developments improve fire fighting efficiency?
7. (a) Discuss the functions of the early mutual fire societies.
   (b) What effect did the formation of fire insurance companies have on these early mutual fire societies?
8. In what ways has the job of fire fighting evolved from that of a mere "water thrower" to a profession that includes many specialized fields?
9. Most of the major fires in early times were caused by inadequate fire protection regulations and fire fighting equipment. Yet despite today's numerous fire protection regulations and advanced fire fighting equipment, major fires are still a threat. Write a brief summary statement explaining why you believe major fires have not disappeared. Compare and discuss your statement with those of your classmates.
10. Independently or with a group of your classmates, trace the development of organized fire fighting efforts throughout history, beginning with the Roman *Corps of Vigiles* and ending with today's fire department. Then write a summary statement of the developments in fire department growth in one of the following areas:
    (a) Fireground organization.
    (b) Fire apparatus.
    (c) Salvage operations.
    (d) Paid fire departments.
    (e) Fire fighting equipment.

## Bibliography

[1]Haessler, Walter M., *The Extinguishment of Fire,* Rev. Ed., NFPA, Boston, 1974, p. 1.

[2]Blackstone, G. V., *A History of the British Fire Service,* Routledge and Kegan, London, 1957.

[3]Braidwood, James, *On the Construction of Fire Engines and Apparatus, the Training of Firemen and the Method of Proceeding in Cases of Fire,* Bell and Bradfute and Oliver and Boyd, Edinburgh, 1830.

[4]Bryan, John L., *Automatic Sprinkler & Standpipe Systems,* NFPA, Boston, 1976, p. 52.

[5]Dana, Gorham, *Automatic Sprinkler Protection,* 2nd Ed., Wiley, New York, 1919.

[6]Freeman, J. R., *Transactions of the American Society of Civil Engineers,* Vol. XII, 1883; and Vol. XXIV, 1891.

[7]*Fire Protection Handbook,* 14th Ed., NFPA, Boston, 1976, p. 9-4.

*Chapter Two*

# Firesafety for People and Property

*Firesafety is an important aspect of building design; firesafe buildings protect the occupants of the buildings and the buildings themselves. Because dwellings, hospitals, schools, industrial plants, stores, and business establishments house and serve people in particular ways, their design must take into account all the features needed to ensure life safety.*

## ASSESSING LIFE SAFETY

Fire is a potential threat to every human being and to practically all property. Despite improvements in building construction, consumer products, and fire protection methods, deaths from fire are averaging about 8,800 a year in the United States. Further, estimates show that each year at least 108,000 people are burned or otherwise injured in fires.

Complete, accurate statistical data on fires and fire losses are not available as yet in the United States or any other country; however, estimates made by the NFPA's Fire Analysis Department show that there are approximately three million outbreaks of fires attended to by the public fire service in the United States in any one-year period. It has been estimated that direct property loss from these fires will exceed three billion dollars annually, and indirect losses (in the form of lost productivity, insurance costs, and hospital and medical costs for injuries) will amount to approximately seven billion dollars.

Estimates of such enormous scope, however, are difficult to comprehend. Although many individuals feel that the tragedy of fire "will happen to someone else," these same estimates indicate that during an average 24-hr period approximately 24 deaths and 296 injuries will result from fires. It is further estimated that property losses from fire during this same time period will total more than ten million dollars, and some 1,800 residential properties will be destroyed or damaged.

## Life Loss and Injury from Fire

On the average, persons that die in fires usually die from inhalation of smoke and toxic gases or from burns when clothing is ignited. Fire deaths can sometimes be attributed to entrapment in a burning house when an individual wakes up too late for escape. Often the victim is a child or an elderly person. In other age groups, fire deaths are more evenly distributed; however, teenagers have the lowest percentage, probably because, due to their agility, their chances of escaping from a burning building are better than average.

In our day-to-day living, we come into contact with ignitable materials, such as combustible furniture, upholstery, and appliances that present dangers through both normal wear and misuse. Additionally, careless storage of combustible and flammable products presents a grave fire hazard, and nonexistent, inadequate, or blocked exits in buildings can result in entrapment and death should a fire occur. Misuse and carelessness with smoking materials, matches, fireworks, and other fire-producing elements and devices can injure or maim both children and adults.

Fire injuries are the tragic and painful symbols of fire's destructiveness, although advances in medical techniques have substantially improved the chances of recovery from serious fire injuries. Nonfatal fire injuries are principally due to burns and to the inhalation of carbon monoxide and other gaseous products of combustion. Added to these physical injuries are the psychological scars left by the loss of loved ones, possessions, or pets. One burn injury in a family unit can affect the entire family, whose reactions, paralleling those of the victim, proceed from shock, to dread, to fear for the burned relative's life, and then to the chronic apprehension and concern about how the patient will look and what the different medical procedures will entail. There is a steady burden upon the entire family.[1]

## Property Loss from Fire

The United States has a greater direct property loss from fire than other countries. The National Commission on Fire Prevention and Control attempted to estimate the annual financial costs of fire waste and reported the figures in "America Burning," issued in 1973 (see Table 2.1).

The figures shown in Table 2.1 can be considered conservative, for it is impossible to estimate all of the indirect economic losses that occur when a business is destroyed by fire. Some firms never recover, and many subsequently go out of business. Others lose employees and customers, and still others may lose the confidence of stockholders. Many find that the cost of replacing buildings and equipment may be substantially greater than the value of those destroyed, especially when essential records are lost. A small community that experiences the loss of one of its businesses may also experience a large tax loss and increased welfare costs.

Table 2.1  Estimated Annual U.S. Fire Costs*

| | |
|---|---:|
| Property loss | $2,700,000,000 |
| Productivity loss | 3,300,000,000 |
| Fire department operations | 2,500,000,000 |
| Insurance costs | 1,900,000,000 |
| Burn injury treatment | 1,000,000,000 |
| Total | $11,400,000,000 |

*From "America Burning," May 1973, National Commission on Fire Prevention and Control, Washington, DC, p. 2.

In some cases, a single fire can seriously hamper production in an entire industry. A 1954 fire in an automotive transmission plant in Livonia, Michigan, halted automobile production for several months because the transmissions produced by the plant were used in six makes of automobiles. The unavailability of transmissions led to sharply decreased sales for five major United States automobile makers. Indirect losses from the fire were never accurately estimated.

However, not all fire losses are true losses; sometimes there are possible "gains" from fire. If, in an obsolete building, outmoded equipment or outdated stock is burned, there is no real economic loss except in damage done to the premises. Also, recapitalization by insurance may be more than could be funded from individual resources. In large-scale fires, the value of funds and services provided by insurance and government disaster-assistance programs may sometimes be greater than the value of the ravaged property. Because of this, some fires are deliberately set to obtain new capital from insurance.

Table 2.2 details estimated building fire causes in the United States. Unfortunately, many of these causes are based on human error or action. Clearly, public firesafety education to eliminate careless smoking and rubbish storage will prevent many fires from starting. Public education might well be the first step towards making properties as firesafe as possible.

Measures can be taken to protect property against the possibility of fire caused by natural acts. For example, the use of lightning protection can help safeguard a property. Other causes of property loss from fire can be eliminated by following recommended codes and standards, by utilizing firesafe building design, and by instituting safer product design.

The building and fire codes and standards that are formulated by many organizations and committees provide recommendations for making properties firesafe by setting forth building design and safety standards. The safety standards of the Occupational Safety and Health Administration (OSHA), other federal government agencies, and private organizations such as the NFPA, cover construction, protection, and occupancy features relative to life safety.

Many of the products that we use in our daily lives need design improvements in order to help reduce fire loss. As shown in Table 2.2, some of the major causes of fires are from heating and cooking equipment, smoking and matches, and electrical appliances. Although manufacturers and standards-writing organizations have generally helped to improve the design and safety features of consumer products, many hazards remain that have not been adequately considered. For example, when the controls for kitchen stoves were located on the fronts of the stoves, they were a safety hazard to children; now that many stoves have controls on the backs, the possibility of ignited clothing or burns to the people using the stoves has increased. Studies of similar potentially hazardous product designs are continually being made by the Consumer Product Safety Commission.

Fire, no matter what its cause or degree, will always be a threat to life and property; however, by helping to protect life through fire protection methods, life loss and property loss can be decreased.

## People Factors

There is surprisingly little research and scientific evaluation available regarding the way people behave when faced with a fire situation. Perhaps this is because no two fire situations are exactly alike. The factors present at the time of ignition (smoke, heat, and flame development; combustibles at hand; drafts; humidity) and many other factors influence the way a fire will behave. The way people behave when they discover or encounter a fire will be influenced by the circumstances of the fire as well as by the "people factors" involved. The reaction to suddenly being awakened in a dark, smoke-filled room will

Table 2.2 Estimated* U.S. Building Fire Causes**

|  | Percent of fires | Percent of dollar losses |
|---|---|---|
| Heating and cooking | 16 | 8 |
| Smoking and matches | 12 | 4 |
| Electrical | 16 | 12 |
| Rubbish, ignition source unknown | 3 | 1 |
| Flammable liquid fires and explosion | 7 | 3 |
| Open flames and sparks | 7 | 4 |
| Lightning | 2 | 2 |
| Children and matches | 7 | 3 |
| Exposures | 2 | 2 |
| Incendiary, suspicious | 7 | 10 |
| Spontaneous ignition | 2 | 1 |
| Miscellaneous known causes | 2 | 6 |
| Unknown | 17 | 44 |
| Total | 100 | 100 |

*NFPA estimates.
**From "America Burning," May 1973, National Commission on Fire Prevention and Control, Washington, DC, p. 57.

usually be quite different from the reaction to a fire started in a place of business in broad daylight.

The NFPA's *Fire Protection Handbook* recognizes seven important human characteristics to be taken into consideration in designing buildings for safety from fire:[2]

1. Physical and mental characteristics.
2. Age.
3. Agility.
4. Decision-making capabilities.
5. Awareness.
6. Training.
7. Special knowledge and beliefs.

These seven characteristics indicate the reasons for and basis of most people's actions and reactions when faced with the danger of fire.

Physical, mental, and emotional characteristics of people vary widely. As has been previously mentioned, most of the fire deaths in ordinary dwellings involve the very young and the very old. Age and agility are important factors that dictate people's actions and reactions.

Decision-making capabilities of people also vary widely. Some people react instinctively and logically when faced with danger; others respond slowly; some are inclined to panic.

It is an accepted fact that getting children out of a burning school building is accomplished easier and faster if the children have had regular fire drills. In industries where there is a possibility of explosion or flash fire, fire drills for the employees are essential to avoid panic and disaster. There is little doubt that training in how to react to a fire emergency is a vital factor in life safety.

In "The Behavior of People in Fires," a study conducted in England by Peter G. Wood,[3] data collected from 1,000 fire incidents revealed that people react to fire in three general ways. In order of frequency, people in a fire situation were found to be *concerned* with:

1. Evacuation of the building either by oneself or with others.
2. Fighting or at least containing the fire.
3. Warning or alerting other individuals or the fire department.

According to Wood's study (which is the most comprehensive study available to date), the majority of people reacted properly when faced with a fire emergency. The most frequent (and usually the first) *actions* taken by people are as follows, listed in order of frequency. The majority of people:

1. Attempted some fire fighting action.
2. Contacted the fire department.
3. Investigated the fire.
4. Warned others.
5. Did something to minimize the danger.

6. Evacuated themselves from the building.
7. Evacuated others from the building.

Some other interesting patterns developed from Wood's study. In cases where it was obvious that the fire was serious and had gained some headway, most people thought first of getting out and not of fighting the fire. People were more likely to leave the building when smoke was present. However, if they encountered smoke only, they were more likely to go back into the building. If people had no means of emergency escape or did not know of such means of escape, they would leave the building faster. If the person involved in the fire emergency had been involved in a previous fire experience, that person would be slower to leave the building and more likely to remain and try to fight or contain the fire. Such a person was also more likely to re-enter the building.

Wood's study revealed that people who had never received any type of drill or training for fire emergencies were more likely to leave the building than those who had received training.

Wood's study also cited the presence of smoke as an important factor in people's activities during a fire incident. The greater the smoke density, the greater the likelihood that people would evacuate the building. This factor was heightened if the smoke spread beyond the room of origin. In addition, the greater the smoke spread became, the more likely people were to use other than normal exits.

In fire incidents where smoke was involved, 60 percent of the people tried to move through the smoke. About half of these people moved 10 yd (9.1 m) or more. The fact that people were aware of a means of escape had no bearing on whether or not they moved through smoke. However, those people who had knowledge of a means of escape were inclined to move farther through the smoke — usually for a distance of 15 yd (13.7 m) or more. As might be assumed, the more familiar a person was with the layout of the building, the more inclined that person was to attempt to move through smoke.

Neither previous experience with fire incidents nor previous training and drills for fire emergencies affected whether people attempted to move through smoke or not. However, people in both of these categories tended to move greater distances through smoke.

The study also revealed that more people were inclined to move through smoke if the fire incident occurred during the daylight hours. However, those people who attempted to move through smoke during fire incidents occurring at night tended to move greater distances through the smoke.

## *People Activities*

Proper planning for the safety of people in any building that might become involved in a fire must take into consideration not only the possible number of people in the building (occupant load), but also the kinds of people who might be in the building and the activities in which these people might be engaged.

Planning for safety in a family home in the event of fire would take into consideration the number of people in the family, their ages, and an estimate of the response of each individual when faced with a fire emergency. In addition, planning for firesafety in both residential and institutional occupancies must take into account the possibility that people may be asleep when fire breaks out.

The seven human characteristics listed in the previous section are a basis for firesafety planning that can be adapted to virtually all situations in which people and people activities are involved. Planning safe egress from an office building or a factory must not only be made with regard for the numbers of people involved, but also for the average mental and physical characteristics of the occupants. Decision-making capabilities must also be taken into consideration. For example, if a person is in a building to perform office work, then such a person can be assumed to have mental characteristics, awareness, and decision-making capabilities that probably fall within an "average" range in terms of actions and reactions in a fire situation. However, if a factory employs people who are mentally or physically handicapped, a different set of firesafety considerations must be employed. The kind and severity of a fire in a particular building can be and often is influenced by the kinds of people who are in the building.

When a new building is being planned for a special purpose that will involve specific "people activities," the designer, architect, and builder must plan for the firesafety of the people with respect to the activities of the building occupants. If, for example, the new building is to be a restaurant, the fact that the dining area is attractive and the kitchen area is efficient does not result in an adequately planned building in terms of the firesafety of the people who will use the building. Special consideration must be given to the possibility of overcrowding, blocking of exits with loose chairs and tables, and lack of emergency lighting in the event of power failure. In a fire emergency, such situations could occur in a restaurant. In addition, special equipment is needed to safeguard against fires caused by grease ducts, deep-fat fryers, and other cooking hazards.

Disastrous loss of life has occurred in fires in restaurants, night clubs, dance halls, and similar occupancies because the possibility of fire and resulting panic was not considered when the occupancy was designed.

## *FUNDAMENTALS OF BUILDING DESIGN*

One of the major goals in designing a building is to provide a structure that fulfills both the needs of the owner and the purpose for which the building is intended. A well-designed building should make efficient use of space and should present an aesthetic environment suited to the purpose for which the building is intended. It should be built as economically as possible, and should be efficiently planned to accommodate the activities of the people who will occupy it. Because buildings are designed for certain (and often specialized) types of activities and occupancy, careful consideration of inherent design features is of vital importance in assessing the life safety factor in case of fire.

The design features of the average one-family dwelling can be easily assessed. Planning for life safety from fire in multiple-purpose buildings such as those containing offices, stores, apartments, garages, and restaurants is complicated and difficult. Assessing life safety from fire in multiple-purpose buildings must take into consideration many and varied design factors so that if a fire develops, the possibility of death and/or injury can be minimized.

A shopping mall is a good example of the complexities involved in planning for life safety in the event of fire. In a shopping mall, there are many different stores of varying sizes and designs containing contents of varying combustibility all under one roof — in effect, all in a single building. Each store opens onto the mall. A fire in one store may well vent itself into the entire mall, thereby endangering both the people in the mall and in the other stores. If the store is small, as many stores in shopping malls are, it will probably have only one visible means of egress. This egress will usually be by way of the mall. This situation results in relatively long travel distances to exits from the mall which, in an emergency situation, will increase the hazard to people in the mall area. In addition, separation between stores may be difficult to achieve and maintain and must be extended to the roof space common to all the stores. Some malls are two stories high — a fact that further complicates easy and speedy exit to a safe point outside the mall.

Multiple-purpose high-rise buildings present great problems in planning for life safety from fire. Some other examples of buildings whose design presents varying and complex problems when planning for life safety from fire are as follows:[4]

1. Large, undivided industrial buildings where compartmentation becomes impractical and travel distance to exits can become excessive.
2. School buildings of open and flexible plan layouts where, again, compartmentation is impractical and access to exits may be varied and obstructed.
3. "Astrodome" structures from which large numbers of people must be quickly evacuated in fire emergencies.
4. Hotel exhibition halls that involve heavy fire loading and numerous ignition sources.
5. Windowless buildings that make entrance for fire fighting and rescue difficult.

Unfortunately, many buildings are in the advanced design stage before local firesafety codes are consulted. Firesafety features should be considered and included in the initial designing and planning stages of buildings. It must be remembered that local firesafety codes usually provide for only minimum firesafety requirements and may not deal with the more complex design problems and firesafety considerations necessary to ensure life safety as well as property safety in more complex buildings.

## Concepts of Egress Design

One of the most important considerations in surviving a fire is how to escape from a burning building. Inadequate escape plans for a fire emergency in a home or in other buildings are the indirect cause of many fire deaths and injuries. Every family or individual should have an escape plan in case of fire so that escape routes are known. Alternate escape routes should also be planned. These preventative measures and the use of protective devices in dwellings are two methods that can be used to help ensure life safety in case of fire.

The following recommendations for life safety provisions to be considered in the home are from the NFPA publication titled *Firesafety in the Home:*[5]

> A fundamental requirement for exits from all dwellings is the provision of at least two independent escape paths for use during a fire. In no case should any of these exitways require passage through space outside the householder's control or through doorways liable to be locked from the outside.
>
> For a dwelling as a whole, two doorways to the outside, as far apart as possible, are necessary. For each individual room or occupied space, there should also be two exits, one of which should be a door that provides unobstructed travel to the outside at ground level. Room exit requirements may be satisfied in a number of ways, such as: 1) one interior door and one door to the outside; 2) two interior doors providing separate ways of escape; and 3) one interior door and one outside window suitable for escape.
>
> These requirements apply to all occupied spaces, but are of utmost importance for basement areas, second-story rooms, and all bedrooms. A space that is normally entered only by ladder, trap door, or folding stairs should never be used for living purposes, regardless of whether or not a window is also available for emergency escape.
>
> Doorways intended to provide escape paths should be at least 2 ft wide, and would best open in the direction of travel. Locks requiring a key or special manipulation for opening should never be employed on such doors. Corridors and stairways that are expected to provide exit to the outside should be at least 28 in. wide, with exit access from bedrooms limited to not less than 3 ft in width. Where doors open directly onto stairs, a landing at least as wide as the door should be provided. Stair treads should be at least 9 in. wide, and risers should be no more than 8 in. high. Stairways should have handrails 30 to 34 in. above the stair tread. Throughout the dwelling, doors, halls, and stairways used for emergency exit should have a minimum headroom of 6 ft, 8 in.
>
> Outside windows intended for use as fire escapes should be capable of being easily opened without the use of tools and should be of sufficient size for the unimpeded passage of an adult. As an absolute minimum, the smallest dimension should be 22 in. and the area 5 sq ft. Escape through a window, even with these minimum dimensions, could be difficult, however, particularly when the vertical dimension is the minimum size and is in upper-story rooms. Much greater utility for escape would be provided by a window with the least dimension greater than 22 in. In any case, the window sill should never be more than 4 ft above the floor; sills within 3 ft of the floor would be better.
>
> As a matter of general good practice — as well as for firesafety — interior

doors with lock mechanisms should be capable of being opened from the outside by using a simple releasing device. All other doors (closets and storeroom doors included) should be capable of being opened from either side at all times.

Multiple-occupancy high-rise buildings present special and difficult egress problems. A fire in any tall building usually has to be fought from inside the building. Reaching fires above 85 ft (25.9 m) is difficult for most fire departments. Also, above this height aerial ladders cannot be used to evacuate people in a fire emergency. It is not considered realistic to expect occupants in a tall building to use the stairway as a practical means of egress. In fact, this procedure can be dangerous if occupants panic and attempt to leave the building by means of the stairway. Time is also a factor: It has been estimated that it would take over two hours to evacuate a 50-story building using a single stairway. An additional consideration is risk of hampering the work of fire fighters who are trying to use the same stairway or stairways. Elevators are not recommended as an acceptable means of egress. Because evacuation is impractical, provision must be made to move the building occupants away from the endangered area to a safe area of the building.

Another factor to be considered when a fire occurs in tall buildings is the phenomenon known as the "stack effect." The movement of smoke is generally more rapid and more dangerous in a tall building than it is in a low building. The natural air movement in a tall building is increased by the differences in temperature between the inside and the outside air, and can be influenced as well by external wind forces and by the air conditioning in the building. A strong draft from the ground floor to the roof can cause smoke to force its way into elevator shafts and stairwells even though the doors of such shafts are closed. Thus the "stack effect" is a significant factor in problems of egress from tall buildings. As buildings become larger and their uses more diversified, the problem of assessing life safety from fire in such buildings becomes more difficult as well as more compelling.

The exit design for a building must take into consideration a variety of factors: aspects of the building provided to resist fire such as walls, stair enclosures, firestopping, door construction, and other similar provisions; the influence of the exit design; and the kinds of built-in fire protection provided, such as sprinklers, standpipes, and alarm systems. In addition, movement of people (or the "flow" characteristics of people) within the means of egress should be accommodated in the designing of means of egress. The design and capacity of passageways, stairways, and other components that make up the total means of egress from a building should be based to a large degree on the physical dimensions of the human body.

The majority of adult males measure less than 20.7 in. (52.6 cm) at the shoulder. The maximum practical standing capacity for one person is estimated at 2.3 sq ft. This is derived from the concept of "body ellipse" on which the major axis measures 24 in. (60.9 cm) and the minor axis 18 in. (45.7 cm). It is necessary to take this factor into account because studies of crowd movement show that people tend to avoid bodily contact with others. Another interesting phenomenon found in the studies of movement of persons is that there is a swaying action that depends upon the type of motion involved — free movement or dense crowds, travel on stairs, etc. Normal sway is 1.5 in. (3.8 cm) left and right; however, when movement is reduced to shuffling in crowds, the sway increases to as much as 4 in. (10.1 cm). Therefore, a width of 30 in. (76.2 cm) is indicated to accommodate a single file of people traveling up or down stairs. There is always the possibility of panic developing when people are crowded together and bodily contact cannot be avoided. The danger of panic is reduced if the crowd can move in a direction toward supposed safety.

*Flow Rates:* The factors involved in the relative rate at which people can be expected to move or flow are also significant considerations in the design of means of egress. Studies conducted by the London Transport Board have produced some helpful figures.[6] On a level walkway, an average walking speed of 250 ft (76.2 m) per minute is attained under free flow conditions and with 25 sq ft (2.35 m$^2$) available per person. A walking speed of 145 ft (44.1 m) per minute indicates a shuffling motion. This can occur when people are crowded and their walking speed as well as their free flow conditions are restricted. A "jam" point is considered to have been reached when the concentration of people is one person to every 2 sq ft. Under fire conditions, space per person of 3 sq ft or less indicates a significant panic potential. Calculations of flow rates of crowds using speed (feet per minute) and density (people per square foot) show that movement increases as the walking area decreases until a point is reached when forward movement becomes restricted and movement slows down. Fairly uniform flow rates were observed over a wide range of conditions making this study even more helpful in making egress design decisions. Figure 2.1, based on the London Transport Board's report, shows the effect of crowding in level passageways.

In addition to determining factors that affect speed in level passageways, the London Transport Board also demonstrated that passageway designs affected the movement of people. The study revealed that for passageways over 4 ft (1.2 m) wide, flow rates are directly proportionate to the width of the passageway. The flow rate in level passages was determined to be 27 people per minute per foot width. Travel down stairways was 21 people per minute per square foot while travel up stairways was reduced to 19 persons. Further conclusions of the study indicated that: (1) minor obstructions in a passageway did not significantly affect the flow of people; (2) corners, bends, and slight grades also appeared not to be a factor in rate flow; and (3) a center handrail can reduce the rate of flow of people.

Two major methods have been employed in order to determine the exit widths required in buildings. These methods are called the "Capacity Method" and the "Flow Method." Projected population characteristics and the special needs or purpose of the structure are used as a basis for determining required

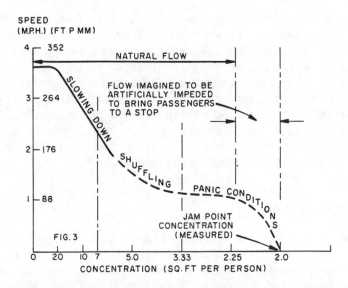

*Fig. 2.1. Walking speed of persons in level passageways. Speeds of less than 145 fpm (feet per minute) are considered to constitute "shuffling." A significant panic exposure exists any time movement is restricted; the problem becomes urgent under fire conditions when concentrations reach one person every 3 sq ft.* (Adapted from Research Report No. 95, London Transport Board, "Second Report of the Operational Research Team on the Capacity of Footways")

exit widths. The "Capacity Method" of determining necessary exit width is based on the theory that adequate stairways should be provided in any building in order to contain all the occupants of that building. This method does not require any flow or movement on the part of the occupants but, instead, assumes that the stairwell provides a safe area of refuge for all the occupants of the building. The occupants may then evacuate the building at a more leisurely pace and in keeping with their physical capabilities, thus avoiding danger and possible panic.

The "Flow Method" utilizes the theory that a building can be evacuated within a specified maximum length of time based on flow rate and density. Most flow rates are set at 45 people per 22-in. (55.8-cm) width per minute through level passageways and doorways. This method is more appropriate in such buildings as schools and theaters where people are assumed to be awake, alert, and of normal physical capability. The "Capacity Method" is more suited to hospitals and nursing homes. There is, of course, more to be considered in the design of safe exits than the flow of people. Good exits permit everyone to leave the fire area quickly, efficiently, and safely. However, in attaining this goal, every factor is of vital importance.

**Numbers of Means of Egress:** Sound exit design demands two separate means of egress which have no common elements as a fundamental safeguard for the proper and safe evacuation of the occupants of a building in a fire emergency. To be considered safe and acceptable, the two exits must be completely separate. Ultimate need to travel through a common space such as the first floor lobby of a building does not fulfill the requirements for two completely separate exits. The only exception to this rule is found when a building or a room is so small that a second exit would not affect the safety factor to any great degree. A large number of people have limited mobility as a result of physical disabilities. In buildings where such people live or work, special exit design features are both desirable and necessary.

## *The* Life Safety Code

The most important and widely used code regarding the safety of people in buildings when fire occurs is NFPA 101, *Code for Safety to Life from Fire in Buildings and Structures* (hereinafter referred to in this text as the *Life Safety Code* or, where clearly identifiable from context, the *Code*).[7] Work on what was to become the *Life Safety Code* began in 1913 when the NFPA Committee on Safety to Life devoted the first few years of its existence to analyzing the causes of loss of life through fire. The efforts of the committee led to the publication of standards for the construction of stairways, fire escapes, fire drills, and for the construction and arrangement of exit facilities for buildings such as factories and schools. The work of this committee formed the basis of the present *Life Safety Code*.

By the early 1940s, increased national attention to fire hazards and loss of life and property through fire resulted in a need and demand for increased legislation regarding firesafety features. Up until this point, the *Code* had been drafted and published to be used as a reference and advisory source. The *Code* was then revised to limit its contents to requirements suitable for legislation. Federal departments and agencies as well as many states and cities have adopted the *Life Safety Code* or referenced it in their own laws and regulations.

The *Life Safety Code* is now on a 3-year revision schedule. Its principal provisions (the present edition is about 250 pages in length) can be summarized in the following twelve requirements:

1. A sufficient number of unobstructed exits of adequate capacity and proper design, with convenient access thereto.
2. Protection of exits against fire and smoke during the length of time they are designed to be in use.
3. Alternate exit and means of travel thereto for use in case one exit is blocked by fire.
4. Subdivision of areas and fire-resistive construction to provide areas of refuge in those occupancies where evacuation is the last resort.
5. Protection of vertical openings to limit fire effects to a single floor.
6. Alarm systems to alert occupants and fire department in case of fire.

7. Adequate lighting of exits and paths of travel to reach them.

8. Signs indicating ways to reach exits where needed.

9. Safeguarding of equipment and of areas of unusual hazard which could produce a fire capable of endangering the safety of persons on the way out.

10. Exit drill procedures to assure orderly exit.

11. Control of psychological factors conducive to panic.

12. Control of interior finish and contents to prevent a fast-spreading fire that could trap occupants.

The *Life Safety Code* recognizes that there are many factors involved in the safety of persons in buildings when fire strikes. The *Code* recognizes that full reliance cannot be placed on any single safeguard. Any item may not function because of mechanical or human failures, so two or more safeguards, any one of which will provide reasonable life safety, must be provided. The *Code* also calls for the safeguarding of fire hazards and specifies various details of automatic sprinkler installation and other protective features. The *Code* differs from building codes in that it makes little distinction between different classes of building construction with the exception that construction types are treated when rapid evacuation of a building is not possible due either to the characteristics of the occupants or the size of the building.

The *Code* recognizes the hazard from smoke. Practically all buildings contain sufficient quantities of combustibles which, when burning, are capable of producing lethal quantities of smoke that can endanger the life and safety of the people in the building. The *Code* faces the problem of existing buildings. For example, it is often argued that because a building complied with all of the legal requirements when it was erected many years ago, its owner should not have to make changes today. A building's age is not acceptable as an excuse for subjecting its occupants to unnecessary peril from fire. Neither is the argument that the cost of making the building safe would be prohibitive. If the building is unsafe for the people in it, then its use should be changed or prohibited. Figure 2.2 illustrates the principles of exit safety.

## THE HAZARDS OF OCCUPANCIES

The vast majority of loss of life and property from fire occurs in buildings. Flaws in building design, improper or poor building maintenance, inadequate firesafety precautions, combustible and easily ignitable materials, human carelessness, and lack of life safety procedures once a fire breaks out are but some of the reasons for the injuries, deaths, and property losses that occur every year because of fire. Table 2.3 indicates the 1976 losses that resulted from fires in buildings.

When a fire occurs in a building, the first consideration is the safety of the lives of the people in the building. Major research has been, and is still being,

36    *Principles of Fire Protection*

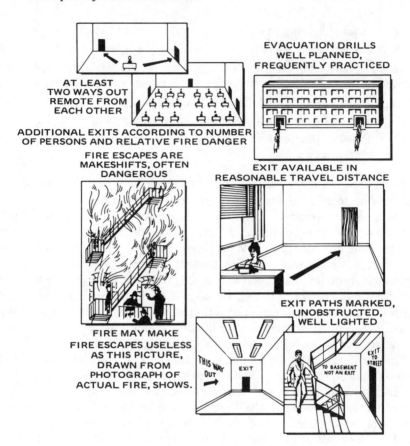

Fig. 2.2.  *Principles of exit safety.* (From NFPA *Fire Protection Handbook*)

Table 2.3   Estimated United States Fire and Property Loss for Selected Property Uses, 1976*

| Property Use | Number of Fires | Property Loss |
|---|---|---|
| Public assembly | 37,700 | $ 188,900,000 |
| Educational | 23,500 | 159,700,000 |
| Institutional | 24,100 | 25,300,000 |
| Residential | 665,400 | 1,433,000,000 |
| Stores and offices | 76,300 | 508,000,000 |
| Industry, utility, defense** | 50,000 | 289,400,000 |
| Storage** | 42,300 | 235,400,000 |

*Since this table does not present results for all property uses, the statistics presented should not be totaled to obtain estimates of overall US fire experience.

**Since some incidents for these property uses are handled only by private fire brigades or fixed suppression systems and are not reported to the NFPA, the results presented represent only a portion of US fire experience.

These estimates are based on data reported to the NFPA by the public fire service. No adjustments were made for unreported fires and losses. Dollar figures represent direct property loss only and were not adjusted for monetary inflation.

conducted into ways and means that can be effectively utilized to ensure rapid and safe evacuation from any and all types of buildings for every kind and type of person who may be in the building. Many lives have been lost in fires in buildings because designers, builders, and/or owners of properties failed to consider the life safety of occupants in the event of fire.

## Firesafe Building Design

High-rise building construction has resulted in some designs that contribute to the hazards of fire. For example, elevators can be potential death traps, sealed windows can cause heat to build up, building height can present problems during rescue operations, and the possibility of overcrowded exitways can cause panic and slowed-down evacuation procedures.

High-rise building design, however, is not the only hindrance to firesafety today. Fires in mobile homes can result in such homes becoming tombs for family members, and in many residential occupancies open stairwells guide fire and smoke upwards into sleeping areas. The creation of codes specifically prepared to cope with the problems imposed by modern building design and the frequent inspection of properties to ensure firesafety can help to protect building occupants against the ravages of fire.

## Types of Occupancies

In addition to the detailed general provisions for means of egress, the *Life Safety Code* presents specific requirements for various occupancies. Sequentially, the *Code* treats the following types of occupancies: places of assembly, educational occupancies (including child day-care centers), institutional occupancies (hospitals, nursing homes, prisons), residential occupancies (hotels, apartments, dormitories, rooming houses, one- and two-family dwellings), mercantile occupancies, business occupancies, industrial occupancies, and storage occupancies.*

The type of occupancy where safe exits are most needed but least provided is the one- and two-family dwelling. Many deaths from fire in these dwellings occur because the people are asleep when the fire starts. Three life safety measures are of prime importance to provide for fire emergencies occurring at night. First, a basic smoke detector should be installed to protect each separate sleeping area in a dwelling. Second, there should be a window in each sleeping room that can be used as a means of escape. If necessary, a collapsible ladder should be provided. Third, the occupants should have an escape plan worked out in advance for evacuating the dwelling in case of fire.

---

*Detailed information regarding the life safety requirements for each of these classes of occupancies is provided in Section 8 of the NFPA *Fire Protection Handbook,* 14th Ed., NFPA, Boston, 1976.

Dwellings are the only type of building for which the *Life Safety Code* permits reliance on windows as a means of escape. Other residential occupancies such as hotels, apartments, dormitories, and rooming houses require two safe means of egress that are completely separate from each other. For hotels the *Code* now requires that all doors between guest rooms and corridors be self closing. This requirement is the result of the fact that, often, when a fire occurs in a guest room, the occupant will rush out and leave the door open, thus endangering the other guests in nearby rooms. Hotels and similar occupancies must provide exit illumination, marking, and emergency lighting, protection of vertical openings to restrict spread of fire, alarm systems, and special protection of hazardous areas such as boiler, storage, and laundry rooms.

In buildings where rapid escape in case of fire is not possible (hospitals, nursing homes, prisons, and high-rise buildings), the occupants need to be protected by being moved from an unsafe to a safe area in the building, by insuring that attendants or employees are properly trained, by providing early alarm systems, and by installing automatic sprinkler systems for rapid extinguishment. The helpless patients in a hospital or nursing home most certainly require special attention and special provisions to protect their lives from fire. The *Life Safety Code* provisions for such people exceed, therefore, the protection requirements for other occupancies.

**Residential Occupancies:** A residential occupancy is defined as one in which sleeping accommodations are provided "for normal residential purposes." Health care facilities and penal institutions are not classified as residential occupancies because sleeping accommodations in them are not considered to be provided "for normal residential purposes." Usually, residential occupancies are divided into five classes: (1) hotels, (2) apartment buildings, (3) dormitories, (4) lodging or rooming houses, and (5) one- and two-family dwellings.

Is it important to determine the major aspects that each type of occupancy has in common in a given classification so that the degree and nature of the potential fire hazards in the various types of occupancies can be assessed. Residential occupancies are defined as those which provide sleeping accommodations for "normal residential purposes." The word "sleeping" is the key word in this definition for two primary reasons: (1) residential occupancies house virtually every individual in this country and, (2) because of the nature of these occupancies, much of the time spent in residential occupancies is spent sleeping. This key factor requires that consideration be given to firesafety and life safety provisions that protect occupants from fire hazards during the times when they might be sleeping and would not be aware of danger.

1. *One- and Two-family Dwellings.* The *Life Safety Code* can adequately define and detail firesafety and life safety requirements for residential occupancies, such as hotels, which are considered in the public sector and are therefore subject to codes and laws regarding public safety. Codes and laws in this area are relatively stringent. However, residential occupancies in the private sector, namely one- and two-family dwellings, are not subject to the

stringent regulations of residences in the public sector. Therefore, they present special problems and contain a higher degree of life safety and firesafety hazards than buildings in the public sector.

As Table 2.3 indicates, most residential occupancy fires occur in one- and two-family dwellings with a resultant higher rate of loss of life and property. The majority of these fires occur at night during the sleeping hours. Such fires tend to start in the living areas occupied by people during the waking hours (family playrooms, kitchens, and basements).

2. *Hotels.* Hotels are generally considered to include buildings or groups of buildings under the control of the same management which provide fifteen or more sleeping accommodations for hire. Motels, inns, clubs, and the like are all considered to be hotels. In addition to sleeping accommodations, many hotels contain ballrooms, exhibition halls, meeting rooms, restaurants, and various shops. These areas for special occupancy and "people actions" require special firesafety and life safety considerations.

3. *Apartment Buildings.* Apartment buildings are considered to be those buildings that contain three or more living units and that provide independent cooking and bathroom facilities. People inhabiting apartment buildings are often considered to be of a more transient nature than homeowners, as are the inhabitants of hotels, although the length of occupancy in apartment buildings is usually measured in months and years rather than in days. Because of the transient nature of the occupants, apartment buildings can be considered as hotels for fire protection purposes, but as a minimum requirement they should meet the requirements for one- and two-family dwellings.

4. *Dormitories.* Dormitories can be designed apartment-style or hotel-style, and are defined as buildings where group sleeping accommodations are provided in one room or in a series of closely associated rooms for persons who are not members of the same family group but who live together under joint occupancy or a single management. College dormitories, ski lodges, fraternities, sororities, and military barracks are considered dormitories.

The particular design of a dormitory occupancy will dictate whether it should meet the minimum life safety and firesafety requirements of a hotel or an apartment building. If a dormitory consists of suites of rooms with one or more bedrooms opening into a living room or study that in turn opens into a common corridor to other such suites of rooms, then that dormitory can be classed as an apartment.

5. *Lodging or Rooming Houses.* Lodging or rooming houses are considered to be buildings that contain sleeping accommodations for fifteen or fewer occupants on either a transient or permanent basis. This type of occupancy is distinguished from apartment buildings in that separate cooking facilities are usually not provided for each occupant. Lodging and rooming houses, because of their relatively low occupancy capacity, are more similar to one- and two-family dwellings than to dormitories or hotels. Therefore, minimum standards for one- and two-family dwellings are required for this type of occupancy. However, because the number of occupants is larger than normally found in

one- and two-family dwellings, additional requirements for lodging or rooming houses are contained in the *Life Safety Code*. These requirements, although higher than the requirements for one-and two-family dwellings, are far less stringent than those required for hotels, apartments, and dormitories.

6. *Mobile Homes*. Mobile homes, although not included in the "technical" classifications of types of residences, require mention at this point. Mobile homes are classified as a special form of residential occupancy. Because of the special construction features of mobile homes, NFPA 501B, *Standard for Mobile Homes*,[8] was developed, which contains special requirements for the construction of such residential occupancies. The *Standard for Mobile Homes* is basically a performance standard. This standard was so constructed to permit innovation in the designs of various mobile homes, although many very specific health and safety requirements are included in this standard. The uniqueness of the *Standard for Mobile Homes* lies in the fact that the building codes, plumbing codes, heating codes, and electrical codes that are normally considered separately in the construction of a residence are all incorporated into the one *Standard for Mobile Homes*.*

***Health Care Occupancies:*** Firesafety and life safety considerations must acknowledge the special human characteristics, listed earlier in this chapter, when fire protection assessment is made of health care facilities. The "people factor" is important in determining fire protection needs for these facilities.

The ability of occupants to escape during a fire emergency has led to classifying health care facilities into two categories: (1) buildings where occupants are "mobile" and can escape rapidly, and (2) buildings where occupants are physically restrained or so physically and/or mentally impaired that rapid escape is impossible and the occupants must be "defended in place."

Generally health care facilities house occupants who are not capable of self-preservation because of age and physical or mental disability. In such cases, the key to life safety and firesafety is the proper and adequate training of the personnel of the health care facility.

A fire hazard that has grown increasingly in this type of occupancy is the problem of combustible loading. Disposable items such as gloves, gowns, tubing, instruments, and drapes have increased in volume and availability at a remarkable rate. Storage area required for such items has not kept pace with the volume of items that are increasingly flooding the market. Also, when these items are disposed of, the result is volumes of highly combustible trash.

---

*The National Mobile Home Construction and Safety Standards Act (Title VI, Public Law 93-383, 42 U.S.C. 5401 *et seq.)* states that "no state or political subdivision of a state shall have any authority either to establish, or to continue in effect, with respect to any mobile home, any safety standard applicable to the same aspect of performance of such mobile home which is not identical to the Federal Standard." The federal standards are identified as the Mobile Home Construction and Safety Standards and are promulgated as Part 280 of the *Code of Federal Regulations*, Title 24 — Housing and Urban Development. The Mobile Home Procedural and Enforcement Regulations are contained in Part 3282 of Title 24 of the *Code of Federal Regulations*. While those standards are in substantial measure based on NFPA 501B, those concerned or required to comply with the federal standards are referred to them.

*Educational and Assembly Occupancies:* Schools and other places of public assembly, such as theaters, restaurants, churches, and exhibition halls, are so varied in character, size, construction, and design that a plan for life safety must be developed almost on an individual basis. Most schools where daily attendance is constant hold regular fire drills for the pupils. Such drills are specified in the *Life Safety Code.* The principal fire hazards endangering life in places of public assembly are: (1) overcrowding, (2) blocking, impairing, or locking exits, (3) storing combustibles in dangerous locations, (4) using open flame without proper precautions, and (5) using combustible decorations.

Life safety hazards in educational occupancies vary according to the physical properties and use for which a building is intended, and also according to the age of the students who occupy the building. The hazards of elementary schools are relatively low. Two major factors for keeping these hazards at their low level are: (1) regular fire drills, and (2) the placing of the younger children (kindergarten and first grade) on the floor of direct discharge, as provided for in the *Life Safety Code.* In addition, rooms set aside for special educational purposes generally do not contain the hazardous materials that are often found in junior high and high school.

A recent problem of increasing concern with regard to life safety provisions has, however, emerged. An increasing number of states have required that handicapped students be educated with those who are not handicapped. Special consideration must be given to evacuation problems in such situations.

Fire hazards and life safety and firesafety considerations increase as the grade level in educational occupancies increases. Shops, laboratories, home economics areas, and the like present special fire hazards, although these facilities are required to be separated from other academic areas.

Some newer school buildings are designed to provide flexible-plan and open-plan buildings. The flexible plan is designed so that walls may be rearranged to accommodate varying educational needs. However, such plans often may not comply with *Life Safety Code* standards in that the movement of walls may result in unprotected corridors. Although the life safety hazard is often increased by a flexible school design instead of the conventional school design, such design permits a view of the entire area so that a fire, if one erupts, may be detected more quickly.

Day-care facilities present special problems for fire protection consideration. The ages of the children involved are, of course, among the first considerations that come to mind. However, because such facilities are increasing at a rapid rate, legislation, firesafety, and life safety requirements constantly need to keep pace with the varying types of day-care occupancies. It must also be remembered that day-care facilities may also provide night-care facilities for dependents of people who work at night, thus increasing life safety hazards.

The *Life Safety Code* classifies day-care facilities according to the number of people cared for in the facility: a facility that provides care for twelve or more occupants is termed a day-care center; a facility that provides care for seven to twelve occupants is termed a group day-care home; and a facility providing care for six or fewer occupants is called a family day-care home. Based on the ages of the children involved and on the type of facility providing care, life safety and firesafety hazards can vary greatly. Following are some of the requirements established for day-care centers by the *Life Safety Code:*

1. Minimum ratios of staff to care for recipients.
2. Adequate fire evacuation plans.
3. Regular fire drills for staff and recipients (if applicable).
4. Frequent firesafety inspections by day-care personnel.
5. Installation and regular inspection of smoke detectors in day-care facilities with sleeping provisions.
6. Proper separation between day-care facilities and other occupancies located in the same building.

***Mercantile Occupancies:*** Mercantile occupancies are those occupancies that are involved with the display and sale of merchandise. Examples of mercantile occupancies are supermarkets, department stores, and shopping malls. The life safety and firesafety considerations involving shopping malls have already been presented in this chapter.

Some major considerations regarding mercantile occupancies that must be taken into account: (1) people who are shopping can be assumed to be ambulatory and/or mobile and, as such, can be considered to have the ability to evacuate a building quickly in the case of a fire emergency; (2) shoppers, being generally unfamiliar with every aspect of the store or building in which they are shopping, may have trouble locating the various means of egress in a store or building; and (3) smaller stores may have only one "main entrance," which could result in panic and disaster in a fire emergency.

***Business Occupancies:*** Business occupancies are considered to be those occupancies that are involved in conducting the financial, managerial, and technical work related to the management of a business operation. Business facilities can be contrasted with manufacturing, servicing, transporting, and warehousing facilities. Examples of business occupancies are: computer and data processing centers, editorial offices, and banks.

Business occupancies are generally considered to be low hazard types of occupancies. A major factor in the relatively low fire hazard risk in business occupancies is the usually high professional or material value of the contents in such a facility. Businesses often go to great lengths to protect the contents of business facilities from fire.

Risk hazards in such occupancies are generally associated with common ignition sources such as heat, light, and power. The presence of eating facilities, parking facilities, and retail facilities in a business occupancy can increase the hazard. Other considerations in a business facility are the number of files which may be stored or kept in an open or partially open filing system and — a relatively new hazard — the increased use of plastic furniture. These two hazards increase with the amount of material involved.

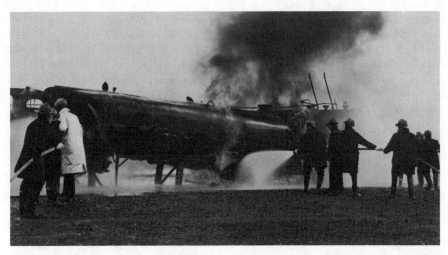

*Fig. 2.3. Members of industrial fire brigades often receive in-plant training by practicing fire suppression techniques under controlled conditions while being supervised by skilled instructors.*

**Industrial Occupancies:** Fire records indicate that the majority of industrial fires with fire deaths are the result of either flash fires or explosions. Generally, although property loss is high in fires in industrial occupancies, loss of life is relatively low. The relatively low loss of life through fires in industrial occupancies can be attributed to many positive actions for firesafety and life safety in industrial occupancies. Good exit design, daily attention to industrial safety, the wide use of automatic sprinkler systems, the training of fire brigades, and regular fire drills all contribute to successful efforts in preventing loss of life from fire.

## Industrial Fire Brigades

Many industrial facilities find it necessary to establish their own private or industrial fire brigades for a variety of reasons. (See Fig. 2.3.) Often industrial facilities are beyond the "jurisdiction" of the local fire department or are involved in producing materials which, if involved in fire, require immediate or specialized attention to contain. In instances such as these, industrial fire brigades perform a vital function until the fire department arrives. The life safety factor is heightened considerably because of industrial fire brigades. In many instances, because of the specialized materials involved in many industrial facilities, members of the industrial fire brigade continue to work in conjunction with the fire department once it arrives at the fire scene.

**The Industrial Fire Risk Manager:** The fire risk manager in an industrial facility (usually the plant manager or the shift superintendent) has the primary responsibility for correlating activities with the public fire department so that agreement is reached on the mutual responsibilities in a fire emergency. Once the public fire department arrives, it is the responsibility of the fire risk manager to make sure the fire chief understands the situation and is made aware of steps already taken. If more than one public fire department is involved, the fire risk manager should communicate with the chief of the fire department having jurisdiction over the property. That chief will then have the responsibility of communicating with the other fire departments.

*Industrial Fire Brigade Organization:* The fire brigade is made up of selected building, operating, maintenance, and security personnel. The fire brigade responds to all fire alarms, reports to the scene of the alarm, establishes communication with the appropriate personnel, takes appropriate action consistent with existing conditions to evacuate the area, and fights the fire with available equipment. Generally, upon the arrival of the fire department, the fire brigade will let the fire department take charge; members of the fire brigade then stand ready to assist as requested.[9]

The organization of an industrial fire brigade varies to suit the needs of the industrial facility it serves. Generally, the fire risk manager determines the nature of the brigade and its organizational structure. The fire brigade may consist of selected personnel or groups of persons functioning as teams.

The availability and proximity of the appropriate public fire department also can affect decisions on the nature and organization of an industrial fire brigade. No matter what circumstances dictate the nature and organization of an industrial fire brigade, provisions should be made in advance so that a fire brigade is on duty on each working shift and during the periods when the plant is shut down.

Operating units or teams might consist of one or two members of the brigade assigned to operate specific equipment, or a larger group to perform complicated operations — if such are considered necessary — for adequate fire protection in a particular industrial facility. A leader should be assigned to each unit or team, and a chief should be appointed as leader of the brigade.

The functions and duties of the fire risk manager and the brigade chief are different. Basically, a fire risk manager has management responsibilities such as determining the size and structure of the fire brigades, ensuring that the brigades are adequately staffed and trained, providing equipment and supplies, and selecting the fire brigade chief.

A fire brigade chief's responsibilities are administrative and supervisory in nature. These responsibilities usually include periodic evaluation and maintenance of equipment, periodic review of the fire brigade personnel, and preparing training and fire attack plans.

Members of fire brigades should meet minimum physical requirements, should be able to be available to answer alarms and attend meetings, should complete a required course of instruction, and should be prepared to attend monthly training sessions.

## Traffic and Exit Drills

When an industrial fire emergency occurs, both ingress and egress traffic problems arise. Fire fighting personnel must be able to get to the location of the fire. At the same time those people not involved in fighting the fire must be evacuated. Traffic control is divided into two categories: (1) external traffic control on public streets and highways that both surround and lead to the industrial structure, and (2) internal traffic control inside the property involved in the fire situation.

***External Traffic Control:*** Preplanning for external traffic control in the event of an industrial fire involves working with the law enforcement agency in which the building or buildings are located. If law enforcement staffing cannot adequately handle external traffic control, a facility may wish to plan for proper external traffic control by providing its own security guards. This type of traffic control plan should be designed in conjunction with the appropriate law enforcement agency.

It is wise to provide personnel involved in the emergency fire control with clear identification that has been approved by the proper law enforcement agency and the Civil Defense Agency so that the fire fighting personnel can quickly get to the scene of the emergency. In addition, alternate routes should be planned so that, in the event of an emergency, fire fighting equipment and emergency vehicles can quickly get to the scene.

***Internal Traffic Control:*** Internal traffic control responsibilities for both pedestrians and vehicles are generally assigned to the private security personnel. The major responsibilities for such personnel are to clear the involved area of unnecessary personnel and to direct the fire department and emergency forces to the emergency scene.

Federal law requires that industrial facilities properly identify emergency exits and access routes to them. Exit drills are essential in preplanning for a possible fire emergency in an industrial facility.

The fire risk manager or plant manager generally has the responsibility for evacuation drills. A recommended procedure for such drills is:

1. Employees should become familiar with the evacuation signal.
2. Employees should immediately turn off all equipment.
3. Employees should immediately proceed along the predetermined exit route. (They should be trained to use alternate routes as well, if alternate exit routes are assigned.)
4. Employees should report to a predetermined assembly point so that it can be determined if all personnel are present.

In addition, daily inspection of emergency routes and exits should be conducted. The frequency of such evacuation drills should be dependent upon both the degree of potential hazard in the facility and the complexity of the facility layout.

## PUBLIC EDUCATION AND COMMUNITY RELATIONS

Public firesafety education is one of the major fire prevention methods that can be used to reduce life loss from fires. Deuel Richardson, in an article titled "The Public and Fire Protection," remarks:[10]

> ... a significant factor contributing to the cause or spread of fire is human failure — failure to recognize hazards and take adequate preventive measures, failure to act intelligently at the outbreak of the fire, failure to take action which would limit damage.

Loss data from the NFPA in 1972 showed that of 826,000 fires, 72.8 percent were attributed to human action.[11] Deaths that occurred from these fires were also attributed to human action either as a result of starting the fire, or as a result of trying to escape the fire. By educating the public in good firesafety methods, human actions that cause fires can be minimized, and safe methods to escape from a fire can be learned.

### Fire Prevention Programs

Various projects in fire prevention education have demonstrated time and time again that increased public awareness of the role the public can play in reducing the hazards of fire can be a significant factor in reducing loss of life and property through fire. The savings in lives, injuries, and property losses from a firesafety education program will exceed the cost of the program.

The report of the National Commission on Fire Prevention and Control acknowledges the importance of fire prevention education as follows:[12]

> Among the many measures that can be taken to reduce fire losses, perhaps none is more important than educating people about fire. Americans must be made aware of the magnitude of fire's toll and its threat to them personally. They must know how to minimize the risk of fire in their daily surroundings. They must know how to cope with fire, quickly and effectively, once it has started. Public education about fire has been cited by many Commission witnesses and others as the single activity with the greatest potential for reducing losses. . . .
> 
> The prevention of fires due to human carelessness is not all that fire safety education can hope to accomplish. Many fires caused by faulty equipment rather than carelessness could be prevented if people were trained to recognize hazards. And many injuries and deaths could be prevented if people knew how to react to a fire, whatever its cause.

Fire prevention programs, in addition to helping to reduce loss through fire, can result in another important benefit to members of the fire service. The increased contact between the public and members of the fire service can result in a heightened awareness of the role of the fire service in public protection

Fig. 2.4. Part of the public education program of the Hayward Fire Department, Hayward, CA, involves in-home counseling of a juvenile firesetter and his family.

and of the fact that the fire service is not an isolated area of employment but actually works in partnership with the public for the benefit and safety of the public itself. In addition:

> ... A good fire prevention program can bring sympathy for large fire department budgets and create an environment in which inspectors are seen as public servants who perform a most useful service, rather than as intruders. This, in turn, makes fire prevention assignments much more rewarding for fire fighters.

## Methods of Public Education

One of the major goals of public education programs is to increase public awareness, support, and involvement in fire prevention. Local fire departments make significant contributions to public education through periodic inspections of dwellings and commercial establishments, through distribution of materials on firesafety, through programs such as the "Junior Fire Marshal Program," and through educational programs conducted in schools.

Periodic inspections of homes and buildings increase the public's awareness of fire hazards and provide good opportunities for the fire inspector to suggest to occupants plans that can be implemented in case of fire. A good publicity campaign by a fire department can also result in home and building owners calling the fire prevention bureau and requesting that their homes or buildings be inspected.

Publicity campaigns (such as the NFPA's "Learn Not to Burn" campaign) conducted through the media, help make the public aware of fire hazards, good fire prevention techniques, and the services provided by the fire prevention bureau. To be effective, publicity programs should disseminate information at a steady, continuous rate. This information should be based on timeliness. For example, special events or seasons of the year present good opportunities for public education regarding specific fire hazards. Articles in newspapers and television news reports can emphasize whether good fire prevention methods were used. A program, such as an "open house" at the local fire department facility, bringing fire fighting equipment to schools, and safety committee programs in which members of the community are involved are just a few of the programs that can result in effective public education in fire prevention.

## Fire Department Activities in Community Relations

A good relationship between the community and the fire department is an important aspect of public relations for fire departments. Fire departments are supported by public funds and, as such, an understanding by the public of the services it receives from the fire department is vital in continued public financial support for good fire protection in the community. Good community relations also result in a high level of understanding and cooperation on the part of the members of the community in helping to achieve effective fire prevention programs and support.

Press releases and items submitted to the news media help to keep the public informed concerning newsworthy developments and events and about all types of departmental activities and programs. Fire departments located in areas where such items are of local interest only can find outlets for this information in the local or regional newspapers.

Usually public information programs are the responsibility of the head of the fire department. Some of the larger fire departments assign public information officers to be responsible for providing positive and continuing community relations. Whatever the organizational structure may be, establishing and maintaining good community relations are a vital part of fire department activities and, as such, should be continual and effective.

## Summary

Loss of life and property through fire is dependent on many factors. The human characteristics of the people involved in terms of age, agility, decision-making abilities, previous firesafety training, and so forth are a vital life safety factor in assessing fire protection planning and techniques. Life safety considerations are also dependent on hazards that may be involved in buildings and areas designed for specific activities, such as stores, restaurants, dwellings, and theaters.

The design of the building and the type of occupancy are also important factors to be considered. The requirements of the *Life Safety Code* provide specific recommendations to enhance the probability of life safety in all types of occupancies. Most, if not all, of these recommendations have been enacted into law at both the federal and state levels.

Public education in firesafety is a major factor in fire prevention. Fire prevention programs and various methods of public education in fire prevention have proven to be effective deterrents to fire and fire loss. Good community relations between the fire service and the public is an important factor in public support for fire prevention and for the fire service itself.

## Activities

1. There are seven commonly recognized human characteristics that must be considered when determining life safety factors in case of fire. Discuss these factors with your classmates and describe how each of these factors affects life safety considerations.
2. Are there any other human characteristics you would add to the list? Discuss these possible additions with your classmates and decide with them whether your recommendations would be valid and all-inclusive enough to be added to the list.
3. It is a well-known fact that most people react differently in unfamiliar environments. How does this fact affect life safety considerations in buildings or areas designed for specific activities?
4. How does building design affect life safety considerations in the event of fire?
5. Compare fire protection considerations with regard to one- and two-story residences and high-rise buildings.
6. How do evacuation plans differ between one- and two-story residences and high-rise apartments? List recommended evacuation and life safety procedures for both types of occupancies.
7. The *Life Safety Code* classifies occupancies into six types. Describe: (1) the six types of occupancies, (2) what makes them different, and (3) the firesafety and life safety requirements for each.
8. Why are industrial fire brigades necessary? Discuss possible organizational structures of fire brigades and the responsibilities of the members of the fire brigades.
9. Design a hypothetical industrial facility. Develop a fire emergency plan for this facility, and include in your plan the following: (1) fire brigade responsibilities, (2) internal and external traffic control, (3) evacuation procedures, and (4) preplanning procedures with the appropriate fire service, law, and emergency agencies.
10. With your classmates, develop a fire prevention education program that will extend for a period of one year in a real or imaginary community.

## Bibliography

[1] *Fire Protection Handbook,* 14th Ed., NFPA, Boston, 1976, p. 1-2.
[2] _____, p. 8-2.

[3]Wood, P. G., "The Behavior of People in Fires," Fire Research Note No. 953, Nov. 1972, Department of the Environment, Building Research Establishment, Fire Research Station, Borehamwood, Herts, England.

[4]*Fire Protection Handbook,* 14th Ed., NFPA, Boston, 1976, p. 8-4.

[5]*Firesafety in the Home,* NFPA, Boston, 1976, p. 22.

[6]London Transport Board, "Second Report of the Operational Research Team on the Capacity of Footways," Research Report No. 95, Aug. 1958, London.

[7]NFPA 101, *Code for Safety to Life from Fire in Buildings and Structures,* NFPA, Boston, 1976.

[8]NFPA 501B, *Standard for Mobile Homes,* NFPA, Boston, 1977.

[9]*Operation Skyline,* NFPA, Boston, 1975, p. 16.

[10]Richardson, Deuel, "The Public and Fire Protection," *NFPA Quarterly,* Vol. 56, No. 1, July 1962, pp. 41-42.

[11]Loss data from "Fires and Fire Losses Classified," *Fire Journal,* Vol. 66, No. 5, Sept. 1972, pp. 65-69.

[12]*"America Burning,"* May 1973, The National Commission on Fire Prevention and Control, Washington, DC, p. 105.

[13]Didactic Systems, Inc., *Management in the Fire Service,* NFPA, Boston, 1977, p. 216.

*Chapter Three*

# Characteristics and Behavior of Fire

*It is virtually impossible to predict exactly when a fire will occur and, upon its inception, the extent of its destructive potential. However, through scientific knowledge of ignition, the combustibility of solids, liquids, and gases, and the products of combustion, effective ways to control the dangers of fire and explosion can be determined.*

## THE UNPREDICTABILITY OF FIRE

It is almost foolhardy to attempt to completely describe in one brief chapter the chemical and physical reactions that take place during a fire. Such an attempt is primarily hindered by the fact that the variables involving the essentials for a fire — something to burn, a source of ignition, and the oxygen necessary to maintain combustion — are almost infinite. Although our technical knowledge about flame, heat, and smoke continues to grow, and although additional information continues to be acquired concerning the ignition, flammability, and flame propagation of various combustible solids, liquids, and gases, it is still not possible to predict with any degree of accuracy the probability of fire initiation or the consequences of such initiation. Thus, while the study of controlled fires in laboratory situations provides much useful information, most unwanted fires happen and develop under widely varying conditions, making it virtually impossible to compile complete bodies of information from actual unwanted fire situations. This fact is further complicated because the progress of any unwanted fire varies from the time of discovery to the time when control measures are applied.

In addition, the combustion process, which is a very complicated chemical reaction in the first place, is affected by many other variables. For example, the size and shape of a solid will influence the way combustion reacts. A solid block of wood is hard to burn. However, if the wood is in the form of shavings it will burn readily; if the wood is finely divided into dust, it will explode violently when a source of ignition is present.

## Principles of Fire

Dr. Richard L. Tuve, in *Principles of Fire Protection Chemistry*, provides the following simple definition of fire:[1]

> Fire is a rapid, self-sustaining oxidation process accompanied by the evolution of heat and light of varying intensities.

Although this definition is complete and almost poetic in its simplicity, some of the basic principles underlying this definition should be considered at this point. As previously mentioned in this chapter, three elements are essential for the initiation of a fire: something to burn (fuel); a source of ignition (heat); and oxygen to maintain combustion. Things that burn will be presented in Chapter 4, "Fire Hazards of Materials." Sources of ignition will be described later in this chapter. The most common sources of ignition are the heat produced by a chemical reaction (such as the striking of a match), by electrical energy, or by mechanical heat energy (such as friction).

A fire is maintained by a supply of oxygen, which undergoes the oxidation process. This process is often thought of solely in connection with fire. However, oxidation is all around us and does not always result in the chemical reaction that produces fire. For example, our bodies are continually undergoing a process of oxidation. Very simply stated, our bodies take in oxygen that "burns up" our food thus releasing energy needed to sustain our bodies. This simplified example serves to demonstrate the fact that oxygen, when it reacts on a substance, results in changing that substance and, in changing the substance, releases resultant energy.

Oxidation in the body cannot, of course, be called a fire. The difference between oxidation in a body and oxidation in a fire is the speed at which oxida-

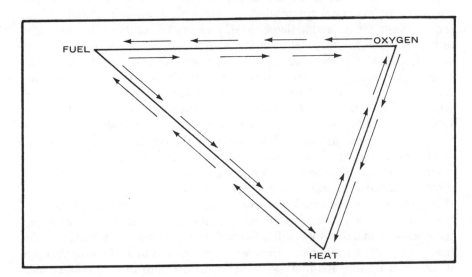

*Fig. 3.1. The fire triangle (circa 1920).*

*Characteristics and Behavior of Fire* 53

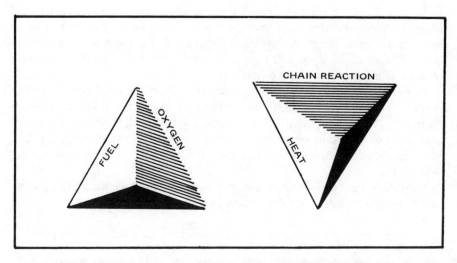

*Fig. 3.2. The fire tetrahedron, which expanded upon the fire triangle by the addition of the fourth element of fire.*

tion occurs. In a fire oxidation is very rapid, resulting in the emission of intense heat and light. Looking again at Dr. Tuve's definition of a fire, the words "rapid" and "oxidation" can be seen to be closely related components of a fire. Oxidation reaction time is an important element of fire.

The most common combustible solids and flammable liquids and gases contain large percentages of carbon and hydrogen and can be readily oxidized. The most common oxidizing material is the oxygen in the air.

Fire was originally thought to require the three basic components already mentioned: something to burn (fuel), source of ignition (heat), and oxygen. To extinguish a fire, only one of these components needed to be removed. This concept, termed the fire triangle, is illustrated in Figure 3.1.

Further research into the nature and principles of fire indicated that a fourth element was required: the *reaction time of oxidation* is a vital and integral factor in a fire. The addition of this important component resulted in changing the concept of the fire triangle to the concept of a fire tetrahedron. The fire tetrahedron is represented in Figure 3.2.

As the heat of a fire increases, another chemical phenomenon takes place and decomposes the matter involved. This action is called "pyrolysis."

Using wood as an example, when the wood decomposes upon being heated, some gases (including water vapor) evolve. The combustible components of these gases increase during the early stages of pyrolysis. The wood surface is attacked and then, as charring occurs, the reaction moves deeper into the wood. The evolution of gas continues and, if a source of ignition is available, these gases will ignite when the lower flammability limit is reached.

The flammability limit of a gas or vapor is the range in which the gas will burn or explode in air. Below a certain point the mixture of gas and air will not

54   Principles of Fire Protection

burn, and above a certain point the mixture will not burn. Gasoline is a useful example. When the percent of gasoline vapor in air is below 1.4, the mixture is too lean to burn. When the percent is above 7.6, the mixture is too rich to burn. The gasoline vapor will burn or explode, therefore, when the mixture with air is between 1.4 and 7.6 percent.

At the temperature at which ignition occurs, the chemical process that is going on becomes exothermic (heat is given off) and the reactions become self-sustaining. From here the pyrolysis reaction will vary depending on the concentration of the oxidizing agent that is present. For almost all materials there is a minimum oxidizing agent concentration below which combustion will not take place.

The following is a summary statement concerning the science of fire protection from the NFPA's *Fire Protection Handbook:*[2]

> In summary, the science of fire protection rests upon the following principles:
> 1. An oxidizing agent, a combustible material, and an ignition source are essential for combustion.
> 2. The combustible material must be heated to its ignition temperature before it will burn.
> 3. Combustion will continue until:
>    (a) The combustible material is consumed or removed.
>    (b) The oxidizing agent concentration is lowered to below the concentration necessary to support combustion.
>    (c) The combustible material is cooled to below its ignition temperature.
>    (d) Flames are chemically inhibited.

All of the material presented in the *Handbook* for the prevention, control, or extinguishment of fire is based on these principles.

## Heat Measurement

All combustion reactions are exothermic: that is, they give off heat. One of the major determinations a fire fighter must make in a fire situation is the intensity of the heat being emitted from the fire. The estimate of heat intensity is vital in determining the potential danger of nearby combustibles reaching their ignition point from the heat that is being emitted by the materials already involved in the fire.

In *Principles of Fire Protection Chemistry,* Dr. Tuve describes the importance of determining the intensity of heat as follows:[3]

> When a fire fighter senses the temperature of a fire situation by sight, touch, or by the amount and type of smoke or gas odor smelled, such information must be translated into the "extent of heat" that has been evolved by the fire. High temperatures mean that ordinary combustibles not already burning are

rapidly reaching their ignition points; in order to slow up the evolution of heat, the prompt preventative efforts of cooling are necessary. Ordinary carbon-hydrogen substances produce temperatures during burning of 1,100°F to 1,800°F (593°C to 982°C), whereas their ignition points may be only 350°F to 1,000°F (177°C to 538°C).

"Heat" and "temperature" are not the same terms. Heat is a form of energy while temperature is a measure that indicates the extent to which heating has taken place in a material. Simply put, temperature is measurement of the result and amount of the heating of a given material. Following are brief descriptions of temperature and heat units as defined by the NFPA *Fire Protection Handbook:*[4]

TEMPERATURE UNITS

*Celsius Degree (also called Centigrade):* A Celsius (or Centigrade) degree (°C) is 1/100 the difference between the temperature of melting ice and boiling water at 1 atmosphere pressure. On the Celsius scale, zero is the melting point of ice; 100 is the boiling point of water.

*Fahrenheit Degree:* A Fahrenheit degree (°F) is 1/180 the difference between the temperature of melting ice and boiling water at 1 atmosphere pressure. On the Fahrenheit scale, 32 is the melting point of ice; 212 is the boiling point of water.

*Kelvin Degree:* A Kelvin degree (°K) is the same size as the Celsius degree. On the Kelvin scale (sometimes called Celsius Absolute), zero is minus 273.15°C.

*Rankine Degree:* A Rankine degree (°R) is the same size as the Fahrenheit degree. On the Rankine scale (sometimes called Fahrenheit Absolute), zero is minus 459.67°F.

HEAT UNITS

*British thermal unit (Btu):* The amount of heat required to raise the temperature of one pound of water one degree Fahrenheit (measured at 60°F) is called the British thermal unit.

*Calorie:* The amount of heat required to raise the temperature of one gram of water one degree Celsius (measured at 15°C) is called a calorie. One Btu = 252 calories.

As heat can be measured in any unit, it is sometimes convenient to convert Btu to joules, one Btu equaling 1,055 joules; or to horsepower, one Btu per minute equaling 0.0236 hp.

## *Heat Transfer*

When heat causes a material to approach its ignition point, preventive cooling measures are necessary. The heat must be transferred to another material to prevent its ignition. For example, when water is poured on a material that is reaching its ignition point, the heat that is endangering the combustible material is transferred to the water, thus lowering the temperature of the endangered material. The transfer of heat is an important and, in many cases, the determining factor in the ignition and extinguishment of most fires. Heat can be transferred by one or more of the following three methods: (1) conduction, (2) radiation, or (3) convection.

***Conduction:*** Heat energy flows unimpeded in all directions from the heat generation point. The flow of heat cannot be stopped by a barrier or an obstruction in the way that the flow of water can be halted. For example, wood, when in direct contact with a steam pipe, will eventually char and burn. The ignition time depends on the degree and constancy of the heat in the pipe, and may require as long as weeks or months before burning occurs.

Substances vary in their ability to conduct heat. Gases such as air conduct heat very slowly. As a result, an air space between a hot object and a nearby combustible object will greatly reduce the chances of ignition.

Some solids conduct heat rapidly while heat conduction in others is very slow. However, solids cannot entirely stop the flow of heat. Heat insulation usually depends on using a material that does not conduct heat readily in combination with air space or some additional way of transferring heat.

***Radiation:*** Heat energy can travel through space with the result that combustible matter not in direct contact with a hot or burning material can ignite by radiation. Heat energy travels in waves or rays. The quality and quantity of heat radiation depends on the temperature of the radiating body and the size of the radiating surface. When two bodies face each other and one body is hotter than the other, radiant energy will flow from the hotter body to the cooler body until both attain the same temperature. The ability to absorb the radiated heat depends on the kind of surface of the cooler body and the area of the radiating surface of the hotter body. If the receiving surface is black or dark colored, it will absorb heat readily. If the surface is light in color or shiny and polished, it will reflect much of the heat away. This property of reflection is utilized in the design of the clothing worn by crews on crash trucks fighting aircraft fires. The shiny metal-coated fabrics reflect the heat of the fire away from the fire fighter.

Radiant heat travels in a straight line, which is why the extent of the radiating surface of the hot body is important. Heat radiation will pass through air, glass, water, and transparent plastics with only a small absorption of energy that only slightly heats the transparent substance. Heat rays can be reflected and can be concentrated in much the same way that light rays are reflected and concentrated.

When large amounts of heat radiation are produced in large fires, the fire will spread rapidly to nearby combustible materials unless water in the form of hose streams conducts the heat away. The development of fog nozzles for fire fighting is most useful in blocking heat radiation. The water fog reflects the heat rays and breaks up the straight line path of the heat radiation.

*Convection:* Heat energy can also be transferred by circulation of a gas such as air or a liquid. Hot gases, vapors, and liquids rise and progressively increase the temperature of the area and the material in the area. Most homes are heated by a furnace or heater supplying warm air near the floor level. This air circulates upward by convection and mixes with the cooler air that travels downward. In a large fire the hot burning gases from the fire flow upward and will transfer heat energy up through the building.

## Sources of Ignition

As previously mentioned, a source of ignition (heat) is needed to maintain a fire. Heat energy may be produced from many sources and circumstances, as described by Dr. Tuve:[5]

> Unwanted or accidental fires are caused by many circumstances, any one of which might be singled out as the source from which runaway combustion begins. It is lengthy and difficult to catalog all such possible situations; such cataloging can usually be found in studies of the science and art of arson investigation. However, it is important that certain basic categories of energy sources should be recognized as primary origins by which heat is developed — heat which, if it is not controlled in some way, can result in fires of a disastrous nature.

There are four major sources of heat energy: (1) chemical, (2) electrical, (3) mechanical, and (4) nuclear.

*Chemical:* Fire is, essentially, a chemical reaction. One of the basic components of a fire is the chemical reaction caused by oxygen.

The chemical process known as oxidation (reaction of a solid, liquid, or gas with the oxygen in the air) usually produces heat. The oxidation process will develop heat whether the oxidation is incomplete or total.

Air is the primary source of oxygen. Oxidation is always limited by the air supply, which affects the amount of heat produced in a fire or in spontaneous heating. Spontaneous heating occurs when the temperature of a material increases without drawing heat from its surroundings. If exposed to the air, practically all organic substances capable of combination with oxygen will oxidize at some critical temperature with a resultant evolution of heat. The rate of oxidation at normal temperatures is usually so slow that the released heat is transferred to surroundings as rapidly as formed with the result that there is no temperature increase in the combustible material being oxidized. When the heat developed cannot be rapidly released, spontaneous combustion may occur. Common examples of spontaneous heating are oily rags in a confined space and wet hay in a barn loft.

*Electrical:* Another common source of ignition is the heat produced by electrical energy. The energy required to move electric current through a substance appears in the form of heat. When electric current flows through a wire or another conductor, some resistance is set up. If the conductor is a good conductor, such as copper or silver, the resistance is low with the result that not much heat will be produced. The electrical resistance of any substance depends on the characteristics of the atoms and molecules that make up the substance.

Dr. Tuve describes the effect of electrical energy on everyday lives, and the relationship between electrical energy and fire in the following manner:[6]

> Because it is used in many ways, electrical energy is a common cause of unwanted fires. In home or industrial oil burners, electrical sparks are used to start and to sustain the ignition of the heating oil. In the kitchen electrical resistance heating elements or other forms of electric heating energy are needed to supply cooking temperatures. In certain living or working areas, electric energy is used to provide comfortable temperatures by means of resistance heaters. In our various appliances electricity is used in many ways, each of which is a potential fire hazard to nearby combustible materials. Unless electrical energy is used in an efficient manner, it will almost always produce heat as an unwanted by-product. Another factor that must continually be remembered is that the fire hazards of electrical energy are almost always concealed. In efforts to beautify and hide electrical wires and equipment, they are often installed in walls and partitions or enclosures so that malfunctions or dangerous conditions that can initiate a fire are difficult to detect. In addition, rather than installing additional wall outlets when they are needed, the tendency is to overload existing ones.

There are five basic forms of electrical heating energy that can result in potential fire hazards. These five forms are: (1) resistance, (2) arcing, (3) sparking, (4) static, and (5) lightning.

1. *Resistance.* Overloading electrical conductors can result in resistance heating. Electrical wiring will heat up if it carries an overload. Portable electric room heaters produce substantial heat and will ignite combustible materials which are placed too close to the heater. Overloading electrical circuits by plugging in too many lights and appliances is a common cause of fires. Most electrical circuit wires are concealed in walls and partitions, which increases the chances of undetected ignitions. Bare wires can carry more current than insulated wires, and the heat generated by bare wires is dissipated more readily. However, most wires are insulated, and if a breakdown occurs, the insulation will contain the heat but the insulation may then burn.

2. *Arcing.* Ignition can also be affected by arcing. When a good electrical connection is not made in a switch or fuse block or in some appliance, the electric energy will jump or arc across the gap. The spark may be hot enough to ignite insulation or nearby combustible material.

3. *Sparking.* Sparking may occur only once. This is in contrast to arcing, which may be continual. Sparking may also result when a voltage discharge is too high for a low energy output.

Heating through sparking does not result in a large or sustained heating effect. Danger can result, however, when sparking occurs in a flammable atmosphere.

4. *Static Sparking.* Static sparks can occur when an electrical charge accumulates on the surfaces of two materials that have been brought together and then separated. One surface becomes charged positively, the other surface becomes charged negatively. If the materials are not bonded or grounded, they will eventually accumulate sufficient electrical charge so that a static spark may occur.

Static sparks do not produce enough heat to ignite ordinary combustibles such as paper but they may ignite flammable vapors, gases, and dusts. Gasoline or oil flowing through a pipe can generate enough static to ignite the flammable vapor.

5. *Lightning.* A powerful, but infrequent source of electrical energy is lightning. A properly installed metallic rod that will conduct the lightning into the ground is the safeguard for preventing fires caused by lightning.

**Mechanical:** Mechanical heat energy is a frequent cause of fire. Mechanical heat is developed when two solids are rubbed together. The friction transforms the energy into heat; unless this heat is rapidly dispersed, fire can result. The procedure for starting a fire by rubbing sticks together is an example of ignition by mechanical heating. Friction is the cause of many fires. For example, in a cotton mill the friction heat of a slipping belt against a pulley or the friction sparks generated when a piece of metal gets in the cotton shredder will readily ignite the cotton particles. Friction sparks cause many industrial fires. Sparks can even be generated from the impact of shoe nails on a concrete floor or from the contact of a steel tool on a concrete floor. The size of the particle involved has a pronounced effect on spark ignition. Small particles of metal can develop sparks that will ignite dust or explosive materials. Larger pieces of metal will usually not develop enough friction heat to reach dangerous temperatures.

Mechanical heat energy can develop when a gas is compressed. For example, in diesel engines the heat of compression of the gas eliminates the need for a spark ignition system. Air is first compressed in the cylinder of the diesel engine, after which an oil spray is injected into the compressed air. The heat that is released when the air is compressed in the cylinder is sufficient to ignite the oil spray.

**Nuclear:** Nuclear energy is heat energy released from the nucleus of an atom through a process known as nuclear fission. Nuclear material such as uranium or plutonium is composed of atoms held together by tremendous forces that can be released when the nucleus is bombarded by energized particles. This energy is released in the form of heat, pressure, and nuclear radiation. Nuclear energy may be a million times greater than the energy released by an ordinary chemical reaction. Controlled nuclear energy is now widely used to generate electric power.

## EXPLOSIONS

The basic difference between an explosion and a fire is the rate at which energy is released. In its broadest sense, an explosion can be said to be the result of the sudden and violent expansion of gases. These gases may already exist, or may be formed at the time of the explosion. The rapid expansion of energy in an explosion may also be accompanied by shock waves and/or the disruption of enclosing materials or structures. Explosions often occur as the result of poor operating and maintenance procedures, or as the result of poor or faulty design. Explosions may be chemical, mechanical, atomic, or thermal.

### Chemical Explosions

Most chemical explosions are the result of chemical changes such as the detonation of an explosive or the combustion of a flammable gas-air mixture. Similar to combustible gas concentrations in the air is the danger of the combination of certain dust particles with the air which, under varying circumstances, can cause what is known as a dust explosion.

When some combustible solids are ground or rubbed into minute particles, the particles tend to mix with the air in much the same way that vapor or gas mix with the air. The finer the dust particle, the more completely it will mix with the air and remain suspended in the air. Although dust particles from all combustible solids do not result in potentially explosive dust particles, a large number of combustible solids can yield explosive dust particles. Table 3.1 lists some common substances that can have potential explosive properties.

Table 3.1* Common Combustible Solid Dusts Generating Severe Explosions**

| Type of Dust | Maximum Explosion Pressure (psig) (bar) | Maximum Rate of Pressure Rise (psig/sec) (bar/sec) |
|---|---|---|
| Corn (processing) | 95 (6.55) | 6,000 (413.7) |
| Cornstarch | 115 (7.93) | 9,000 (620.5) |
| Potato starch | 97 (6.89) | 8,000 (551.6) |
| Sugar (processing) | 91 (6.27) | 5,000 (344.7) |
| Wheat starch | 105 (7.24) | 8,500 (586.0) |
| Ethyl cellulose plastic molding compound | 102 (7.03) | 6,000 (413.7) |
| Wood flour filler | 110 (7.58) | 5,500 (379.2) |
| Natural resin | 87 (6.0) | 10,000 (689.5) |
| Aluminum (powder) | 100 (6.9) | 10,000 (689.5) |
| Magnesium (powder) | 94 (6.48) | 10,000 (689.5) |
| Silicon (powder) | 106 (7.31) | 10,000 (689.5) |
| Titanium (powder) | 80 (5.52) | 10,000 (689.5) |
| Aluminum magnesium alloy (powder) | 90 (6.20) | 10,000 (689.5) |

*From *Principles of Fire Protection Chemistry*, by Richard L. Tuve.
**Extracted from Bureau of Mines Investigations and Reports, No. 5753, RI 5971, RI 6516.

The following excerpt contains a further warning concerning explosive dust particles, from Dr. Tuve's book *Principles of Fire Protection Chemistry:*[7]

> Obviously, dust explosions take place only when a source of ignition is present in, or at the edge of, a mixture of combustible dust and air. Open flames, electric arcs, suddenly broken electric light bulbs, friction sparks, and even static sparks can initiate dust explosions.

## Mechanical Explosions

Mechanical or physical explosions, such as the bursting of a boiler, account for many explosions and resultant loss of life and property. Because of their large size, boilers used in industry can be particularly devastating when they explode. Trapped-steam explosions such as those that occur at foundries, and rupture of pressurized equipment such as pressurized tanks and processing equipment are other fairly common sources of mechanical explosions. Adequate precautions and proper maintenance as well as the installation of adequate pressure relief devices could reduce the danger of loss from explosions of this type.

## Atomic Explosions

A nuclear or atomic explosion is the result of the redistribution of protons and neutrons within the interacting nuclei. Atomic explosions are produced by two processes: fission and fusion. Fission involves the use of uranium 235 and plutonium 239, while fusion uses deuterium. The resulting energy yield is generally divided as follows: blast and shock, 50 percent; thermal energy, 35 percent; nuclear radiation, 15 percent.

## Thermal Explosions

A thermal explosion occurs when a confined unstable material decomposes, with the resulting production of self heat. If the heat in the reaction cannot readily escape, the process of decomposition accelerates the self-heating process and an explosion occurs. One factor that greatly affects the explosion potential of a material is the shape of the vessel or container in which the material is confined or stored.

**Deflagration:** A deflagration is a heat-producing reaction that propagates from burning gases to the unreacted material by conduction, convection, and radiation. In this reaction, the combustion progresses through the material at a rate that is less than the velocity of sound.

The term "deflagration" cannot be used interchangeably with the term "explosion." In an explosion, deflagration may not occur.

***Detonation:*** A detonation is a heat-producing reaction accompanied by a shock wave in the material that establishes and maintains the reaction. (Again, this term is not synonymous with the term "explosion," although it is fairly common for these terms to be used interchangeably.) This reaction propagates at a rate greater than the velocity of sound. The release of chemical energy in a detonation energizes the shock wave.

The burning of combustible dust particles in the air is an example of deflagration of dusts. Often detonation is not present. A further example: if a vessel containing a flammable vapor-air mixture is ignited, resulting in a pressure buildup that ruptures the vessel, an explosion is said to have occurred. If, however, deflagration and/or detonation does not rupture the vessel, then no explosion is said to have occurred.

## PRODUCTS OF COMBUSTION

There are four major products of combustion: (1) fire gases, (2) flame, (3) heat, and (4) smoke. All of these products are produced in varying degrees by fire. However, the material or materials that are involved in the fire and the resulting chemical reactions produced by the fire result in many variables that must be considered in fire protection.

### Fire Gases

Most people think a fire death or injury results from contact with flame or heat. Actually, the primary cause of loss of life in fires is the inhalation of heated, toxic, and oxygen-deficient gases and smoke. The amount and kind of fire gases present during and after a fire vary widely with the chemical composition of the material burning, the amount of available oxygen, and the temperature. The effect of toxic gas and smoke on people will depend on the time of exposure, the concentration of the gases in air, and also, to a large degree, on the physical condition of the individual.

There are usually several gases present during a fire. Those that are most lethal are carbon monoxide, carbon dioxide, hydrogen sulfide, sulfur dioxide, ammonia, hydrogen cyanide, hydrogen chloride, nitrogen dioxide, acrolein, and phosgene.

***Carbon Monoxide:*** Although carbon monoxide is not the most toxic of fire gases, it is one of the more frequently produced fire gases. If a fire burns with a good air supply, the carbon in most organic combustible materials is converted into carbon dioxide. However, most fires develop under conditions where the supply of air is less than that sufficient for complete combustion, and carbon monoxide is produced. This gas in any mixture of fire gases is the chief danger to persons because it rapidly robs the blood of the oxygen needed to support life. Low concentrations of invisible and odorless carbon monoxide cannot be tolerated for very long. Exposure to as little as 1.3 percent will cause unconsciousness in two or three breaths; death will result in a few minutes. Another dangerous quality of carbon monoxide is its effect upon coordination and judgment when it is inhaled.

*Carbon Dioxide:* Carbon dioxide is not toxic like carbon monoxide. However, large quantities of carbon dioxide are usually produced in a fire. Inhaling above-average amounts of this gas increases the speed and depth of breathing. Two percent carbon dioxide can increase breathing by about 50 percent. If the concentration of this gas nears 10 percent, carbon dioxide can cause death in a few minutes. Since high concentrations of carbon dioxide increase the breathing rate, danger to life is further increased as other toxic gases developed by the fire may be inhaled more readily.

*Hydrogen Sulfide:* Organic materials that contain sulfur, such as rubber, wool, hides, and meat, when subjected to fire and incomplete combustion, will produce hydrogen sulfide. Breathing as little as 0.04 percent for more than half an hour is dangerous, and concentrations in the air of more than 0.08 percent is acutely poisonous. This gas smells like rotten eggs, which makes it easily detectable. However, above a concentration of 0.02 percent, the average human sense of smell fades rapidly, and hydrogen sulfide gas cannot be detected after a few breaths.

*Sulfur Dioxide:* Sulfur dioxide is released when materials containing sulfur are oxidized completely by fire. This gas has an extremely irritating effect on the eyes and on the respiratory tract and, as such, can be quickly detected. Concentrations in the order of 0.05 percent are dangerous for even a short period of time.

*Ammonia:* Ammonia is toxic and extremely irritating to the eyes, nose, throat, and lungs. As a result, breathing ammonia is so disagreeable that people are quick to react and try to escape from the fumes. Ammonia is widely used as a refrigerant in the larger commercial and industrial refrigeration systems; the accidental release of the gas during a fire in these systems presents a great danger. Ammonia is formed during the burning of combustible material containing nitrogen, *e.g.,* resins, wool, silk, and some plastics. Any prolonged breathing of ammonia fumes will cause serious injury or death.

*Hydrogen Cyanide:* Hydrogen cyanide is highly toxic. Exposure to 0.3 percent is fatal. While this gas is, fortunately, not present in dangerous quantities in most fires, it is widely used to fumigate buildings. Therefore, fire fighting in buildings being treated with hydrogen cyanide can be dangerous. Hydrogen cyanide has an odor of bitter almonds.

*Hydrogen Chloride:* The wide use of polyvinylchloride, which releases hydrogen chloride as a product of combustion, makes this gas of concern to fire fighters. Like ammonia, the irritating effects of breathing this gas will drive people away automatically. In Europe a number of cases of metal corrosion from the combustion of polyvinylchloride have been reported. The hydrogen chloride developed in such fires has attacked the steel reinforcing rods in concrete buildings.

*Nitrogen Dioxide:* Nitrogen dioxide (commonly referred to as peroxide) is very toxic. Brief exposures to from 200 to 700 parts per million can be fatal. Nitrogen dioxide is formed during decomposition and combustion of cellulose nitrate, and also in fires involving other inorganic nitrates such as ammonium nitrate. This gas is also formed when nitric acid comes in contact with metals or combustible material. An insidious feature of this gas is that breathing it will anesthetize the throat so that its presence may not be recognized. In moderate exposures a delayed reaction can occur, and the effects may not be felt for as much as 8 hours.

*Acrolein:* When petroleum products, fats, oils, and many other common materials burn, acrolein is produced. Acrolein is highly irritating and extremely toxic. A concentration of one part per million is intolerable. However, acrolein is not too frequently encountered in quantities that can be considered dangerous in fire gas environments.

*Phosgene:* Phosgene is not usually present in the products of combustion of ordinary combustible material. While it is highly toxic, it does not produce a serious threat to health except when the ventilation is poor or when large quantities of chlorinated vapors are produced from fires involving polyvinylchloride or other chlorinated solvents. There have been many instances where phosgene was reported as the cause of death when carbon tetrachloride was used as an extinguishing agent in contact with hot metal.

## *Flame*

The burning of combustible materials in the air is almost always accompanied by visible flame. Serious burns can be caused by direct contact with flames. Burns can, also, be caused by direct radiation of heat from flames.

## *Heat*

In *Principles of Fire Protection Chemistry,* the characteristic of heat as a product of fire is explained by Dr. Tuve:[8]

> One of the basic characteristics of a fire is the emission of heat. All combustion reactions, or oxidation reactions, are exothermic (exo = out, thermo = heat), meaning they give off heat. The rate and the extent to which this heat is

given off is highly variable and depends upon many factors. In the case of most fires, it is difficult — if not impossible — to identify the fire's rate and extent. Because of this, these factors can only be determined in a general way.

Exposure to heat from a fire will affect persons exposed to it in proportion to the length of exposure and the temperature of the heat. The dangers of exposure to heat from fire range from minor injury to death. Exposure to heated air increases the heart rate, causes dehydration, exhaustion, blockage of the respiratory tract, and burns. Fire fighters should not enter atmospheres exceeding 120°F to 130°F (48.8°C to 54.4°C) without special protective clothing and masks. The maximum survivable breathing level of heat from fire in a dry atmosphere for a short period has been estimated at 300°F (148.8°C). Any moisture present in the air greatly increases the danger and sharply reduces the time of survival.

## *Burns*

Burns are usually classified as first, second, or third degree. First degree burns affect the outer layer of skin. They are painful but not as serious as second or third degree burns. Second degree burns penetrate more deeply into the skin. They form blisters, and a considerable amount of fluid usually accumulates under the skin. Third degree burns penetrate still further, and are the most serious. Third degree burns are not painful like first and second degree burns because the nerve endings have been made inactive.

Any doctor specializing in the treatment of burns will testify that serious burns are among the most painful of injuries, among the most long-lasting and difficult to treat, and among the most costly for the patient.

Persons exposed to excessive heat may die if the heat is conducted into the lungs rapidly enough. This may occur without visible signs of burning. Blood pressure will decline and circulation of the blood will fail. Exposure to excessive heat may raise the body temperature sufficiently to damage the nerve centers of the brain.

## *Smoke*

Smoke, like flame, is generally a visible sign of fire. Under the usual conditions of insufficient oxygen for complete combustion, wood, petroleum oil, paper, and other common combustibles give off particles of carbon that appear as smoke.

Some experts believe that smoke, including the invisible poisonous gases that are in it, is the major killer in fires, responsible for 50 to 75 percent of deaths in fires. Smoke often creates dangerous conditions for life safety before the temperature in the building reaches dangerous levels.

Smoke irritates the eyes and the lungs, and frequently causes panic conditions. Fire gases created under conditions of incomplete combustion, such as methane, formaldehyde, and acetic acid, may condense on the smoke particles and be carried into the lungs with lethal results.

## Oxygen Deficiency

Another effect of the combustion process dangerous to life is a drop in the oxygen level. The usual level of oxygen in the air is about 21 percent. If the level drops to 15 percent, anoxia (diminished muscular control) develops. If the oxygen drops lower (10 to 14 percent), a person can remain conscious but judgment will be impaired and the person will tire easily. At an oxygen content range of 6 to 10 percent, a person will collapse and can only be revived by fresh air or oxygen.

## FIRE AND EXPLOSION CONTROL

The combustion process occurs in two modes: (1) the flaming mode, and (2) the flameless surface mode. The flaming mode includes explosions. An example of a flameless surface mode is glowing embers. The flaming mode is characterized by relatively high burning rates, which are expressed in terms of the heat energy that is released. Intense and high levels of heat are usually associated with the flaming mode.

The flaming mode is graphically represented by the tetrahedron that was described earlier in this chapter. The four components of a tetrahedron are heat, fuel, oxygen, and unrestrained chain reaction. By contrast, the flameless combustion mode can be graphically represented by the traditional triangle containing the three elements heat, fuel, and oxygen.

The flaming and flameless modes are not mutually exclusive; combustion may involve one or both modes. Often combustion may occur in the flaming mode and gradually make a transition to the flameless mode. At one point in this process, both modes occur simultaneously.

However these modes occur, flaming modes require four considerations for extinguishment and flameless modes require three. Four means of fire and explosion control will be described in the following sections in this chapter. These four means are: (1) water (to reduce the heat), (2) removal of oxygen, (3) removal of fuel, and (4) flame inhibition by chemical means (to remove the unrestrained chain reaction).

## Use of Water

For most common combustibles such as wood, paper, and cloth, the simplest and most effective means of removing the heat of a fire is through the application of water. Water application can be varied and will depend on the fire. Water spray is usually the most effective application for a house fire. For a large fire, a straight stream of water will provide longer range and more powerful drenching action.

Applying water to the burning fuel cools the fuel to the point where the rate of release of combustible vapors and gases is reduced and ultimately stopped. Heat developed by a fire tends to be carried away by radiation, conduction, and convection.

This fact helps to reduce the amount of heat and makes the use of water more effective. Only a relatively small proportion of the heat evolved needs to be cooled by the water in order to extinguish the fire.

Effective use of water spray or solid streams cannot be accomplished if the water cannot reach the burning fuel directly. For this reason, areas where fire fighters cannot readily reach the fire with water streams, such as high-rise buildings, high piled storage areas, and the like, must be provided with automatic sprinklers or other automatic fire protection systems.

When a water spray is used on a fire, the water may turn to steam. At one time it was thought that this steam was helpful in controlling a fire. However, steam diffuses rapidly, and while it helps to dilute the oxygen concentration to some degree in the immediate fire area, the presence of steam is not a major factor in extinguishing a fire.

## *Removal of Oxygen*

The amount of dilution of oxygen necessary to stop the combustion varies greatly with the kind of material that is burning. Ordinary hydrocarbon gases and vapors will not burn when the oxygen level is below 15 percent. (The oxygen in the atmosphere is normally about 21 percent.) Acetylene will continue to burn unless the oxygen concentration is lowered, but will continue to glow on the surface even if the oxygen level is as low as 4 to 5 percent.

A fire in a closed space can extinguish itself by consuming the oxygen. However, incomplete combustion, which takes place when the oxygen is consumed, usually results in considerable generation of flammable gases. Fire fighters should use great caution and guard against violent flashbacks or explosions if such a space is improperly ventilated and a supply of oxygen suddenly rushes in.

A commonly used method of putting a fire out by removing or diluting the oxygen is by flooding the entire fire area with carbon dioxide or with some other inert gas.

Some burning materials react violently with water and can best be extinguished by covering them with a suitable inert material. New ways of shutting off combustion have been and are being developed to control and extinguish fires in solids, liquids, and gases under varying conditions through chemical means.

## Fuel Removal

Fuel removal can be accomplished in a variety of ways. One of the most common examples is the practice of bulldozing a fire break across the path of an advancing forest fire so that the trees and brush are removed and the fire runs out of fuel.

Fires in large coal or wood pulp piles can usually only be controlled by moving the pile out of the fire zone. Fires in large oil storage tanks have been controlled by pumping the oil out of the burning tank into an empty tank. If a gas line is ruptured and the gas ignited, shutting off the supply of gas is the only way to stop the fire.

If it is not practicable to remove the fuel, extinguishment can be accomplished by shutting off the fuel vapors or by covering the burning or glowing fuel. Fire fighting foams and dry powder extinguishers are examples of effective procedures for covering or coating a fire.

Forest fire fighters have effectively used gelling agents in the water to coat burning wood and vegetable materials and also retard water runoff.

## Flame Inhibition

The extinguishment of fires by chemical flame inhibition is a new development and is the subject of much continuing research. There are certain extinguishing agents such as some of the gaseous and liquid halogenated hydrocarbons (halons), some of the alkali metal salts (dry chemicals), and ammonium salts, which will extinguish flames with efficiency and dispatch without the accompanying action of the other traditional methods of flame extinguishment such as cooling, oxygen dilution, or fuel removal.

When the proper amounts of these agents are injected into flames, the agents act as a negative catalyst, and the flame becomes "inhibited," and the fire is extinguished. This fourth method of extinguishment is used for the flammable mode only. The object of this type of extinguishment is to inhibit and halt the unrestrained combustion chain reaction.

## Summary

Although the essential components of a fire are known, and extensive laboratory and field research has been conducted regarding fire and its causes and behavior, fire is still unpredictable.

Combustion, which is commonly referred to as fire, consists of four components: fuel, heat, oxygen, and unrestrained chain reaction. Some fires, however, consist only of fuel, heat, and oxygen. These fires are considered to be fire "in a flameless mode."

The heat of fire, its measurement, and its ability to transfer to other combustible materials, is a vital factor in fire control. There are four primary sources of ignition that produce enough heat to generate a fire: chemical sources, electrical sources, mechanical sources, and nuclear sources. The basic difference between fire and explosions is the rate at which energy is released. Explosions can be caused by chemical, mechanical, atomic, or thermal reactions.

Contrary to popular opinion, the greatest life safety hazard from fire is not flame or heat, but the inhalation of heated, toxic, and oxygen-deficient gases and smoke. The gases formed in any fire depend on many variables such as the chemical composition of the material or materials that are burning, the amount of oxygen available for combustion, and the temperature of the heat that is generated. Fire fighters must be aware of the common fire gases and their lethal properties so that they will have a healthy respect for any hazard potential based on the fire gases they encounter.

Fire and explosion control is based on an understanding of the basic components of fire — whether the fire is in the flaming or flameless mode — and the means required to control the fire situation.

## *Activities*

1. Discuss with your classmates some of the reasons why fire is considered unpredictable. List five major reasons for the unpredictability of fire.
2. Why are the conclusions reached as a result of the study of fire in research laboratories different from conclusions which may be reached about fire in "real life" situations?
3. List five characteristics of fire and the fire protection considerations to be kept in mind with regard to combatting and controlling a fire or fires containing these characteristics.
4. Describe the chemical phenomenon called "pyrolysis."
5. Define and give examples of a flammability limit.
6. Why is determining the intensity of the heat of a fire important? Give reasons for your answer by presenting situations that might occur if the measurement of heat is not considered in fire protection procedures.
7. Describe the three types of heat transfer and their effect on fire ignition.
8. Describe the four major sources of ignition and the circumstances under which they might cause combustion and/or explosion.
9. Give three reasons why fire is considered to be essentially a chemical reaction.
10. What is the difference between an explosion and a fire?
11. Describe the major causes of explosions.
12. Explain why deflagration and/or detonation are not necessarily elements of an explosion.
13. Describe the effect of fire gases in a fire situation.
14. Describe the four major methods of fire and explosion control, and the circumstances under which these methods might be used.

## Bibliography

[1] Tuve, Richard L., *Principles of Fire Protection Chemistry*, NFPA, Boston, 1976, p. 125.

[2] *Fire Protection Handbook*, 14th Ed., NFPA, Boston, 1976, p. 2-6.

[3] Tuve, Richard L., *Principles of Fire Protection Chemistry*, NFPA, Boston, 1976, p. 59.

[4] *Fire Protection Handbook*, 14th Ed., NFPA, Boston, 1976, p. 2-6.

[5] Tuve, Richard L., *Principles of Fire Protection Chemistry*, NFPA, Boston, 1976, p. 71.

[6] _____, p. 72.

[7] _____, pp. 100-101.

[8] _____, p. 58.

*Chapter Four*

# Fire Hazards of Materials

*The world contains a great variety of highly flammable materials, both natural and synthetic, that can present a threat to both life and property. A wealth of knowledge exists concerning these substances, their physical and chemical properties, suitable methods for their handling and storage, and the proper steps to take when a fire involving them occurs.*

## COMBUSTIBLE SOLIDS

The two basic components of which everything in the physical world consists are matter and energy. Energy is the basic force that brings about both chemical and physical changes — it is the result of the interaction of matter under varying circumstances. Energy is "work," "performance," and is most readily described by our senses in terms of what energy "does." Matter, the other basic component of our physical world, has two fundamental characteristics: (1) it has weight, and (2) it occupies space. Although it is easier to weigh a piece of wood than it is to weigh a molecule of oxygen, both still are forms of matter. Matter can be in the form of gas, liquid, or solid. For example, water is a liquid. When it is boiled, it becomes steam — a vapor or gas — and when it is frozen, water becomes a solid. It is "matter" in all its forms with which this chapter is concerned. Matter, both in natural and synthetic form, can, in varying combinations and circumstances, become hazardous in terms of firesafety.

In Chapter Three, "Characteristics and Behavior of Fire," the basic components of fire were described. In the flameless mode, three components are necessary for combustion: heat, fuel, and oxidation. The flaming mode added another characteristic: unrestrained chain reaction. Fuel is a basic component to both modes of combustion. Matter provides this fuel in one or a combination of the three forms of matter: solid, liquid, or gas.

Most people, when asked to name things that burn, will generally name a solid form of matter such as wood. Wood, or the products of wood, such as paper and cardboard, are the most commonly thought of and, in fact, the most common form of combustible solid. Plastics and textiles are other solids that require major consideration from a fire protection point of view.

## *Wood*

It has already been noted in this text that fires in one- and two-family dwellings are responsible for causing the highest percentage of death and injury from fire in the United States. Wood, one of the most commonly used construction materials, is generally the predominant material employed in building one- and two-family dwellings as well as in many other types of occupancies. The wood used for framing, sheathing, flooring, and interior finish as well as the wood products, such as furniture, which are installed once the house is built, provide a ready source of fuel in the event of a fire. When in contact with sufficient heat, all wood or wood-based products will sooner or later ignite, the time of ignition depending on the ignition source and the length of exposure.

Wood or wood-based products can be treated with fire retardant chemicals. When so treated, the flammability of wood and wood-based products is greatly reduced. The flammability of these products can also be reduced when they are used in combination with other materials, such as insulation.

***Physical Properties:*** The physical form of the wood is important from the standpoint of ease of ignition and rate of burning. A log or beam is hard to ignite; a pile of wood shavings will ignite easily and burn rapidly; wood dust can explode violently when ignited by a small spark. The shavings and dust expose the wood particles to air and provide less mass to conduct heat away from the wood particles. Wood is a poor conductor of heat. Heat will not readily penetrate into wood that is in log or beam form. Also, the char that develops on the surface of the wood when attacked by fire results in an insulating effect. In his book titled *Principles of Fire Protection Chemistry,* Dr. Tuve explains the importance of this property of wood with regard to fire protection:[1]

> The low heat conductivity of wood is another important physical property of this combustible material, and is of considerable importance in some fire situations. If one considers the situation of a moderate heat or flame exposure of considerable thicknesses of wood, charring immediately takes place from the exposed surface inward. The char formation possesses an even lower heat conductivity than does wood, and the rate of heat penetration into the wood is progressively lowered by this mechanism. In the case of heavy timber construction or where large, laminated timber trusses are employed, the structural integrity of a building may be preserved for longer lengths of time under fire exposure than if the building had been constructed with steel framing of a type for similar stresses. The higher heat conductivity of the metal would be more quickly vulnerable to strength loss due to rapid penetration by heat attack.

Table 4.1, Thermal Conductivity of Materials, shows how the thermal conductivity of various materials can affect a fire situation.

*Moisture Content:* Wood is primarily made up of carbon, hydrogen, and oxygen. Live wood cells retain considerable moisture. When the wood is dead, air replaces most of the water in the cellular structure of wood.

The fire behavior of wood and other combustible solids of the same size and shape will vary greatly with the moisture content. Obviously, wet wood is harder to ignite and will therefore not burn as fast as dry wood. Thus, burning rate is influenced by the moisture content in materials.

The effects of moisture content can best be illustrated by the action of forest fires. When the forests are wet, the water content in the trees and forest vegetation absorbs much of the heat from the sun or from other sources. The water then evaporates as a result of absorbing the heat. Water has a high specific heat, with the result that a great deal of heat is required to evaporate the moisture. The large quantities of water vapor that are given off dilute the oxygen in the air surrounding the combustible wood and forest duff (the partly decayed organic matter on the forest floor). This process affects both the development of combustible vapors and the ignition or continued burning of the combustible vapors. However, if the forest material is exposed to a long, hot, dry spell, and if high winds develop to further dissipate the remaining moisture, then the forest vegetation dries out and dies. At this point, forest materials can easily be ignited and the ensuing flames can race through the forest area at incredible speed.

Table 4.1 Thermal Conductivity of Materials*

| Material | Rate of Heat Conductivity (k) (in calories per sec per cm per °C) (*Note:* Multiply all values by 1/1,000) |
|---|---|
| Aluminum (metal) | 500.0 |
| Brick (common) | 1.7 |
| Charcoal | 0.21 |
| Concrete | 4.1 |
| Copper (metal) | 910.0 |
| Corkboard | 0.1 |
| Fiberboard ("celotex" type) | 0.14 |
| Glass | 2.3 |
| Iron (metal) | 150.0 |
| Marble | 6.2 |
| Mineral wool (blanket) | 0.1 |
| Paper | 0.3 |
| Plaster | 1.7 |
| Plastics (solid) | 0.45 |
| Vermiculite | 0.14 |
| Wood (oak) | 0.41 |
| Wood (white pine) | 0.29 |
| Wool (loose clothing) | 0.8 |

*From *Principles of Fire Protection Chemistry,* by Richard L. Tuve.

One of the major factors in determining whether "fire weather" is present or can be expected is the measurement of the moisture content that is present in the forest materials. Other important factors are atmospheric humidity, wind, the condition of the vegetation, and the season of the year.

Even when exposed to a relatively high source of heat for a prolonged period of time, ignition is generally difficult when the moisture content of wood (and similar fuels) is above 15 percent. For example, the conditions of high humidity that are often present in the summertime make wood less susceptible to ignition than conditions present in cold climates, where heated buildings cause indoor relative humidities to fall below the 15 percent level. Once ignition and resultant fire have begun, however, heat radiation and the rate of pyrolysis reduce the importance of the moisture factor to such a point that, under some conditions, wood with a moisture content of 50 percent or more will burn.

*Ignition of Wood:* Burning of wood will only occur if it is heated to the point where combustible gases are released from the surface of the wood. At normal temperatures and pressures wood and other similar combustible solids, unlike combustible liquids and gases, do not give off flammable vapors. Wood must be exposed to the heat source for a long enough period of time to permit the wood to give off flammable vapors. The time of contact is a determining factor. Wood may not ignite if a blow torch is held momentarily to the wood surface even though the temperature of the blow torch flame is higher than the ignition temperature of the wood. On the other hand, prolonged contact of wood with a steam pipe having a temperature much lower than the blow torch flame can well raise the heat of the wood to its ignition point.

The temperature at which wood will ignite varies widely, depending on the form, size, and moisture content of the wood. Low density softwoods will ignite at lower temperatures than high density hardwoods. The rate of heating and the length of time that heat is applied also influence the ignition point. The nature of the heat source attacking the wood is another factor that influences the temperature at which wood will ignite. The amount of oxygen available will also affect the ignition temperature.

It is difficult to identify the specific ignition temperatures of wood because of the large number of variables involved. Test results vary greatly. Generally, the average ignition temperature of wood is considered to be about 392°F (200°C). At this temperature combustible vapors are produced and begin to burn. As the burning progresses and the temperature increases, carbon monoxide begins to be emitted. Finally, the burning wood becomes charcoal and ash. The four stages of decomposition of wood and the corresponding temperatures that cause the reaction are described in Table 4.2.[2]

Wood, however, can ignite at temperatures lower than 392°F (200°C). For example, if wood is in constant contact with a temperature source such as steam pipes over a long period of time, the wood may undergo chemical changes that can result in the formulation of charcoal, which can heat spontaneously. Research indicates that 212°F (100°C) — the boiling point of

Table 4.2 The Thermal Degradation of Wood[2]

| Temperature | Reaction |
|---|---|
| 392°F (200°C) | Production of water vapor, carbon dioxide, formic and acetic acids — all noncombustible gases. |
| 392°F — 536°F (200°C — 280°C) | Less water vapor, some carbon monoxide — still primarily endothermic reaction. |
| 536°F — 932°F (280°C — 500°C) | Exothermic reaction with flammable vapors and particulates. Some secondary reaction from charcoal formed. |
| Over 932°F (500°C) | Residue principally charcoal with notable catalytic action. |

water — is the highest temperature to which wood can be continually exposed without the danger of ignition. Dr. Tuve explains as follows:[3]

> Concerning the spontaneous ignition temperature of wood, the influence of the variables involved is even more indeterminate than where an igniting flame for the evolved gases is present. Considering the fact that wood slowly changes its chemical composition under sustained heat attack lesser in amount than is necessary to cause ignition of the combustible gases evolved, the end point of such an exposure over long periods of time is the formation of extremely porous charcoal. Charcoal is a material that is capable of absorbing gases and vapors. If these gaseous materials are combustible and are capable of further slow breakdown with the evolution of heat within the charcoal, a temperature may be reached that is capable of spontaneous ignition of the charcoal and of any wood in contact with it.

*Flashover:* If an area such as a compartment or a room has limited ventilation, flashover conditions may be present. Flashover occurs when the mixture of gases reaches the flammable range or ignition point, at which time the whole area bursts into flame. When flashover occurs, in seconds temperatures can increase as much as fivefold, oxygen is greatly reduced, carbon monoxide gas is generated to lethal levels, and carbon dioxide increases rapidly. Experiments with room fires have repeatedly demonstrated this phenomenon, and fire fighters therefore need to be constantly aware of the dangers presented by flashover.

*Flame Spread:* The way flame spreads along the surface of wood or other combustible solids also affects the severity of a fire. Flame spread is heat transferred by convection, and is the most frequent means of transferring fire in which combustible solids are involved. Combustible wall boards and other interior finish may spread flame so rapidly that life is quickly endangered. Flame spread across a combustible ceiling tile or wall covering is different from "flashover," in which a whole room and its contents become simultaneously involved. The flame spread characteristics of wood and other common building materials have been measured and defined by the "Tunnel

Test" apparatus designed by A. J. Steiner of Underwriters Laboratories Inc. (UL). Further details of this will be developed in Chapter Six, "Firesafe Building Design and Construction."

*Fire Loading:* When considering wood and similar combustible solids, it is possible to estimate the amount of fuel that can be involved in a fire (the fire loading of a building). Wood and similar materials will provide heat of combustion of from 7,000 to 8,000 British thermal units per pound (Btu/lb). Thus, the severity and duration of a fire in the building can be evaluated. This subject will be presented in greater detail in Chapter 6, "Firesafe Building Design and Construction."

*Smoke:* In the early stages of a fire, wood can produce large quantities of smoke. The smoke develops when there is insufficient oxygen for complete combustion and the vapors and gases from the burning wood include droplets of flammable tars (smoke). Research is being conducted to understand more fully the hazards of smoke and to control the use of various building materials to limit the amount of smoke and toxic gases that can be produced in fires. Smoke and toxic gases are serious life safety hazards.

*Storage:* Wood and wood by-products are generally stored outdoors. At sawmills, papermills, pulp mills, and lumberyards, wood is stored in large quantities that may vary in amount throughout the year, but which, at times, may also represent a year's supply of the material.

Logs may be stored in rank piles, *i.e.*, in parallel form, or stored by simply dumping them in a pile called a stack pile. The rank pile method of log storage is preferable in terms of fire protection because the piles usually do not exceed 15 ft (4.5 m) in height, and the logs can be stored in aisles with more adequate provision made for fire fighting than with stacked piles.

Storage of logs in chip form at paper and pulp mills presents considerations based on the physical properties of wood in this form. Chip fires may occur in two different ways: surface fires or internal fires.

Lumber storage can pose serious problems in urban areas. Lumber storage is generally prohibited in congested areas; however, zoning considerations can result in urban lumber storage areas located in or near congested areas. A fire hazard that is present in areas such as these is, of course, the danger of fire exposure to nearby areas.

## *Plastics*

Another common combustible solid is a large and varied group of materials called "plastics." There are thousands of plastic product formulations that are produced in a wide variety of shapes and sizes, such as solid shapes, films and sheets, foams, molded forms, synthetic fibers, pellets, and powders. There are about thirty major groups or classes of plastics and polymers. In addition, most finished products contain additives such as colorants, reinforcing agents, fillers, stabilizers, lubricants, and the like. These additives vary the chemical nature of the product still more.

Most plastics are combustible, but the degree of combustibility varies widely because of the wide range of chemical compositions and combinations in plastics. As a result, it is virtually impossible to assign a fire hazard or flammability limit to any general plastic group. The only method of determining the fire hazard of a particular plastic is to fire test the plastic under exact "end use" conditions.

*Manufacture:* The processing of plastic involves three basic steps: (1) manufacturing the basic plastic; (2) molding, extruding, or casting the basic plastic into the shape of the article itself (processing); and (3) bending, machining, cementing, decorating, and polishing the article (fabricating).

The manufacture of the basic plastic material usually involves flammable ingredients or solvents. Most finished plastic products also contain additives such as colorants, fillers, stabilizers, and reinforcing agents that influence the flammability characteristics of the plastics. The plants that make articles from plastic material by molding or machining do not have the same degree fire hazards that exist in the plants making the basic plastics where large quantities of flammable chemicals are used. The improper handling of the flammable organic solvents used in most plastic plants has resulted in many fires. The installation of vapor removal systems, the use of explosion-proof electrical equipment, and the reduction or elimination of the possibility of static sparks are all steps that can and should be taken to minimize fire hazards.

*Fire Behavior:* The fire behavior of plastic materials is dependent on the chemical composition of the basic plastic, the kind of additives used, the size, shape, and "end use." All of the major plastic materials are combustible and, if they are involved in a fire, will contribute fuel to the fire. The ignition point of plastics will vary widely, as will the rate of flame propagation. It is comparatively easy to predict what wood and wood products will do when exposed to fire, but the same type of product made of plastic will provide a totally different hazard from wood. This hazard will vary depending on the particular type of plastic used.

The smoke and toxic fumes given off by a burning plastic material will again depend on the particular plastic. Some burning plastics will yield quantities of dense smoke while others burn more cleanly and produce less smoke.

The growing use of plastics for structural and finishing materials in buildings has led to flame-retardant treatments for plastics in such use. Such treatments make the material more difficult to ignite, and also reduce the rate of burning.

*Storage:* The chemical composition, the physical form, and the manner and arrangement in which plastics are stored greatly affect the degree of fire hazard that is present. Large quantities of smoke are usually generated when stored plastics are involved in fire, a condition made more or less difficult by the amount of ventilation present or available in a given storage area.

Thermoplastics such as polyethylene and plasticized polyvinylchloride, and thermosets such as polyesters, present severe fire hazards. Plastics in foamed material form present the most severe hazard of all. In a fire, thermoplastics will melt and break down and behave and burn like flammable liquids. Automatic sprinkler systems with high sprinkler discharge densities are necessary for adequate fire protection.

## Textiles

Textiles are a major component in our everyday lives. Clothing, bedding, carpets, upholstery, and many other items with which people come in daily and frequent contact are textiles or textile-based. Virtually all textile fibers are combustible. The chemical composition, the weight of the fabric, the type of weave, and the type of finish and finishing treatment given to textiles are some of the many variables that affect the way a textile burns.

The combustibility of textiles, combined with the high degree of people's involvement with textiles in their daily lives, accounts for the large numbers of textile-related fires and the correspondingly large number of deaths and injuries as a result of textile-related fires. The NFPA Fire Analysis Department estimated from a one-year study that approximately 2,000 reported fires occur in the United States every year in which textile products are the first materials ignited. Of the 2,000 fires, 211 deaths resulted from clothing ignitions; 845 deaths resulted from fires originating in bedding.[4]

Cotton is a chemical composition of carbon, hydrogen, and oxygen. Any small ignition source such as a match or a cigarette can ignite cotton. The ensuing fire will develop and spread rapidly. The same is true of rayon. Cotton and rayon are a very large part of the total clothing fabric materials in common use. According to a 1971 report from the Department of Health, Education, and Welfare, of the people whose clothing catches fire, one out of four is a child under ten years of age; 15 percent of the victims are people over sixty-five years of age.[5] The extent and severity of clothing burns is much greater than burns received on uncovered skin area. Research has shown that clothing burn victims are four times more likely to die than burn victims not involved in clothing fires.

*Natural Fiber Textiles:* There are two basic types of natural fiber: (1) fibers derived from plants, and (2) fibers derived from animals.

Cotton and other plant fibers such as linen and hemp are composed mainly of cellulose. Cotton and other plant fibers are combustible, but when involved in combustion, the plant fibers do not melt. By-products of burning plant fibers are carbon dioxide, carbon monoxide, and water.

Fibers derived from animals are chemically different from plant fibers in that the basic component of animal fibers is protein. Animal fibers such as wool are not as likely to ignite and support combustion as are plant fibers such as cotton. For example, the ignition temperature of cotton is 752°F (400°C) while the ignition temperature of wool is 1112°F (600°C).

***Synthetic Textiles:*** The use of natural fibers in textiles is increasingly being challenged by synthetic textiles. Rugs and clothing are major examples of this fact. Synthetics are fabrics woven entirely or mostly from synthetic fibers. Rayon, for example, is a synthetic material produced by "reconstituting" cellulose made from cotton and wood fibers. Cellulose acetate is made by reacting cellulose with acetic acid. Acetate, which was previously classed as a rayon, is now listed by the U.S. Federal Trade Commission as a separate category. Primarily, rayon and acetate are the only two major synthetic fibers that chemically resemble plant fibers.

Table 4.3 lists the various classes of plastic resins that are presently utilized in the production of synthetic fibers. Also shown are some of the more common fiber trade names associated with each class, and the fire hazard properties, including generalized statements about relative flammability.

The descriptions of burning are based on small-scale tests and for this reason may be misleading. Some of the synthetic fabrics will give the appearance of

Table 4.3  Synthetic Fibers*

| Plastic Resin Class | Trade Names | Fire Hazard Properties |
|---|---|---|
| Acetate | Chromspun, Celaperm, Arnel (Triacetate) | Burns and melts ahead of flame. Ignition temp., 475°C |
| Viscose | Avisco, Avril, Bemberg (Cuprammonium) | Burns about the same as cotton |
| Nylon | Antron, Caprolan | Supports combustion with difficulty. Melts and drips — melting point, 160-260°C. Ignition temp. 425°C and above |
| Polyester | Dacron, Fortrel, Vycron, Kodel, Terylene | Burns readily. Ignition temp., 450-485°C. Softens, 256-292°C, and drips |
| Acrylic | Acrilan, Orlon, Zefchrome, Zefran, Creslan | Burns and melts. Ignition temp., 560°C. Softens, 235-330°C |
| Olefin | Herculon | Burns slowly. Ignition temp., 570°C. Melts and drips |
| Modacrylic | Verel, Dynel | Burns very slowly. Melts |
| Saran | Rovana, Velon | Does not support combustion. Melts |
| Fluorocarbon | Teflon | Does not support combustion. Softens, above 327°C. Ignition temp. above 600°C |
| Spandex | Lycra, Vyrene | Burns and melts. Ignition temp., 415°C. Softens, 230-260°C |
| Rubber | Lastex | Burns |
| Phenolic | Kynol | Burns |

*From NFPA *Fire Protection Handbook.*

being flame retardant when tested with a small flame source, such as a match. However, when the same fabrics are subjected to a larger flame or full-scale test, they may burst into flame and consume themselves while generating quantities of black smoke.[6]

*Flame Retardant Textiles:* Because of the fire hazard of flammable clothing, much attention has been given to flame retardant treatments for cotton and other textiles. So far, only sleepwear up to size 14 for children and highly flammable material such as brushed rayon have been subject to control.

The textile industry resists producing flame retardant materials because of the economic factors and because it feels the public has not yet been aroused enough to demand safer wearing apparel.

Most of the synthetic fibers used in manufacturing clothing are subject to melting when exposed to sufficient heat. This property can aggravate the severity of a burn.

Many fires are caused by ignition of mattresses or overstuffed furniture through the careless use or disposal of a burning cigarette or other smoking material. Some hospitals, nursing homes, and other health care facilities are now flameproofing the mattresses used by patients. Recently, the Consumer Products Safety Commission promulgated regulations requiring the flameproofing of mattresses and mattress pads.

In most areas the law requires that theater scenery, curtains, and draperies in places of public assembly be treated with flame retardants.

Disposable or nonwoven fabrics consist primarily of cellulose fibers and are generally treated with flame retardants.

*Storage:* As has been previously noted, both natural and synthetic textiles have varied combustibility and flammability characteristics. The method of textile packaging for storage is the main factor in determining the level of fire hazard through storage. A solid wood case is preferable to combustible burlap wrappings. The general storage considerations of good aisles, stability of piles, and piling limitations are important factors in minimizing the fire hazards that are generated through storage.

## FLAMMABLE AND COMBUSTIBLE LIQUIDS

The improper storage, handling, and use of flammable and combustible liquids has been the cause of many deaths, injuries, and disastrous fires. It is vital that the physical properties of such liquids, their classifications — particularly the fire and burning characteristics — and the procedures for fire prevention when dealing with storage and handling of dangerous liquids be understood by the industries producing and using such liquids, and by fire fighters and fire inspectors.

In general, liquids can be considered as a midpoint in the physical world of solids, liquids, and gases. Molecules in liquids move more freely than the molecules in solids. For example, most liquids can easily adapt themselves to the shape of the vessel containing them. Although the molecules in liquids move more freely than the molecules in solids, they do not have the tendency to separate themselves and expand indefinitely as the molecules in gases do.

Although most materials are thought of as existing in solid, liquid, or gaseous forms, many materials are capable of existing in any one of these three forms or states. For example, water in its solid state is ice; water in its gaseous state is steam or vapor. Temperature and pressure conditions can change matter from one state to another.

Because their molecules are freer, changes in state are more common between liquids and gases. Liquids will become gases as the temperature increases or the pressure decreases. Conversely, a gas tends to become a liquid as the temperature decreases or the pressure increases. The critical temperature of a material is that temperature above which the material can exist only in a gaseous state.

It is the vapor from the evaporation of a flammable or combustible liquid when exposed to air or under the influence of heat, rather than the liquid itself, which burns or explodes when mixed with air in certain proportions in the presence of some source of ignition. As was noted in a previous chapter, there is a flammable range below which the vapor mixture is too lean to burn or explode, or above which the vapor mixture is too rich to burn or explode. For gasoline, the most common and widely used flammable liquid, the flammable range is between 1.4 and 7.6 percent by volume. When the vapor-air mixture is near either the lower flammable limit (LFL) or upper flammable limit (UFL), the explosion is less intense than when the mixture is in the intermediate range. The violence of the explosion depends on the concentration of the vapor as well as the quantity of vapor-air mixture and the type of container. Thus, storing gasoline or other flammable liquid in the proper type of closed container and minimizing the exposure to air is of fundamental importance in controlling the fire hazard during storage or use. It must be noted, however, that a tank or other container, when exposed to heat from a fire, may rupture with dangerous results if properly designed vents are not provided and if the exposed tank or container is not cooled by hose streams.

Table 4.4 shows the wide variations in flammable and explosive ranges of some common liquids and gases. Those substances with a wide range of flammability limits are, of course, more hazardous than others because of the possibility of confrontation with a flammable mixture that can occur over a wide range of circumstances.[7]

Fighting a fire involving flammable or combustible liquids requires, if possible, the shutting off of the fuel supply, excluding air from the area, cooling the liquid, or a combination of these measures. The principal fire and explosion prevention measures under such circumstances are: (1) exclusion of sources of ignition, (2) exclusion of air, (3) keeping the liquid in a closed container, (4) ventilation to prevent the accumulation of vapor in the flammable range, and (5) use of an atmosphere of inert gas instead of air.

Table 4.4  Flash Points and Flammable Limits of Some Common Liquids and Gases*

| Liquid (or gas at ordinary temps.) | Flash point °F | (°C) | Flammable limits (percent by volume) |
|---|---|---|---|
| Acetylene | (Gas) | | 2.5 to 81.0+ |
| Benzene | 12 | (−11) | 1.3 to 7.1 |
| Ether (ethyl ether) | −49 | (−45) | 1.9 to 36.0 |
| Fuel oil | | | |
| (Domestic, No. 2) | 100 (min.) | (38) | None at ordinary temps. |
| (Heavy, No. 5) | 130 (min.) | (54) | None at ordinary temps. |
| Gasoline (high test) | −36 | (−38) | 1.4 to 7.4 |
| Hydrogen | (Gas) | | 4.0 to 75.0 |
| Jet Fuel (A & A-1) | 110 to 150 | (43 to 65) | None at ordinary temps. |
| Kerosine (Fuel oil, No. 1) | 100 (min.) | (38) | 0.7 to 5.0 |
| LPG (propane-butane) | (Gas) | | 1.9 to 9.5 |
| Lacquer solvent (butyl acetate) | 72 | (22) | 1.7 to 7.6 |
| Methane (natural gas) | (Gas) | | 5.0 to 15.0 |
| Methyl alcohol | 52 | (11) | 6.7 to 36.0 |
| Turpentine | 95 | (35) | 0.8 — (undetermined) |
| Varsol (standard solvent) | 110 | (43) | 0.7 to 5.0 |
| Vegetable oil (cooking, peanut) | 540 | (282) | (Ignition temp. = 833°F) |

*From *Principles of Fire Protection Chemistry,* by Richard L. Tuve.

## Classification by Properties

An arbitrary division between liquids and gases has been defined in NFPA 321, *Standard on Basic Classification of Flammable and Combustible Liquids.*[8] The division established by NFPA 321 places liquids that will burn into three categories: Class I, or flammable liquids, and Class II and Class III, liquids called combustible liquids.

***Flammable Liquids:*** Class I liquids are those flammable liquids with flash points below 100°F (37.8°C). This classification is based on the premise that indoor temperatures could reach 100°F (37.8°C). Liquids in this category are called flammable liquids. Following is the NFPA *Fire Protection Handbook* summary of the classification of flammable liquids:[9]

> Flammable liquids shall mean any liquid having a flash point below 100°F (37.8°C) and having a vapor pressure not exceeding 40 psia (2068.6 mm) at 100°F (37.8°C).
>
> *Class I* liquids shall include those having flash points below 100°F (37.8°C) and may be subdivided as follows:
>
> *Class IA* shall include those having flash points below 73°F (22.8°C) and having a boiling point below 100°F (37.8°C).
>
> *Class IB* shall include those having flash points below 73°F (22.8°C) and having a boiling point at or above 100°F (37.8°C).
>
> *Class IC* shall include those having flash points at or above 73°F (22.8°C) and having a boiling point below 100°F (37.8°C).

***Combustible Liquids:*** A combustible liquid is classified by the NFPA as any liquid with a flash point at or above 100°F (37.8°C). Class II liquids require moderate heating to reach their flash points. The flash point range is from 100°F to 140°F (37.8°C to 60°C). Liquids requiring considerable heating to bring them to their flash points [over 140°F (60°C)] are classified as Class III liquids.

Following is the NFPA classification for combustible liquids:[10]

> Liquids with a flash point at or above 100°F (37.8°C) are referred to as combustible liquids and may be subdivided as follows:
>
> *Class II* liquids shall include those having flash points at or above 100°F (37.8°C) and below 140°F (60°C).
>
> *Class IIIA* liquids shall include those having flash points at or above 140°F (60°C) and below 200°F (93.4°C).
>
> *Class IIIB* liquids shall include those having flash points at or above 200°F (93.4°C).

Some typical liquids would be classed as follows:

| | |
|---|---|
| Denatured Alcohol | Class IB |
| Fuel Oil | Class II |
| Gasoline | Class IB |
| Kerosine | Class II |
| Peanut Oil | Class IIIB |
| Turpentine | Class IC |
| Paraffin Wax | Class IIIB |

Underwriters Laboratories Inc. (UL) has a useful system for grading the relative flammability hazards of liquids, which is based on the following scale:

| | |
|---|---|
| Ether class | 100 |
| Gasoline class | 90–100 |
| Alcohol (ethyl) class | 60–70 |
| Kerosine | 30–40 |
| Paraffin Oil class | 10–20 |

## *Characteristics*

There are many terms used to describe the characteristics or physical properties of liquids. These characteristics can also be thought of as fire characteristics. The following terms describe some characteristics of liquids with relation to their flammability or combustibility potential.

***Flash Point:*** The flash point of a liquid is the lowest point at which the vapor pressure of a liquid will produce a flammable mixture and resultant flame. The flame will not continue to burn at this temperature if the source of ignition is removed. Dr. Tuve further explains:[11]

Because it is an indicator of the degree of safety of a material, the flash point of a liquid is one of the most important fire characteristics of substances. At its flash point, a liquid continuously produces flammable vapors at the right rate and amount (volume) to give a flammable and even explosive atmosphere if a source of ignition should be brought into the mixture.

***Vapor Pressure:*** Molecules escape from the surface of all liquids when the liquids are not enclosed in containers. When a liquid is in a closed container, molecules from the liquid discharge into the space above the liquid level and also condense back into the liquid. When the rate of discharge and the rate of return are in balance or equilibrium, the resulting pressure exerted by the molecules in the space above the liquid is called vapor pressure.

Vapor pressure of a liquid is used to measure the vapor-air mixture above the liquid surface in a closed container. The percentage of vapor is in direct proportion to the relationship between the vapor pressure of the liquid and the total pressure of the mixture. When heat is added, the molecules become more agitated, thus increasing the pressure. The vapor pressures of many liquids have been measured and are listed in various chemical handbooks. The vapor pressure is expressed in pounds per square inch absolute (psia). When the flash point and the vapor pressure of a liquid are known, it is possible to calculate the lower flammable limit of the vapor.

***Boiling Point:*** The boiling point of a liquid is the temperature at which the vapor pressure equals the total pressure on the surface. The normal boiling point is the temperature at which the liquid boils when under normal atmospheric pressure (14.7 psia). The boiling point increases as pressure increases and is dependent on the total pressure.

***Specific Gravity:*** The specific gravity of a liquid is an important consideration in fighting a flammable or combustible liquid fire. The specific gravity of a liquid is the ratio of the weight of the liquid to the weight of an equal volume of water. The specific gravity of water is one. Any liquid with a specific gravity of less than one will float on water (unless it is soluble in water). If a liquid has a specific gravity greater than one, the water will float on the liquid. Gasoline, with a specific gravity of 0.8, will float on water. Glycerine, with a specific gravity of 1.3, will sink to the bottom when mixed with water.

***Evaporation Rate:*** As has been previously stated, molecules will escape from the surface of all liquids. When a liquid is in an open container, molecules in the form of a vapor will escape. The evaporation rate is the rate at which the liquid is converted to vapor at any given temperature and pressure.

Some liquids like asphalt or wax are on the border line between liquids and solids. These liquids are viscous. The viscosity of a liquid can be measured in relation to the time required for the liquid to flow into a container or through an opening. Another common measure for liquids is called the latent heat of vaporization. This is measured by the heat that is absorbed when one gram of liquid is transformed into vapor at the boiling point under one atmosphere pressure (14.7 psia). Latent heat is usually given in British thermal units per pound (Btu/lb) or in calories per gram (cal/g).

*Ignition Temperature:* The ignition temperature of a liquid is the minimum temperature to which the liquid must be heated in order to initiate self-sustained combustion independent of the heating element. The ignition temperature of kerosine is 410°F (210°C), and of motor oil is around 500°F (260°C).

## Storage and Handling

There are certain fundamentals in the proper storage and handling of flammable and combustible liquids that must be observed to prevent fire or explosion. Ventilation to prevent accumulations of flammable vapors is of primary importance. Although liquids are usually exposed to air at some stage of a process, there is always the possibility of breaks or leaks in the storage and handling in a closed system. It is important to eliminate possible sources of ignition in an area where flammable liquids are stored, handled, or used.

Ventilation of an area where flammable liquids are manufactured or used can be accomplished by natural or mechanical means. Wherever possible equipment such as compressors, stills, pumps, and the like should be located in a spacious, open area. Gasoline and most other flammable liquids produce heavier-than-air vapors that flow along the ground or floor and settle in depressions. Unless these vapors are removed at the ground level, they can travel long distances and be ignited and flash back from a point remote from the origin of the vapors. NFPA 30, *Flammable and Combustible Liquids Code*[12] is the accepted national standard for fire protection and prevention involved with the storage, handling, and use of such liquids.

A great deal of attention has been given to the construction, installation, spacing, venting, and diking of aboveground and underground storage tanks, as well as container storage in buildings. NFPA standards also cover loading and unloading practices, safeguards for dispensing the liquids, and standards for transporting the liquids in trucks, ships, or pipelines.

## GASES

As in the case of solids and liquids, there are a great variety and number of materials that exist in the form of gas. A gas is made up of molecules that are in constant motion. The higher the temperature, the more rapid the motion. A solid has shape and volume, a liquid has volume but no shape of its own, while a gas has neither shape nor volume of its own, but will take the shape and occupy the entire volume of whatever enclosure it occupies.

Since all substances can exist as gases if the temperature and pressure are high enough, in general, gases are thought of and described when the substance exists in a gaseous state at normal temperature and pressure [70°F (21.1°C) and 14.7 psia].

## Classification by Properties

Gases can be broadly classified in various ways such as by chemical properties, by physical properties, or by usage. Classification by chemical properties helps to define the hazards of gases to people and in fires.

*Flammable Gases:* Any gas that will burn in the normal concentrations of oxygen in the air is a flammable gas. Like flammable liquid vapors, the burning of this gas in air is in a range of gas-air mixture (the flammable range).

The term "flash point," which is necessary information concerning flammable liquids, has no real bearing with regard to flammable gases. The flash point for flammable liquids is always below the normal boiling point. Normally, a flammable gas in its gaseous state and even in its liquid state exceeds its normal boiling point, and therefore has already exceeded its flash point. The ignition temperature of a gas is the temperature required to initiate combustion.

*Nonflammable Gases:* Nonflammable gases will not burn in air or in any concentration of oxygen. A number of nonflammable gases, however, will support combustion. Such gases are often referred to as "oxidizers" or oxidizing gases. Common oxidizers are oxygen or oxygen in a mixture with other gases.

Nonflammable gases that will not support combustion are usually called "inert gases." Among the most common inert gases are nitrogen, carbon dioxide, and sulfur dioxide.

*Toxic Gases:* Toxic gases are those gases that endanger life when inhaled. Gases such as chlorine, hydrogen sulfide, sulfur dioxide, ammonia, and carbon monoxide are poisonous and/or irritating when inhaled, and are serious considerations when fire fighting involves such gases.

*Reactive Gases:* Reactive gases are gases that will react with other materials or within themselves by a reaction other than burning. When exposed to heat and shock, some reactive gases rearrange themselves chemically. Such gases can produce hazardous quantities of heat or reaction products. Fluorine is a highly reactive gas. At normal temperatures and pressures it will react with most organic and inorganic substances, often fast enough to result in flaming. Other examples of reactive gases are acetylene and vinyl chloride.

## Physical Properties

A second way of classifying gases is by their physical properties. Gases can be compressed. A compressed gas is one that at normal temperatures inside the gas container exists solely in the gaseous state under pressure. Ranges of pressure in gas containers can vary greatly from a lower limit of about 25 pounds per square inch gage (25 psig) to an upper limit that may be 3,000 pounds per square inch gage (3,000 psig). The common portable cylinders holding compressed gases do not contain very large quantities of gas.

When the gas in a container is more concentrated than the normal compressed gas it is known as liquefied gas. This gas exists in a partly liquid and a partly gaseous state at normal temperatures, and is under pressure as long as any liquid remains in the container.

When the liquefied gas in a container is at a temperature far below normal atmospheric temperatures, the gas is classified as a cryogenic gas.

***Compressed Gases:*** As has been previously stated, a compressed gas is one that is at normal temperature inside a gas container and that exists solely in the gaseous state under pressure. Common compressed gases are hydrogen, oxygen, acetylene, and ethylene.

Hydrogen has a high diffusion rate and is extremely flammable. It also has a wide flammability range and requires very low energy sources, such as a friction spark, for ignition.

Oxygen, although a nonflammable gas, can support combustion and accelerate flame and explosive conditions. Because of its ability to drastically alter the combustibility of other substances and materials, oxygen, although not in itself flammable, should be regarded as a dangerous material.

Acetylene is an extremely reactive, flammable gas that must be stored in special containers to control its reactivity.

Ethylene is used as a fruit ripening agent in very low concentrations. Ethylene is highly flammable and reacts quickly with oxidizing gases. When released into the atmosphere, ethylene mixes quickly with air and can quickly produce an explosive mixture.

***Liquefied Gases:*** Liquefied gases are gases that can be liquefied relatively easily and stored at ordinary temperatures at relatively high pressure. When stored, liquefied gas exists in both liquid and gaseous states. At storage pressure, both the liquid and gas in the liquefied gas container are in equilibrium and will remain so as long as any liquid remains in the container.

Liquefied gas is a much more "concentrated" quantity of gas than compressed gas. For example, although liquefied and compressed oxygen will not be stored in the same type of container, for purposes of comparison a cylinder of compressed oxygen will contain about 20 lbs (9.1 kg) of oxygen, while the same size cylinder of liquefied oxygen could contain about 116 lbs (52.6 kg) of oxygen — almost six times more oxygen.

***Cryogenic Gases:*** Cryogenic gases are those gases that are stored in a completely liquid state. These gases must be maintained in their containers as low-temperature liquids at relatively low pressure. Cryogenic gases must be stored

88   Principles of Fire Protection

in special containers that allow the gas from the liquid to escape in order to prevent a pressure buildup caused by the production of the gaseous state within the container, which would result in container failure. The following excerpt is an important warning on the hazards of cryogenic gases from Dr. Tuve's book *Principles of Fire Protection Chemistry:*[13]

> With the exception of acetylene, all of the common gases that we dealt with earlier are transported and are used as cryogenic gases. In terms of their chemical and physiological characteristics, their properties as liquids are similar to those found in the gaseous state. However, emphasis must be directed to the fact that when gases are liquefied they represent a high concentration of the gas. When liquid oxygen is mixed with another liquid combustible, or when it is dispersed in a solid combustible (*e.g.,* carbon), the resulting mass will burn at explosive rates. Similarly, liquid chlorine in a mixture with a combustible will react (oxidize) rapidly, sometimes with only the slightest ignition. Chlorine gas and hydrogen gas can even react at explosive rates by ignition from sunlight alone.

## Classification by Usage

An understanding of gases as they are classified by usage is of great importance to those involved in fire protection as the terms of these classifications are used in codes, standards, and general industrial and medical terminology. Most people think of gas as the natural gas or liquefied petroleum gas used for heating and cooking in the home. This gas is commonly known as fuel gas. There are two other broad classifications of gases based on usage: (1) industrial gases, customarily used for welding and cutting, refrigeration, heat treating and chemical processing, and (2) medical gases, a specialized class of gases used for anesthesia and respiratory therapy.

The following are NFPA definitions of the three major classifications of gases, based on their use:[14]

### Fuel Gases
Fuel gases are flammable gases customarily used for burning with air to produce heat which in turn is used as a source of heat (comfort and process), power, or light. By far the principal and most widely used fuel gases are natural gas and the liquefied petroleum gases, butane and propane.

### Industrial Gases
Industrial gases embrace the entire gamut of gases classified by chemical properties customarily used in industrial processes, for welding and cutting, heat treating, chemical processing, refrigeration, water treatment, etc.

### Medicinal Gases
By far the most specialized usage classification, the medical gases are used for medical purposes such as anesthesia and respiratory therapy. Cyclopropane, oxygen, and nitrous oxide are common medical gases.

## Gas Laws

There are three basic physical laws that predict the behavior of compressed gases under most conditions. These laws are based on three components regarding the physical laws of gases: volume, temperature, and pressure. Boyle's Law states that the volume occupied by a given mass of gas varies inversely with the absolute pressure if the temperature is constant. Charles's Law states that the volume of a given mass of gas is directly proportional to the absolute temperature if the pressure is kept constant. Gay-Lussac's Law states that when gas temperatures are raised and the volume in which they are confined stays the same, pressures are raised proportionate to the change in the absolute temperature of the gas.

Again, in very simple terms, Boyle's Law describes the behavior of gases when the temperature is constant, Charles's Law deals with the behavior of gases under conditions where the pressure is kept constant, and Gay-Lussac's Law describes the behavior of gas if the volume is kept constant. Although the following descriptions of these laws are general, the basic factors involved are important in terms of fire protection considerations and decisions.

*Boyle's Law:* Boyle's Law relates to gases under the pressure that is normally used in compressed gas containers. Boyle's Law states that the volume of a gas varies with the absolute pressure, if the temperature is kept constant. Very simply stated, *if the absolute pressure on a given volume of gas is increased to twice the original pressure, the volume of gas will be compressed to one half its original volume. Conversely, if the pressure on a certain volume of gas is reduced by one half its original pressure, the volume of gas would increase twofold, or to twice its original volume.*

*Charles's Law:* Boyle's experiments dealt with the effect of pressure on the volume of gas if the temperature were kept constant. Jacques Charles's experiments kept the pressure constant, and looked to the effect of temperature on the volume of gases.

Most people today are aware of the fact that gases will expand when heated. Solids and liquids will expand when exposed to heat, but not to the dramatic extent that heat or temperature will affect gases.

Charles's Law is a result of experiments in which Jacques Charles discovered that *if the pressure were kept constant and the temperature became the variable, gases would expand in direct proportion to the temperature to which the gases were exposed. If the absolute temperature were doubled, the volume of gas would double.*

*Gay-Lussac's Law:* Joseph Gay-Lussac experimented with the third aspect of the physical laws regarding gases. In his experiments, Gay-Lussac kept the volume of gases constant and discovered that *when gas temperatures are raised, gas pressures are raised in proportion to the change in temperature.*

## Gas Fires

The gas containers holding liquefied and cryogenic gases will expand with heat, as in the case of compressed gas. In addition, the liquid in the container will expand and the vapor pressure of the liquid will increase as the temperature increases. All of these reactions combine to increase the pressure when the container is heated. To relieve the pressure in the container when it is exposed to fire, most compressed and liquefied gas containers are provided with relief valves or bursting discs or both. The danger is greater from failure of a liquefied gas container than from failure of a compressed gas container because larger quantities of gas are released. Failure of a compressed gas container is more a flying missile hazard than an explosion danger.

## BLEVEs

The term BLEVE refers only to the physical phenomenon that results from combustion involving unreactive liquefied flammable gases. BLEVE is an acronym that stands for Boiling Liquid — Expanding Vapor Explosion. When a container of liquefied flammable gas such as LP-Gas is exposed to fire, great energy is developed as a result of the large liquid-to-vapor expansion. This expansion may cause cracks in the structure, propulsion of pieces of the container through the air (see Fig. 4.1), and may result in the characteristic huge fireball. Heat from the fireball can kill a person who is as far as 250 ft (72.6 m) from the container. Deaths resulting from the propulsion of pieces of the container have occurred to persons who were located as far as 800 ft (243.8 m) from the container.

Cooling action on that portion of the container metal not in internal contact with the liquid in the container is the most ideal fire protection method in order to lower the temperature that causes the gas vapors to expand. However, many factors should be taken into consideration before a decision is made regarding the best action to take. A BLEVE can occur at any time and without any warning. Hazard can be increased by the production of internal heat as a result of chemical reaction and the inability to cool the liquid itself.

## Combustion Explosions

If a flammable gas or vapor from a liquefied flammable gas escapes or leaks from its container, piping, or other equipment, the gas mixes with the air. When the flammable range is reached, the mixture is ignitable and will burn. A gas fire of this sort burns rapidly and produces heat rapidly. Virtually all materials expand when heated. All the materials in the vicinity of such a gas fire absorb heat and expand. Air will double its volume for every 459°F (237.2°C) it is heated. If the heated air cannot disperse, then the pressure will rise. If the room or structure is not strong enough to withstand the increasing

*Fig. 4.1. Aftermath of a BLEVE. Arrow 1: position where the largest portion of BLEVEd tank came to rest; Arrow 2: another portion of the tank; Arrow 3: tank portion that remained in the approximate BLEVE location; Arrow 4: tank head that was punctured; Arrow 5: covered hopper tank car containing plastic pellets; Arrow 6: ethylene glycol tank car.* (Dallas Fire Department Photo Unit, Dallas, TX)

pressure from the heated air, an explosion will result. This type of explosion is called a "combustion explosion."

Most structures can withstand pressures of one pound per square inch (psi) or less. A flammable gas-air mixture can develop pressures of from 60 to 110 psi (41.370 kPa to 758.540 kPa) with the result that a room or enclosed structure could blow apart even if 25 percent or less of the enclosed area is occupied by the flammable vapors.

The basic safeguards against a combustion explosion are the use of emergency flow control devices and the reduction of the possibility of leakage by using strong containers and equipment.

## *Odorizing*

Natural and LP-Gas is odorized so that a normal person can detect gas concentrations in air not exceeding one-fifth of the lower limit of flammability.

## Gas Standards

LP-Gas, acetylene, oxygen, and other common gases are contained in cylinders for transportation purposes and during use. The cylinders are built and used subject to regulations of the U.S. Department of Transportation. Large volumes of gases are transported by pipelines. This method is also regulated by the Department of Transportation. Piping of natural and LP-Gas into homes is safeguarded by NFPA 54, *National Fuel Gas Code,*[15] which has been adopted by most state and local utility authorities.

## Comprehensive Reference

A reference with which all students of the fire hazard of materials should be familiar is the Table of Flammable Liquids, Gases, and Volatile Solids sponsored by the NFPA Committee on Flammable Liquids and published in NFPA 325, *Fire-Hazard Properties of Flammable Liquids, Gases, and Volatile Solids.*[16] This table provides data such as flash point, ignition temperature, flammable limits, specific gravity, vapor density, boiling point, water solubility, method of extinguishment, and health, flammability, and reactivity ratings for more than one thousand substances. These substances are listed alphabetically.

# HAZARDOUS MATERIALS

Of growing concern to members of the fire protection community is the category of materials known as hazardous materials. Their unique properties require a knowledge on the part of members of the fire service of how such materials will react in emergency situations and the dangers they can present to life safety. Their increased use and the fact that they are transported on roads, in rail cars, aboard ships, and in planes has created the possibility of hazardous materials emergencies occurring in nearly every type of community.

Although this text is not intended as a course of instruction in fire fighting methodology, this section has been included to give a brief overview of the diversity and complications involved in fighting fires in hazardous materials.

## Corrosive Chemicals

Corrosive chemicals are usually strong oxidizing agents that can heighten fire hazards and affect fire protection decisions. Caustics, which are classified as water- and air-reactive chemicals, are also corrosive. All chemicals should be treated with caution and care unless they are *known* to be harmless.

*Inorganic Acids:* Concentrated aqueous solutions of inorganic acids are not in themselves combustible. However, in addition to their corrosive and therefore destructive effect on living tissue, their chief fire hazard results from the possibility of these materials mixing with combustible materials or other chemicals; such a situation could result in fire or explosion. The fact that almost all corrosive chemicals are strong oxidizing agents accounts for their potential as a serious component of a fire hazard.

1. *Hydrochloric acid.* The chief hazard in hydrochloric acid is its ability to react with metals such as tin, iron, zinc, aluminum, and magnesium. Reaction with such metals results in the formulation of hydrogen gas. The combination of the two corrosive chemicals — hydrochloric acid and nitric acid — produces chlorine and nitrous oxide.

2. *Hydrofluoric acid.* This acid is noncombustible and does not generate combustion of other combustible materials. But hydrofluoric acid attacks metals, with the resultant production of hydrogen gas. Hydrofluoric acid is highly toxic and can cause serious eye irritation as well as severe skin burns.

3. *Nitric acid.* Under certain conditions, nitric acid will react with cellulose material such as wood. Nitric acid or its vapor, when it reacts with cellulose material, drastically increases the ease-of-ignition potential of wood. If the acid solution is strong enough, mixture with cellulose material can result in spontaneous heating.

The oxides of nitrogen that are given off when dilute nitric acid reacts with organic materials produce a concentration of gases which, if inhaled in even very small amounts, can cause serious illness or death.

4. *Perchloric acid.* At the normal commercial strength of 72 percent, heated perchloric acid is a strong oxidizing and dehydrating agent. This and other advantageous properties have resulted in the wide use of perchloric acid in analytical laboratories.

However, if not used properly or if the concentration level of perchloric acid exceeds 72 percent, this chemical can be extremely hazardous. Perchloric acid greatly increases the burning rate of organic materials and can cause explosions as a result of prolonged exposure to its vapors. Dehydrating agents should never be mixed with perchloric acid. Strong dehydrating agents (such as concentrated sulfuric acid) convert perchloric acid into anhydrous perchloric acid which, even at room temperature, can explode with tremendous violence.

5. *Sulfuric acid.* Sulfuric acid absorbs water from any organic material with which it may come in contact, with the result that sufficient heat might be evolved to cause ignition.

**Halogens:** Halogens are salt-producing chemicals that are very active. Halogens are noncombustible but will support combustion. For example, the presence of halogens causes turpentine, phosphorus, and finely divided metals to ignite spontaneously. Halogen fumes are poisonous and corrosive.

Halogens have similar chemical properties and differ basically in the chemical activity they generate. The following halogens are listed in terms of greater-to-lesser degree of activity with regard to fire hazards: fluorine, chlorine, bromine, and iodine.

1. *Fluorine.* Fluorine is a greenish-yellow gas and is one of the most reactive elements known. Under varying conditions, fluorine will react almost immediately with virtually all known elements and compounds. Fluorine reacts violently with hydrogen as well as with many organic substances, and attacks glass and most metals. Also, it reacts explosively with water vapor.

2. *Chlorine.* This heavy, greenish-yellow gas is poisonous and, although not flammable in itself, can react with other materials to cause fires and explosions.

3. *Bromine.* Bromine is a dark reddish-brown corrosive liquid that may cause fire when it comes in contact with combustible materials.

4. *Iodine.* Iodine's chemical form is usually in purplish-black crystals that are highly volatile as well as corrosive. Iodine can be explosive when mixed with chemicals such as ammonia and turpentine.

**Storage and Fire Protection for Corrosive Chemicals:** Storage of corrosive chemicals should be provided with two major considerations in mind: (1) protection against the damaging effect of corrosive chemicals on living tissue, and (2) guarding against any fire and explosion hazard that might be associated with the corrosive chemical.

1. *Inorganic acids.* Inorganic acids should be stored in cool, well-ventilated areas that are not exposed to the sun or other chemical and waste materials. Inorganic acids should be protected from freezing temperatures as well.

Generally, water in spray form is the recommended procedure for fighting fires in inorganic acid storage areas. However, in fires involving perchloric acid, extra care should be taken as this acid may mix with other organic material and result in an explosion. Fire fighters should, of course, avoid any contact with spilled acid and, particularly, should avoid inhaling any of the toxic fumes.

2. *Halogens.* The halogens fluorine and chlorine should be stored in special containers. Fluorine may be safely stored in nickel or monel cylinders. Although fluorine reacts with these two metals, the reaction results in the formation of a protective nickel fluoride layer that prevents further action. However, impurities or moisture in the cylinder may cause an explosive reaction. Chlorine, a serious inhalation hazard, should be stored in areas where ventilation is a prime consideration. Areas where chlorine leakage is suspected should not be entered without self-contained breathing apparatus. Fluorine, in addition to requiring protection by the wearing of self-contained breathing apparatus, should not be approached without special protective clothing to guard against severe flesh burns.

## Corrosive Vapors

Very often, corrosive vapors are a by-product of an industrial or chemical process. Ducts must be used to carry these vapors safely from the area. The type of duct utilized is determined by the corrosive properties and potential of the vapor. Often a heavier gage metal may be sufficient to protect against the corrosive effects of a vapor. Other circumstances may require a protective coating or even a special lining in the ducts. Stainless steel, asbestos cement, and plastic linings have been used with success depending on the corrosive vapor involved.

## Radioactive Materials

The process by which an atom of an element emits radiant energy because its nucleus is unstable is called radiation. The emission of radiant energy in the form of alpha ($\alpha$), beta ($\beta$), or gamma ($\gamma$) radiation from an element is the element's radioactivity.[17]

Whether an element or compound is radioactive or not has no real bearing on that material's fire or explosion hazard. The major difference to be considered in fire protection is the damage that can be caused by radioactive materials to living tissue. Under fire conditions involving radioactive materials, radioactive dust and vapors may form which could contaminate neighboring areas as well as the area involved.

***Characteristics of Radioactive Materials:*** The presence of radioactive dusts and vapors is not detectable by any of the human senses. Special instruments are required to detect and measure the presence of radioactivity.

Radioactivity can cause injury and loss of life. In addition, the presence of radioactivity can hamper fire fighting and salvage efforts as well as cause extended loss because of contaminated buildings and materials.

Usually, relatively small amounts of radioactive materials are involved in the manufacture and production of certain items. The manufacture of luminous watch dials and tracers for chemical and biological reactions are examples of processes that require limited use of radioactive materials. Nuclear reactors and radiation machines used in research, on the other hand, require the use of significant amounts of radioactive materials.

***Handling and Storage of Radioactive Materials:*** Because of the possibility of accidental release of radioactive materials in case of fire or explosion, great care must be taken in the handling and storage of radioactive materials. NFPA 801, *Recommended Fire Protection Practice for Facilities Handling Radioactive Materials,*[18] presents detailed guidelines on this subject.

Because radioactive materials present little or no additional fire and explosion hazards, the main emphasis in fire protection of buildings and areas in which such materials are stored is to ensure that the buildings, or areas themselves, are as firesafe as possible so that leakage or escape of radioactive materials will not result if a fire occurs.

Fire hazards of buildings or laboratories in which radioactive materials are stored can be determined by applying the combustibility considerations of buildings or laboratories not involved with radioactive materials, *i.e.,* nature and design of the building, furnishings, and nonradioactive chemicals.

One aspect of radiological laboratories or areas that deal with radioactive materials is the duct system that is used to safely dispose of radioactive vapors, gases, and dusts. The duct system must be designed and constructed so that in the event of fire, contaminated materials are not released to other parts of the building, area, or laboratory.

Immediate control of any fire in an area handling radioactive materials is vital to prevent any leakage or escape of the contaminated matter. Automatic sprinkler protection or specially designed piped water spray systems should be considered mandatory equipment in such areas.

*Fire Protection for Radioactive Materials:* The main concern in fire protection involving radioactive materials is to prevent the release (or at least control the release) of these materials during fire extinguishment. Although fire protection operations are similar to those used when nonradioactive materials are involved, fire involving buildings or areas containing radioactive materials presents two additional main factors to consider: (1) the presence of harmful radioactive materials might necessitate that normal fire fighting procedures be changed; and (2) because of the presence of radioactive matter, delay in salvage and resumption of normal operations may occur.

The dangers and difficulties inherent in fires involving radioactive materials require thorough hazard analysis and careful emergency prefire planning. In buildings, plants, laboratories, or areas where radioactive materials are involved, a fire emergency cannot be handled by simply calling the public fire department. Decisions must be made well in advance by the appropriate personnel with regard to the type of fire emergency that can be expected. Safety measures such as shutting down or isolating parts of a plant or individual equipment, and the special procedures necessary for immediate protective action must be planned and designed in advance of any fire emergency. Preplanning should involve working out proper and effective coordination between personnel and public fire fighting teams.

Once an area is contaminated, new considerations are involved. Large supplies of water help, of course, in fire control, but can also be helpful in decontamination operations. However, safe disposal and/or storage of contaminated water that results must be prearranged.

## *Transportation*

The safe transportation of chemicals and radioactive materials depends on three major factors: (1) knowledge of the hazardous properties of the chemicals, (2) the normal and abnormal conditions to which the materials may be exposed during shipment, and (3) guarding against the possibility of accidental release or reaction of the chemicals.

State regulations govern intrastate transportation of chemical and radioactive materials. Interstate transportation of such materials is governed by the

*Fire Hazards of Materials* 97

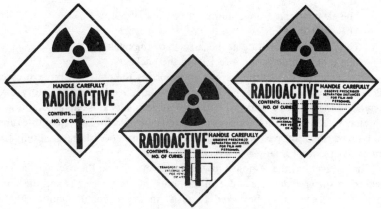

*Fig. 4.2. Labels for radioactive shipment.*

U.S. Department of Transportation (DOT). Both state and federal regulations include regulations concerning container construction, method of packaging of materials, marking, labeling, and placarding, and regulations concerning the mode of transportation: air, sea, rail, highway (see Fig. 4.2). The NFPA *Fire Protection Handbook* lists the major principles concerning the safe transportation of chemical and radioactive materials:[19]

1. Make container of material that will not react with or be decomposed by the chemical.

2. Exclude chemicals that can react dangerously with each other from the same container.

3. Package toxic and radioactive chemicals so that they will present no health hazard during normal transportation conditions and will not be released in an accident or under other abnormal conditions.

4. Provide sufficient outage for maximum expansion of liquids under conditions to be expected during transportation.

5. Limit the amount of chemical that can be released by container breakage or leakage by limiting maximum size of individual containers.

6. Cushion containers to minimize possibility of breakage.

The rules and regulations that govern the transportation of radioactive materials cover the responsibilities of the three agents handling such materials: the shipper, the carrier, and the receiver. The following description of the responsibilities of these three agents are described in *Handling Radiation Emergencies* as follows:[20]

The shipper must package the material properly and label it correctly. The package must survive and contain the contents under any credible accident. The shipping documents must give the following information:

The element and atomic number (*e.g.,* cobalt 60).
The number of curies.
The radiation dose rate at the surface of the package.
The radiation dose rate 3 feet from the package.

The carrier must take due care in transporting the material. In most cases the carrier must label the package as "radioactive," properly store and segregate the material, and promptly report accidents and damaged and lost shipments. Rail and truck transportation must be used for large quantities and long-lived radioisotopes, while common-carrier airplanes can be used only for short-lived isotopes, largely for medical use.

The receiver has the responsibility for determining that the package has not leaked or suffered damage and, if it has, must notify the carrier and the shipper. The receiver may be asked to assist in any decontamination that is necessary, and must notify the proper authorities of lost shipments or shipments that may have gone astray.

Shipping containers used to transport radioactive materials are also subject to stringent regulations. Such containers are classified as Type A containers or Type B containers. Regulations for Type A containers require that, under normal transportation conditions, loss of radioactive materials and/or container shielding will not occur. Type B containers are required to be so constructed that loss of radioactive materials and shielding will not occur and, in case of an accident, there will be no loss of radioactive materials and only limited loss of shielding will occur. Table 4.5 shows the kinds of testing packages used for shipment of radioactive materials must undergo.

## FIRE FIGHTING AND CONTROL

When hazardous materials are (or may be) involved in a fire incident, special precautions and attack procedures are necessary. Knowledge of the specific materials involved and their reaction potential is, of course, mandatory. Specialized fire fighting procedures should be planned in advance. This is particularly difficult when a fire emergency occurs as a result of an accident involving the transportation of hazardous materials. Incidents involving hazardous materials that are in transit present many additional variables that must be considered in fire protection decisions.

### Radioactive Emergencies

Professional radiation monitors are required in fires involving radioactive materials. However, fire fighters involved in such fires must have enough knowledge and skill regarding handling radioactive fire emergencies so that they may take whatever steps necessary and possible to limit danger and protect themselves from the dangers of radiation. Because fire fighters are often the first emergency personnel to arrive at a fire emergency involving radioactive materials, the responsibility of life safety, control of the fire, and initial steps to reduce contamination requires knowledge of and skill in dealing with fire emergencies involving radioactive materials (see Fig. 4.3).

Following are some precautions that are necessary when handling fire emergencies involving radioactive materials. These precautionary suggestions are adapted from Charles Bahme's book titled *Fire Officer's Guide to Emergency Action*.[21]

1. Using the instruments carried on the apparatus, evaluate the radiation level. If no equipment is available, call for such equipment.

2. Do not enter a placarded area unless accompanied by an authorized monitoring official.

3. Until the arrival of the monitoring equipment, attempt to confine the fire to the area or room involved.

4. If radiation is present, all personnel not directly contributing to the rescue or fire fighting operations, including personnel in nearby buildings, spectators, reporters, etc., should be kept at least 500 feet from the storage building or location.

5. Wear breathing apparatus, preferably of the self-contained type, helmet, coat, rubber boots, and gloves in any area where radioactive material may be involved to minimize the danger of contamination.

Table 4.5 Tests for Packages for Shipment of Radioactive Materials*

| Tests | Type A | Type B |
|---|---|---|
| Water spray that keeps the package wet for 30 minutes and after 1½ to 2½ hours, a 4-foot free drop onto an unyielding surface. | X | X |
| A 1-foot free drop onto each corner. (Applies only to wood or fiberboard packages of 110 pounds gross or less.) | X | X |
| Impact of the hemispherical end of a 13-pound steel cylinder, 1½-inch diameter, dropped 40 inches onto the vulnerable side of the package. | X | X |
| A comprehensive load of either 5 times the weight of the package or 2 lb/in.² times the maximum horizontal cross-section, whichever is greater. | | |
| Direct sunlight at ambient temperature of 130°F. Ambient temperature of −40°F in still air and shade. | X | X |
| Reduced pressure equal to 0.5 atmospheres. | X | X |
| Vibration normally incident to transportation. | X | X |
| A free drop of 30 feet onto a flat unyielding surface so as to cause the most damage. | | X |
| A drop of 40 inches onto the end of a rigidly mounted 6-inch diameter steel shaft. | | X |
| Whole surface thermal exposure at 1475°F for 30 minutes and not cooled artificially until at least 3 hours after the test. | | X |
| Water immersion under 3 feet of water for 8 hours. (For fissile materials packaging only.) | | X |

*From *Handling Radiation Emergencies*, by Robert G. Purington and Wade Patterson.

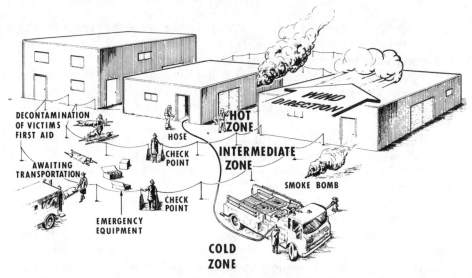

*Fig. 4.3. An emergency situation involving a radioactive hazard. Barricades and checkpoints separate "hot" and "intermediate" zones from "cold" zones. Fire fighters working in the "hot" zone wear air masks and extra protective clothing. The smoke bomb indicates wind direction.*

6. Use low-velocity fog, unless $CO_2$ or dry chemical is indicated, owing to the presence of water-reactive chemicals.

7. Fight fires from upwind as far as possible, keeping out of any smoke, fumes, or blowing dust that might result from the accident. Treat the fire as a fire involving toxic chemicals.

8. Conduct no salvage or overhauling operations except in the presence of a monitoring official, who should first determine that the area is safe to enter for limited periods of time.

9. Follow department procedure regarding notation of exposures to personnel, decontamination of clothing and skin, medical records, etc.

***Extinguishment of Fire:*** The choice of extinguishing agent is determined by the chemical properties of the materials involved, not by the fact that the material is or may be radioactive. There are, however, a few situations where the use of a special extinguishing agent may be necessary.

Water is usually the best extinguishing agent. However, if alkali metals such as sodium are present, water should not be used, because such metals react violently with water. In such a case, combustible metal extinguishing agents should be used.

Whatever extinguishing agent is used, it should be used as minimally as possible or feasible so that cleanup operations can be facilitated. Water that is used should be contained in runoff so the cleanup of possibly contaminated water can be complete.

*Overhaul:* The following recommendations can be found in *Handling Radiation Emergencies:*[22]

1. Perform only enough overhauling to confirm that the fire is completely out and that there is no possibility of a rekindle. Leave the general cleanup to trained personnel at the facility.

2. Once the fire has been completely extinguished, seal off the area by closing all doors and windows and placing plastic or plywood over openings made in the roof. Tape all openings shut so that the cracks are sealed. This ensures that contaminated material is not transported by the air to the outside.

4. Tools and equipment that are highly contaminated should be left inside for cleanup by facility personnel.

5. All gear and personnel should be monitored and decontaminated before leaving the hot zone.

6. Run-off water should be contained if possible. Keep it from spreading to the street or getting into the sewer system. This can be done with dikes made of earth, sand, sawdust, etc.

## Chemical Emergencies

When a chemical emergency arises, it is vital to determine the nature of the chemicals involved and the potential hazards these chemicals possess. The following are preliminary steps to be taken in a chemical emergency:[23]

1. **Try to find out what chemicals are involved.** Check labels, placards, shipping documents, etc. Do not accept the guesses of watchmen, truck drivers, train brakemen, etc. Trade names of products are not sufficient unless you can determine the chemical involved.

2. **If you can't find out what chemical is involved** from any source (shipper, receiver, warehousemen, driver, etc.) assume that it should have a "DANGEROUS" placard, and may be highly toxic, violently reactive, or explosive, and act accordingly.

3. **If a fire is about to reach chemicals** placarded "DANGEROUS," or "EXPLOSIVES," or bears an NFPA 704 label with a "3" or "4" in the right hand corner, evacuate the area and prepare to protect exposures from fires which may result from an explosion.

4. **If you find out what the chemical is,** determine the hazard from references you carry . . . or have the dispatcher look them up in NFPA references, Chemcards, Sax, etc., or call CHEMTREC (Chemical Transportation Emergency Center), 800-424-9300, toll free.

5. **If you find out what the hazards are,** take the proper action to safeguard lives and property, but never jeopardize the lives of your personnel merely to save someone's property. . . . you will probably need to:

    a. Block off the area.

    b. Approach from upwind with hose lines or other extinguishing equipment.

    c. Wear protective clothing and self-contained breathing apparatus.

d. Rescue injured or overcome victims.

e. Depending upon the chemical and the nature of the emergency, either make a quick retreat or attempt to handle the situation through containment, dispersal, absorption, neutralization, control or extinguishment.

***Extinguishment of Fire:*** Once the chemical materials and their potential hazards have been identified and the appropriate preliminary action taken, the following procedures are recommended:[24]

**6. Use no water, fog, foam, or soda acid extinguishers,** on chemicals with which water produces a dangerous reaction, including:

a. Sodium, potassium, lithium, rubidium. (Water produces a violent reaction on these chemicals, producing enough heat to ignite the hydrogen which is liberated.)

b. Concentrated acids or hydroxides, sulfur chloride, phosphorus oxychloride, and similar anhydrous liquid chlorides. (Water produces violent spattering and sufficient heat to ignite combustible materials; fumes are very corrosive and toxic.) EXCEPTION: In an area where safe to do so, small spills may be flooded from a safe distance. . . .

c. Sodium peroxide, potassium peroxide. (Water produces oxygen and sufficient heat to ignite combustibles; the reaction is not violent, however, and small spills may be flooded, using care to avoid contact with the caustic fumes and solutions produced.)

d. Carbides, phosphides, and silicides, of various light metals. (Water produces a flammable, and sometimes toxic, gas; small outdoor spills can be flooded, using care to use enough water to dissipate the heat sufficiently to prevent spontaneous ignition of the gases.)

e. Baths of molten salts, lead, zinc, etc. (Water will produce a violent steam explosion.)

**7. Use large volumes of water on chemicals which**

a. *Explode* when involved in a fire in a confined condition, such as ammonium nitrate in the hold of a ship, drums of sodium chlorate, drums of chromic acid flakes, packages of benzoyl peroxide.

b. *Intensify a fire* by supplying oxygen when heated, such as sodium nitrate, potassium bromate, and similar oxidizing salts.

c. *Rupture their containers* when exposed to sudden high temperatures, such as drums or carboys of acids and caustics, and cylinders of gases, including:
1. flammable gases, such as acetylene, hydrogen, ammonia, hydrogen sulfide, butane, propane, methyl chloride, etc.
2. combustion-supporting gases, such as nitrous oxide, oxygen, chlorine.
3. nonflammable gases, such as sulfur dioxide, nitrogen, carbon dioxide, freon, etc.

d. *Are extinguished by water,* such as yellow or white phosphorus.

e. *Are controlled by water application,* such as magnesium castings and fires in nonmiscible flammable liquids and gases.

f. *Are dissolved by water application,* such as ammonia and sulfur dioxide gases, and soluble solvents, such as alcohol, acetone, etc.

## Fire Hazards of Materials 103

**8. Do not attempt to fight a fire** that is about to reach:

a. *High explosives,* such as dynamite, TNT, etc.

b. *Low explosives,* such as black powder, gun cotton (cellulose nitrate over 12 percent nitration), smokeless powder, etc.

c. *Chemicals* that burn like smokeless powder, such as dry benzoyl peroxide, lauroyl peroxide, and acetyl peroxide.

d. *Containers of perchloric acid.*

**9. Wear breathing apparatus wherever:**

a. Toxic gases may be released, such as chlorine, ammonia, sulfur dioxide, nitric oxide, methyl bromide, hydrocyanic acid, ethylene oxide.

b. Toxic vapors may be released, such as benzene, xylene, toluene, carbon tetrachloride, carbon disulfide.

c. Toxic gases may be generated from the action of acids or water on cyanides, sulfides, phosphides.

d. Toxic gases may be formed as products of combustion, such as carbon monoxide, sulfur dioxide, hydrogen cyanide.

e. Toxic gases may be formed when halogenated hydrocarbons are exposed to high temperatures, as when freon refrigerants are decomposed to form hydrochloric acid, chlorine, phosgene, etc.

Note: A few gases are also poisonous through skin absorption, such as hydrogen cyanide and methyl bromide, both of which have practically no perceptible odor capable of giving adequate warning of their presence in lethal quantities.

**10. During overhauling operations,** be careful to:

a. Avoid mixing incompatible chemicals which may result in forming an explosive or causing an explosion, such as chlorine and acetylene, nitrates and sulfur (or other combustible materials); chromic acid flakes and alcohol or glycerine; acids and solid oxidizers, such as sulfuric acid and potassium chlorate; nitric acid and flammable liquids; hydrogen peroxide and flammable liquids. Keep packages of chemicals bearing DOT "White," "Yellow," or "Red" labels separate from each other.

b. Wear available protective clothing, such as boots, rubber aprons, rubber gloves, goggles (or breathing apparatus), and avoid ingestion, inhalation, or contact with chemicals.

**11. Make entry in personal record** of any fire fighter who has suffered ill effects, giving length of exposure, effects, etc.

**12. Make any required special report.**

## *Transportation Emergencies Involving Hazardous Materials*

Transportation accidents in which hazardous materials are involved, such as derailed tank cars or overturned tank trucks, present special life safety hazards (see Fig. 4.4). Because the vehicles carrying hazardous materials are mobile, such an accident can occur virtually in any area into which trains or trucks can travel. An accident may occur in an area that is highly congested or in an area that is far-removed from adequate fire fighting equipment and facilities. In addition, unlike buildings, factories, or plants where hazardous materials might become involved in fire but where the personnel may be expected to know how to react in such emergencies, transportation accidents, for the most part, involve people who may be unaware of the potential hazards of the materials involved and may be untrained in how to react in the face of such an emergency.

*Fig. 4.4. This overturned tank trailer containing 9,000 gal (34 m³) of liquefied propane gas could not be righted without transferring the propane into two empty tankers. The tanker was then filled with water, and air bags were used to raise the vehicle. However, the tanker shifted on the bags and trapped two workmen, as shown in the photo. Both men were rescued, and a major disaster was averted. (Paul Ockrassa,* St. Louis Globe-Democrat, *St. Louis, MO)*

***Life Safety Hazards:*** People are generally unaware that cryogenic liquids such as liquid oxygen, liquid hydrogen, liquid nitrogen, and liquid fluorine can cause severe frostbite, or skin burns; that a tank truck of such a liquid can, upon evaporation, create the equivalent of 600 to 800 tank trucks of gas; that even static energy can ignite evaporated cryogenic liquids; that wind can carry toxic gases that can contaminate clothing and buildings at distances that seem to be far removed from the scene of the emergency.

In a transportation accident involving hazardous materials, relief valves, refrigeration systems, insulation, and the tank itself may be damaged. If the material is exposed to fire, the evaporation rate can increase the internal pressure to the point where the tank ruptures and the resultant explosion and flying pieces of metal present even greater hazards. Leaks and spills and the possibility of contamination through movement of dangerous vapors through the air are other potential hazards.

***Action at Emergencies:*** Two types of emergency action are major considerations in incidents involving hazardous materials transportation accidents: (1) ensuring the safety of people and personnel, and (2) containing or controlling the emergency.

Following are some precautions and suggestions that involve people and personnel safety and may apply in a given transportation emergency. These suggestions and precautions are not all-inclusive, but serve to identify precautions and procedures that must be taken into consideration during a transportation emergency.

*In the case of cryogenic liquids such as liquid oxygen and liquid hydrogen:*

- Avoid getting any of the liquid on clothing, as it can be readily absorbed. Splashed clothing should be removed immediately.
- Do not touch any cold metal with bare hands, as it may stick to your flesh and rip off your skin when you pull away. Remember also that cryogenic liquids give off extremely cold vapors which, in the absence of special clothing, can cause frostbite damage to feet and hands.
- If a large spill has occurred, lay lines to protect exposures, and keep apparatus and people out of the vapor area. Remember that in the case of any cryogenic liquid, the visible cloud will not define the danger limits; it can extend much farther, even where no visible cloud can be seen. It takes very little energy to ignite hydrogen — even static energy can do it.
- In the case of large outdoor spills confined in dikes, let them stay where they are and divert vapors after they leave the spill in the safest direction by means of high velocity water spray. Do not ignite large spills. If the spills are already burning, do not apply hose streams, as the heat of the water will only increase the rate of burning and evaporation. Do not try to extinguish a large spill fire, but protect exposures.
- Do not stand over cracks in pavement where liquid oxygen has penetrated and water has flowed in on top. Past experience has shown that the expanding

gas, trapped by the inverted wedge-shaped ice plug in the crack, can blow up the concrete.
• Water is not recommended on large pools of liquid oxygen where the heat of the water might produce such large quantities of gaseous oxygen as to endanger persons or property in the vicinity — particularly to the leeward.

*In the case of toxic chemicals:*

• Keep cooling streams of water on drums as well as on cylinders that are exposed to heat to prevent flame impingement, fusing of the safety devices, or rupture of the cylinders that have no safety devices.
• Use water spray streams to flush away spills, disperse vapors, and to protect personnel attempting to shut off a valve to stop the flow, or who are trying to plug a leak.
• Be careful to avoid touching cyanides and other poisons with your bare hands. Keep your hands, which may have contacted poisonous solutions, away from your eyes and mouth.
• On combustible oils such as aniline and cresol, use water spray, dry chemical, or carbon dioxide; foam can also be used, but in the case of aniline, a higher rate of application than normal will be required because of the initial foam breakdown caused by aniline.
• Do not use carbon dioxide on chemicals that give off poison gas in the presence of acids, such as cyanides, because the carbonic acid formed by the carbon dioxide in moist air will tend to accelerate the evolution of hydrocyanic acid.
• Wear full protective clothing (including respiratory protection) during fire fighting and overhauling operations.
• Do not drive apparatus on an asphalt road that has liquid oxygen on it, nor drop couplings, spanners, or other tools on it — casualty-producing explosions can happen.

The following are suggestions to be considered in handling the emergency itself:

*For cryogenic liquids:*

• Approach the scene from upwind and keep apparatus far enough back to stay out of the vapor cloud (which is not always visible).
• Evacuate the immediate area of people and automobiles; divert traffic if necessary, using police assistance.
• Radio for assistance from military or industrial experts if needed.
• Try to stop the leak or shut off the flow if possible.
• If unsuccessful in controlling the leak, or in case of a large spill, ask your dispatcher to notify the nearest plant or agency from which the product has come or is intended for delivery; obtain advice and special emergency gear if available.

- Stop the flow of leaking oxygen if possible. Where valves are ineffective, massive ice caps made by covering the leak with sheet polyethylene and wet rags or roller cloth towels soaked with fine water spray may be effective.
- Where possible, have the contents of the leaking vessel transferred into another vessel to end the hazard as soon as possible.
- Small spills can be evaporated sooner by use of hose streams.
- Mixtures of liquid oxygen and fuels are shock-sensitive and sensitive to detonation. Where such mixtures have accidentally occurred in the absence of fire, shut off the supply to the mixture and, if the fuel is water soluble, dilute it with water. Where large amounts of insoluble fuel have mixed with large amounts of liquid oxygen, evacuate the area and protect exposures from barricaded positions while the oxygen evaporates.
- In the case of small spills in an area where there are no exposures, prompt ignition will result in rapid boiling, and elimination of the hazard. Should it become necessary to extinguish a liquid hydrogen or methane fire, first extinguish fires in all surrounding materials to eliminate the source for reignition; water spray will generally be effective for this. If only a very small quantity of cryogenic is involved, potassium bicarbonate can be used successfully. For a fire in small amounts of hydrogen-air mixture, bromotrifluoromethane (Halon 1301) will effect extinguishment.
- Where the fire involves liquid oxygen flowing into flammable solids, first stop the flow if possible and simultaneously use water to cool adjacent tanks or other exposures. Remember that almost *anything* will burn in the presence of oxygen, including steel, as such fires are extremely hot and fast burning. If the fuel is finely divided, such as sawdust or similar organic material, a shock-sensitive high explosive is created when soaked with liquid oxygen; in such a case, protect exposures from a shielded position and from a "safe" distance, keeping in mind the story of the man being chased by the bull who yelled at the farmer, "Is that bull safe?" and the farmer replied, "He's safer than you are!"
- If the fire is burning in liquid oxygen that has flowed into a water-soluble fuel, the application of water may dilute the fuel eventually to a point where the fire can be put out.
- If the fire is in a mixture of nonmiscible fuel and liquid oxygen, and the flow of oxygen can be stopped from going into the fuel (or vice versa), then the fire can be extinguished with appropriate agents for the type of fuel involved. If either the flow of oxygen or the flow of fuel cannot be stopped to prevent their commingling, then take cover and protect exposures from a shielded location.
- If a flammable liquid or other contaminant should flow into liquid nitrogen and catch fire, use the type of extinguishing agent appropriate for the material involved; follow normal procedures and safety precautions the same as for any cryogenic liquid. A serious hazard with liquid nitrogen is anoxia, or permanent damage to body tissue from a deficiency of oxygen.
- If fluorine vessels or tank trailers are well involved in fire upon arrival,

## 108   Principles of Fire Protection

evacuate the surrounding area as quickly as possible, and take cover at a safe distance until the fluorine has been consumed; then fight any remaining fires. The fluorine will be consumed in the fire very quickly, so direct your efforts toward covering exposures and secondary fires that may continue after the fluorine has gone.

• Fight any fire that exposes liquid fluorine vessels or tank trailers by means of unstaffed wagon batteries, portable monitors, ladder pipes, or deluge guns. Do not direct water on the fluorine, as it will produce an extremely violent reaction.

• Use neoprene gloves, garments, and boots to protect against both fluorine and the hydrofluoric acid which it forms in moist air. Use only self-contained breathing apparatus with a full face piece.

### *Summary*

The three basic types of materials that provide fuel for all fires are: (1) combustible solids, (2) flammable and combustible liquids, and (3) flammable gases. The number and variety of these materials amounts to thousands.

Wood is still the most "basic" fuel for fires. However, even the combustible properties of wood are not a simple subject. The physical properties of wood, the moisture content, and the smoke factor involved in burning wood are but a few of the necessary considerations in fire protection preplanning and fire control decisions. Plastics and textiles, which are also combustible solids, present special fire hazards because of their chemical or synthetic nature and properties.

The characteristics of flammable and combustible liquids are many and varied. Temperature and pressure have a marked effect on the fire hazards that might be presented by flammable and combustible liquids. In addition the special characteristics of these liquids, such as ignition temperature and evaporation rate, affect all fire protection considerations and decisions.

Gases can be classified by their physical properties — whether they are flammable, nonflammable, toxic, and/or reactive — as well as by their physical state — whether they are compressed, liquefied, or cryogenic. These properties or states of gases have direct bearing on fire protection considerations and decisions. In addition, gases can be classified by usage — fuel, industrial, or medicinal. Knowledge of the physical laws regarding the behavior of gases as well as the chemical reactivity and behavior of gases is a vital decision-making factor for all fire fighting personnel.

Hazardous materials such as corrosive chemicals and vapors and radioactive materials require special knowledge and consideration because of the high life safety factors involved. Inorganic acids, halogens, and nuclear materials all require special handling, storage, and fire fighting procedures.

Fire fighting and control with regard to fire hazards and materials is dependent on knowledge about the materials involved. In addition, more specialized knowledge and information may be required regarding those materials that may be stored or used in buildings or plants in a particular fire protection district. Preplanning is a necessary factor where hazardous materials are known to be used in relatively large quantities in a given area. In addition, members of the fire service should be aware of the kinds of hazardous materials that may be transported through a given area so that preplanning and the necessary firesafety and life safety procedures that may be required may be planned well in advance of any emergency.

## *Activities*

1. Discuss the following characteristics of liquids in relation to their flammability:
   (a) Vapor pressure.
   (b) Flash point.
   (c) Evaporation rate.
   (d) Specific gravity.
2. (a) What does Boyle's Law predict about the volume occupied by a given mass of gas?
   (b) Both Charles's Law and Gay-Lussac's Law deal with the temperature and volume of a gas. Discuss the differences that these characteristics play in the two laws.
3. Radioactive materials do not necessarily present fire and explosion hazards greater than other materials, but fire protection in areas where these materials are stored is of utmost importance.
   (a) What are some of the reasons for this?
   (b) What are the difficulties inherent in fighting a fire involving radioactive materials?
4. Liquids themselves do not burn, yet some are termed "flammable" or "combustible." Explain what happens before a fire or an explosion involving a flammable or combustible liquid can take place.
5. Evacuation of an area endangered by a hazardous materials emergency may sometimes appear necessary. However, evacuation can often create as many problems as it attempts to solve. With a group of classmates discuss the advantages and disadvantages of an evacuation procedure during a hazardous materials emergency.
6. (a) Why do natural fiber textiles generally create less of a fire hazard than synthetic textiles?
   (b) Why is it difficult to assign a fire hazard or flammability limit to a general group of plastics?
7. (a) What are the dangers of flashover?
   (b) How does flashover differ from flame spread?
8. (a) Explain how a piece of wood may not burn even though its surface is subjected to a heat source higher than its ignition point.

110    Principles of Fire Protection

    (b) Conversely, tell how a piece of wood might ignite while in contact with a heat source whose temperature is lower than its ignition point.
9. (a) To what phenomenon does the term BLEVE (Boiling Liquid — Expanding Vapor Explosion) refer?
    (b) What are some of the hazards of a BLEVE?
10. What are the responsibilities of the following agents in the transportation of hazardous materials?
    (a) The shipper.
    (b) The carrier.
    (c) The receiver.

## Bibliography

[1] Tuve, Richard L., *Principles of Fire Protection Chemistry*, NFPA, Boston, 1976, p. 81.

[2] Beall, F. C. and Eichner, H. W., "Thermal Degradation of Wood Components: A Review of the Literature," Report No. 130, May 1970, Forest Products Laboratory, Madison, WI (from NFPA *Fire Protection Handbook*, p. 3-4).

[3] Tuve, Richard L., *Principles of Fire Protection Chemistry*, NFPA, Boston, 1976, p. 81.

[4] *Fire Protection Handbook*, 14th Ed., NFPA, Boston, 1976, p. 3-8.

[5] "America Burning," May 1973, The National Commission on Fire Prevention and Control, Washington, DC, p. 67.

[6] *Fire Protection Handbook*, 14th Ed., NFPA, Boston, 1976, p. 3-8.

[7] Tuve, Richard L., *Principles of Fire Protection Chemistry*, NFPA, Boston, 1976, p. 106.

[8] NFPA 321, *Standard on Basic Classification of Flammable and Combustible Liquids*, NFPA, Boston, 1976.

[9] *Fire Protection Handbook*, 14th Ed., NFPA, Boston, 1976, p. 3-19.

[10] _____, p. 3-19.

[11] Tuve, Richard L., *Principles of Fire Protection Chemistry*, NFPA, Boston, 1976, p. 50.

[12] NFPA 30, *Flammable and Combustible Liquids Code*, NFPA, Boston, 1976.

[13] Tuve, Richard L., *Principles of Fire Protection Chemistry*, NFPA, Boston, 1976, p. 120.

[14] *Fire Protection Handbook*, 14th Ed., NFPA, Boston, 1976, p. 3-41.

[15] NFPA 54, *National Fuel Gas Code*, NFPA, Boston, 1974.

[16] NFPA 325M, *Fire-Hazard Properties of Flammable Liquids, Gases, and Volatile Solids*, NFPA, Boston, 1977.

[17] Purington, Robert G. and Patterson, Wade, *Handling Radiation Emergencies*, NFPA, Boston, 1977, pp. 30-31.

[18] NFPA 801, *Recommended Fire Protection for Facilities Handling*

*Radioactive Materials,* NFPA, Boston, 1975.

[19]*Fire Protection Handbook,* 14th Ed., NFPA, Boston, 1976, p. 3-66.

[20]Purington, Robert G. and Patterson, Wade, *Handling Radiation Emergencies,* NFPA, Boston, 1977, p. 55.

[21]Bahme, Charles W., *Fire Officer's Guide to Emergency Action,* 3rd Ed., NFPA, Boston, 1976, pp. 108, 109.

[22]Purington, Robert G. and Patterson, Wade, *Handling Radiation Emergencies,* NFPA, Boston, 1977, p. 111.

[23]Bahme, Charles W., *Fire Officer's Guide to Emergency Action,* 3rd Ed., NFPA, Boston, 1976, pp. 30-31.

[24]_____, pp. 32-35.

*Chapter Five*

# Investigating the Fire Loss Problem

*Very seldom can the cause of fire be determined without some degree of investigation. Our present regulatory codes, standards, and inspection and suppression procedures were developed mainly from the investigation of fires and from the careful analysis of the important information that these investigations yielded.*

## THE NEED FOR INVESTIGATIONS

Fire has always been a major concern to society. Although originally fire was often thought of as a supernatural act, its destructive qualities and the tragedies it caused soon made people realize the need for more advanced knowledge concerning how to protect themselves from fires. In order to better understand fire protection needs, investigations were made to determine the causes of fires that had occurred and to discover ways to eliminate similar occurrences in the future.

Although it is not known exactly when the first fire investigation took place, it can be speculated that investigations into the causes of fires may have resulted from the curiosity of early humans. Following the fires that sprang up in their dwellings, early humans often returned to the fire scene to sift through the rubble in search of salvageable articles. As they searched, they probably thought about how and why the fires started. The earliest recorded instances of what can be considered as organized fire investigation date from the days of the Roman Empire when the *Quarstionarius* was assigned to investigate the causes of all destructive fires.

As early civilization expanded and the use of various materials for building construction became more common, some of the reasons for the simultaneously increasing number of fires became more obvious. For example, overcrowded building conditions, thatched roofs, and wood shingles were soon recognized as being directly related to fire occurrences. Still later, medical research revealed that specific by-products of fire such as toxic gases were the actual cause of many deaths from fire. Such information led to the more in-depth testing of the burning characteristics of manufactured materials.

## The Purpose of Fire Investigations

Because local fire departments are present at most fires, they have the opportunity to compile firsthand reports of fire incidents. Thus, fire investigations are an important function of every fire department. According to the NFPA *Fire Protection Handbook,* fire investigations by the public fire service serve the following three basic purposes:[1]

> 1. *To determine what happened, so that preventative measures can be taken in the future.* Too often fire investigations are conducted strictly to affix blame. A fire occurrence generally is a failure of either a code enforcement program or a public education program, except in the case where criminal activity is involved. Fire investigations can lead to better prefire activities that exact better control over the fire ignition sequence, particularly in those areas where the fire investigation shows shortcomings on the part of the fire service.
>
> 2. *To ascertain whether there was any criminal activity involved.* The rate of incendiarism is on the increase, and it is only through proper fire investigation that incendiarism will be detected and the proper evidence secured for conviction of arsonists.
>
> 3. *To provide accurate information for the fire report.* (By definition, the fire report is the legal record of a fire department incident.) The fire service must maintain an accurate report of each fire occurrence, the circumstances surrounding it, and the damage and/or casualties resulting from it. Information for the fire report can only come from an investigation of the fire.

## The Scope of the Investigation

The three most significant areas in fire investigation are: (1) fire ignition sequence, (2) fire development, and (3) fire casualties.

*Fire Ignition Sequence:* It is important to determine the location within the property where the fire started, and the ignition sequence that caused the fire to start. This ignition sequence consists of identifying three factors: a heat source, a kindling fuel, and an event, human action, or natural act that combines the heat source with the kindling fuel to start the fire. Each of these three factors must be identified separately in order to fully explain the ignition sequence. Information provided by fire scene witnesses can be helpful in determining the fire ignition sequence. However, there are occasions when the fire ignition sequence cannot be determined so easily. In such instances, the point of origin needs to be ascertained as closely as possible. Points of origin can be easily established by first reconstructing the layout of the room or area of origin. Information provided by surviving occupants, or by individuals who are familiar with the interior of the property, can be of value in locating the position of furniture and the types of combustible materials that were in the area. When the fire ignition sequence has been established, investigators can begin to determine the fire's development.

*Fire Development:* Once a fire starts, its growth or development is based on a number of factors, each of which is important in understanding why a fire grew as it did or remained as small as it did. The contents of a building often provide fuel to the fire, thus causing the fire to spread or to remain concentrated in a particular area. These materials should be identified, and their role in the overall fire development should be evaluated. From a reconstruction of the room or area of origin, an examination of the contents of the room can be used to determine the level of origin of a fire, and the direction of burning or fire travel. Additionally, contents of the room may have had an effect on fire fighting conditions or may have resulted in fire casualties because of gases or smoke they created.

Another factor directly involved in fire development is the compartmentation, or other subdivision, of a structure. These physical barriers play an important part in limiting the development of fire and smoke conditions, and their performance should be evaluated after the fire. If fire walls, doors, or other smoke- or fire-limiting devices were present and did not perform as they should have, the reasons for unsatisfactory performance should be explored.

The time the fire burned prior to its detection or discovery, and the time between the detection and the transmission of an alarm to the fire department, are important factors in the ultimate extent of damage from the fire. The effect that automatic fire detection and extinguishing equipment had on the extent of damage from the fire or on the survivors from the fire should also be analyzed. If there were delays in the transmission of alarms from automatic detection equipment, the reasons for these delays should be evaluated as part of the overall fire development sequence.

Fire department tactics at the fire scene can affect the outcome of a fire. To help determine whether or not a fire department had the available resources to properly tackle a fire or if it properly utilized available resources, and to help determine whether any part of the fire's development or spread resulted from inadequate tactical procedures, a thorough evaluation of fire scene tactics should be made after each major fire.

*Fire Casualties:* Each injury and fatality associated with a fire incident should be thoroughly investigated for cause and for any information that may lead to measures for preventing future casualties. Casualties should be followed up for a period of time to determine their outcome. To evaluate the reasons for the casualties, an understanding of the fire development sequence may be necessary. Finally, the fire department's public education program should be studied to determine its effect upon the reasons for the survival of occupants, or for the lack of knowledge of those who suffered casualties.

## Conducting the Investigation

Fire departments should maintain an accurate report of each fire, the circumstances surrounding it, and the casualties and damage resulting from it. Such a procedure depends upon the investigation following each fire. This investigation may only take a few minutes if the fire is small, or it may take many days if the fire is large or complex and the details of its ignition and development are difficult to ascertain. The amount of time spent on an investigation, however, should be governed by how long it takes to fully understand the facts about the fire.

The NFPA *Fire Protection Handbook* recommends the following as a fairly effective general guide for investigation of the fire scene:[2]

1. Review structural exterior, fire suppression, and timing.
2. Reconstruct as much as possible.
3. Establish approximate burning time and temperatures.
4. Determine path of heat travel and point of origin.
5. Evaluate combustion characteristics of all materials involved.
6. Compare similar materials and situations.
7. Fit known facts to various possibilities.
8. Corroborate information from occupants and witnesses.

A good fire investigation will include all of these factors and any others that have a bearing on the way the fire behaved. The information obtained from a thorough investigation can increase firesafety awareness and thus benefit the inspections that fire departments conduct in their communities. Knowledge of such information as the causes of fires and the potential hazards of building contents can help a community to face its fire problems by instilling in it a greater awareness of firesafety.

***Fire Scene Evidence:*** In investigations of large-loss fires the entire fire scene can be considered evidence, although items at or near the point of origin will have greater value than others. These articles supply information such as the initial materials ignited and the source of ignition, and may also yield data on time-temperature factors.

Before removal, all fire ignition sequence evidence should be photographed and tagged or marked with the owner's name or initials, date, type of evidence, and its location at the fire scene. If the evidence is of any material of intrinsic value, a receipt should be given to the owner.

***The Preliminary Investigation:*** There are many conditions that should be noted both on the way to a large-loss fire and at the fire scene itself. These conditions include: the automobiles and people in the area of the fire, the color and intensity of the fire and smoke, and the direction of fire travel. Although most fire fighters usually note direct fire conditions automatically, others might well need to be trained to remember indirect fire conditions such as vehicles and witnesses in the fire area, as often most vehicles and witnesses will leave the area before the fire is extinguished. Names of all persons present should be obtained because witnesses can provide information on factors such as point of origin, time of burning, and actions of occupants in both accidental and arson-related fires.

Fire fighters should also note whether doors and windows were locked, unlocked, or open at the fire scene. If windows were forced open by fire fighters, the condition of the windows or doors should be noted. After the fire has been extinguished, the information obtained from the preliminary investigation will be valuable for conducting the actual investigation.

*Fire Scene Analysis:* Once a large-loss fire is out, guards should be posted. Unnecessary personnel should be excluded from the area until either the fire cause has been established or the investigator has taken charge. Additionally, overhauling or disarranging the premises should not be permitted unless it is needed to control or black out the fire. Such action may disturb or destroy essential evidence.

The fire investigation should include a discussion with eyewitnesses concerning what they have observed, and a survey of debris remaining at the fire scene. Sometimes a physical or chemical laboratory analysis of the remaining debris must be made in order to determine the condition of materials prior to the fire. Scientific experts may be able to provide fire departments with a fairly accurate sequence of events and to evaluate the effects of the fire on certain pieces of debris.

In many American states and all Canadian provinces, the primary responsibility for the investigation of incendiary fires is assigned to the office of the state or provincial fire marshal. In states that have no such office, or where the fire marshal's office does not have the responsibility for arson investigation, such investigation is conducted by the local police and fire departments.

Ideally, a fire department should have a team of specialists available for assistance when investigation circumstances are too complex to handle, or when the situation is too time-consuming for the fire officer. In any event, the fire officer present during the extinguishment of the fire should remain involved because of the support that can be given to the investigative team in reconstructing the scene and identifying the time sequence of certain events.

Loss-of-life fires, big fires, and arson fires are likely to be investigated by other agencies, as well as by local fire departments. For example, in addition to the state fire marshals who are often called in by the fire department in these cases, insurance companies and insurance bureaus often investigate fires of special interest to them. Also, manufacturers of products that were involved in the fire may send investigators to report on the performance of their products.

In turn, the fire reports that provide records of fire investigations will assist a fire department by showing trends in both fire losses and fire prevention and suppression effectiveness. For many years NFPA's Fire Analysis Department published information on large-loss fires (fires causing a loss of $250,000 or more). These annual studies have provided a great deal of useful information on fire protection and the most frequently encountered types of inadequate firesafety awareness of individuals.

The purpose of these reports and the types of fire reports that are available are discussed in the following section.

## RECORDING THE FIRE PROBLEM

All the information gathered during the fire investigation should be properly recorded into a fire report. The fire report is the legal record of a fire incident, and includes information on the time of the incident, response to the incident, action taken, details of the fire, and damage or casualties resulting from the incident. Some fire incident reports may be brief, while others may be extensive reports that include photographs, physical evidence, and laboratory test results. Some departments, such as those who use the *UFIRS* system (*Uniform Fire Incident Reporting System*), may have various types of standardized forms. Whatever its length, scope, or form, the fire report should be in the words of the fire officer present at the incident, and should be complete and clear so that another fire service member not present at the incident can understand what happened. This section of this chapter will explore the purposes of fire reports and fire reporting systems, their application to fire investigations, and how these reporting methods are used.

### The Purpose of Fire Reports

In his book *Fire and Arson Investigation,* John Kennedy discusses three major purposes of a fire investigation report for fires dealing with arson:[3]

1. To communicate the important information of the investigation to a supervisor, prosecuting attorney, or other interested person who will be utilizing the information obtained in the investigation.
2. To provide a permanent record.
3. To provide guidance in the preparation or conduct of litigation.

Within these three main purposes are more specific purposes that have a direct relationship to a fire department. The fire report, as well as being a legal record of the fire's occurrence, is also the fire department's legal record that the department took action to extinguish the fire and to protect building contents or occupants. Specifically, the fire report records both the actions that the fire department took and the origin and cause of the fire; it also assesses the damage and reports the casualties that resulted from the fire. The fire report also provides information on the performance of the fire fighting unit. From an assessment of the fireground tactics, the fire fighting unit and its superiors can judge individual unit performance and, if necessary, revise strategy. The fire report also provides a fire department with data on the community's fire problem. This data can be used to track trends, measure the effectiveness of fire prevention and fire suppression procedures, evaluate the impact of new methods, and indicate those areas that may require further attention. The information can also be fed into state and national fire data banks for use in evaluating the scope of the fire problem on a broader basis.

## *Types of Fire Reports*

Two reports that are of importance to fire departments are the "Basic Incident Report" and the "Basic Casualty Report." These reports were developed in NFPA 902M, *Fire Reporting Field Incident Manual:*[4]

> ... to allow a fire department to collect and summarize basic details about all incidents to which it responds and to use that information in making decisions affecting the fire protection of the community.

**Basic Incident Report:** Figure 5.1 shows an example of the Basic Incident Report. The first block is designed to collect information on:

- Time and date of incident.
- Location of the incident.
- Property owner or responsible party.
- The incident itself.
- Fire department actions.
- Fire department equipment used.

The second block collects information on the number of casualties and type of property involved in the fire; the third block identifies why a fire started and the location of fire origin. Fire damage and factors for fire and smoke travel are recorded in the fourth block. Fire suppression information and mobile property records are collected in blocks five and six, respectively.

**Basic Casualty Report:** In any incident that results in death or injury, casualty information should be recorded in the Basic Casualty Report (see Fig. 5.2). Each report contains space for up to three casualties. Because of the necessity to follow up all hospitalized injuries, revised reports should be submitted. All reports should reflect uniform definitions of fire injury and fire death as recommended by NFPA 902M, *Fire Reporting Field Incident Manual:*[5]

> **Injury.** Physical damage to a person suffered as the result of an incident that requires (or should require) treatment by a practitioner of medicine within one year of the incident (regardless of whether treatment was actually received), or results in at least one day of restricted activity immediately following the incident.
>
> **Death.** An injury which is fatal or becomes fatal within one year of the incident.

*Investigating the Fire Loss Problem* 119

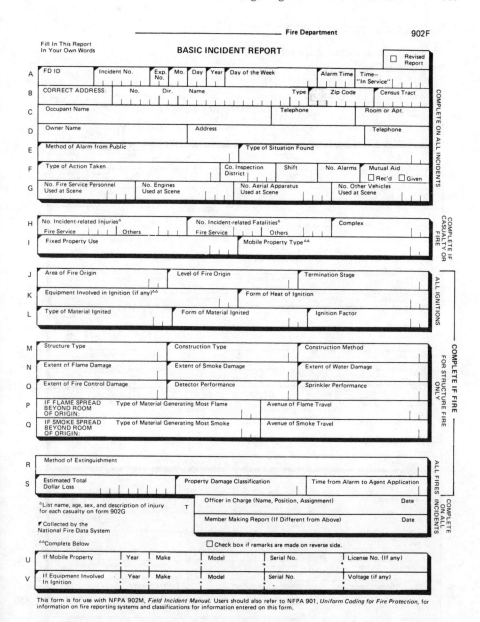

Fig. 5.1. *Example of the Basic Incident Report. This report, which is divided into six blocks, should be completed by fire investigators.*

## Fire Reporting Systems and Their Objectives

A good fire incident reporting system is valuable to a local fire department for several reasons. According to the NFPA *Fire Protection Handbook,* the data can be collected and analyzed to:[6]

- Describe a community's fire problem.
- Support budget requests.

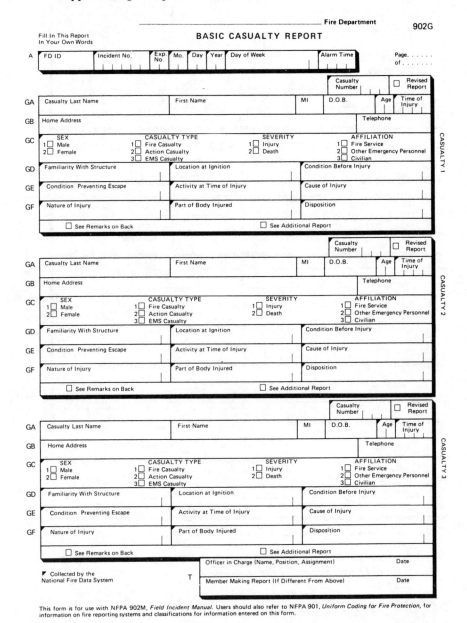

Fig. 5.2. *Example of the Basic Casualty Report for reporting incidents that result in death or injury. This report should be completed by a member of a fire department.*

- Support code changes.
- Evaluate code enforcement programs.
- Evaluate and guide fire prevention educational programs.
- Plan future fire protection needs.
- Improve allocation of resources.
- Schedule nonemergency activities to avoid peak response periods.

Until recently there has been little uniformity in fire reports; within a department, terminology, details, and reporting procedures have been left to an individual officer's choice. As a result, department uniformity in reporting has been lacking and has made it difficult for fire data collection at the state and national level.

Recognizing this problem, NFPA's Committee on Fire Reporting has been working to improve fire reporting uniformity. Their report — NFPA 901, *Uniform Coding for Fire Protection* (hereinafter referred to as the NFPA *Uniform Coding Standard*) — establishes basic definitions and terminology for use in fire reporting systems, and also sets forth a means of coding data so that it can be aggregated either manually or by data processing equipment. The stated purpose of the NFPA *Uniform Coding Standard* is to assist a fire department that wishes to build its own fire reporting systems, and to help a department recognize the objectives of collecting and using uniform fire incident information.[7]

According to the NFPA *Uniform Coding Standard,* a common fire reporting language and a method for systematic information collection, processing, and use comprise a fire reporting system, including the following objectives:[8]

- To provide for the routine collection of information required by a local fire service . . .
- To provide local fire service management with information to indicate trends, to measure the effectiveness of fire prevention and fire suppression measures presently in practice, to evaluate the impact of new methods, and to indicate those areas that may require further attention.
- To provide a local prefire inventory of all property in a fire service district so that future needs may be anticipated . . .
- To provide uniform data to regional, national, and international fire control organizations in order: to make the full extent of the fire problem known; to reveal facts that require action on these levels; to guide the effective development and administration of codes and standards; and to guide fire prevention and fire protection research.

In order to be complete, a fire reporting system has three elements that are necessary for the successful aggregation of information. (See Table 5.1.) These elements are: (1) fact finding, (2) fact processing, and (3) fact use.

*Fact Finding:* To serve as input to a fire reporting system, an incident report must be clearly structured and must use uniform definitions and terminology. The collection of information for an incident report requires some type of

Table 5.1  The Three Elements of a Fire Reporting System*

|  | | |
|---|---|---|
| **Element I** Fact Finding | A. | Obtain information |
| | B. | Complete report form |
| | C. | Send completed report to processing |
| **Element II** Fact Processing | D. | Receive completed reports |
| | E. | Edit (and code) reports |
| | F. | Enter facts |
| | G. | Process facts |
| | H. | Update fact file |
| **Element III** Fact Use | I. | Report periodically |
| | J. | Analyze these reports |
| | K. | Request special report (if needed) |
| | L. | Decide specific action |
| | M. | Act |
| | N. | Analyze results of that action |
| | O. | Return to J and repeat |

*From NFPA 901, *Uniform Coding for Fire Protection,* NFPA, Boston, 1976, p. 11.

form on which to record the information, instructions to ensure uniformity in completing the form, and procedures for forwarding a completed form to a central point.

*Fact Processing:* Once data have been received, they must be processed into a record that is useful for legal, statistical, planning, and management purposes. The first step in processing involves checking the incident reports for accuracy and completeness, then collecting information for compilation into a composite record.

The second step involves creating a file consisting of every available record of the reported incidents. This "Fire Fact File" serves as a compilation of the basic sources of information about past incidents, and its use will determine the facts that must be recorded on an incident report.

*Fact Use:* Once an incident file has been started, it may have many potential uses. It should first meet the informational needs of all sectors of the local fire service. Such needs include both information required from a legal standpoint, and information necessary for periodic statistical reports. Specifically, the file can be used to provide company officers with data on their designated part of the protected community. A more general use would be to spot trends in fire incidence and provide data for program evaluation and fire-related research.

Through each incident report and through the Fire Fact File, the company officer and the chief of the department can work to "manage" the local fire problem. State and federal authorities can "manage" their interests, and on a broader scale, industry, educators, architects, research scientists, and fire protection engineers can work as a team to reduce the fire problem as defined by the basic local Fire Fact File.

The National Fire Prevention and Control Administration is currently at work on a national data file that is based on the NFPA *Uniform Coding Standard*. As more local fire departments accept this standard, and as more states collect fire information in conformity with the standard, a more accurate and useful picture of the total fire problem will emerge. From this picture, an analysis of fire losses can be made.

## ANALYZING FIRE LOSSES

Any analysis of fire losses includes all the information about a fire, especially the fire ignition sequence, and all other factors related to its spread, such as structural integrity and the performance of fire protection equipment. Three basic types of fire analysis — large-loss fire analysis, fire ignition sequence analysis, and "group fire" analysis — are of major importance, as are the steps that can be taken to eliminate, or at least decrease, the loss of life and property from fire.

### Large-loss Fire Analysis

As previously stated, fires resulting in individual losses of $250,000 or more are termed "large-loss fires" by the NFPA Fire Analysis Department. The following three factors are usually the cause of large-loss fires:[9]
1. Structural defects.
2. Improper handling and storage of combustible contents.
3. Private fire protection weaknesses.

Table 5.2 shows each of these three factors, and how each factor is distributed according to occupancy.

A large-loss fire is usually more thoroughly investigated than a small-loss fire, and thus can often provide greater information for use as the basis for an accurate fire analysis. An accurate analysis can provide a clear indication of what fire protection weaknesses and human weaknesses are most frequently encountered at fires. Such information can be useful in preventing loss in other types of fires since the factors influencing fire spread are very often the same, irrespective of property type and value.

**Structural Defects:** Table 5.2 shows that structural defects are the major factor in the large-loss fire experience. The most prevalent structural fault is the absence of walls to subdivide large areas. Even when division walls do exist, they can be ineffective as fire barriers when there are no fire doors or other protection for openings in the walls, or when fire doors are blocked open.

The principal structural defect responsible for vertical fire spread is the absence of fire cutoffs at openings between floors. Open stairways and elevator shafts are the most frequent paths of vertical fire spread. Combustible

### Table 5.2  Factors Responsible for Fire Spread Involving Property Damage of $250,000 or More to Buildings*

The items detailed in this table reflect some of the significant factors responsible for the spread of fire in buildings which resulted in loss to the structure or its contents of $250,000 or more. The data are based on studies of selected fire experiences as reported and analyzed by the NFPA Fire Analysis Department. In any given fire a number of contributing factors might be involved. The purpose of the information presented is to assist in evaluating those features of structures or their contents, private fire protection weaknesses, or public protection handicap which have proven to be significant in these fires. The numbers reflect only the recorded number of instances as reported in the fires studied and should not be taken to reflect the relative importance of each factor in any given property.

| | Public Assembly and Office Facilities | Mercantile Facilities | Industrial Facilities | Manufacturing Facilities | Storage Facilities |
|---|---|---|---|---|---|
| **A. Principal Structural Defects Influencing Fire Spread** | | | | | |
| Vertical Spread | | | | | |
| 1. Stairways and elevators not enclosed by fire walls or partitions | 27 | 9 | 4 | 4 | 3 |
| 2. Nonfire stopped walls | 14 | 8 | 1 | 4 | 4 |
| Horizontal Spread | | | | | |
| 1. Nonfire stopped areas including floors and concealed spaces above or below floors and ceilings | 73 | 60 | 3 | 51 | 53 |
| 2. Interior wall openings unprotected | 8 | 6 | 2 | 10 | 5 |
| 3. Exterior finish | 4 | 1 | 2 | 7 | 15 |
| Combustible Framing/Finishes | | | | | |
| 1. Structure or framing | 62 | 41 | 7 | 56 | 58 |
| 2. Ceilings, walls, floors | 8 | 6 | 1 | 2 | 4 |
| **B. Contents Features Influencing Fire Spread** | | | | | |
| 1. Products in storage | 3 | 19 | 1 | 24 | 53 |
| 2. Flammable liquids, gas not properly contained | 3 | 3 | 1 | 12 | 9 |
| **C. Principal Fire Protection Defects Influencing Fire Spread** | | | | | |
| Automatic Sprinkler Performance | | | | | |
| 1. Extinguished/controlled fire | 4 | 3 | 2 | 17 | 9 |
| 2. Did not control/extinguish fire | 12 | 4 | 2 | 37 | 34 |
| 3. Standpipe/hand extinguisher helped control fire | 6 | 1 | 10 | 11 | Not reported |
| 4. Standpipe/hand extinguisher did not help control fire | 22 | 2 | 6 | 10 | Not reported |

*From *Fire Protection Handbook,* 14th Ed., NFPA, Boston, 1976, p. 1-25.

construction materials such as fiberboard wall panels and plywood, poorly constructed party walls, and interconnecting passageways also contribute to large-loss fires.

***Building Contents:*** As noted in Table 5.2, improper storage of building contents and flammable and combustible liquids contribute to the fire spread. When not stored properly, flammable and combustible liquid vapors can escape from their receptacles and intensify a fire. The improper storage of combustible materials (such as oil-soaked rags) is usually a result of either poor housekeeping practices or the lack of firesafety awareness.

***Fire Protection Weaknesses:*** Only a small percentage of large-loss fires have occurred in buildings with complete or partial sprinkler systems. However, in those large-loss fires in which automatic sprinkler systems were unsuccessful, the principal reasons were: (1) closed water control valves, (2) obstructions to sprinkler distribution, and (3) only partial protection of occupancies by sprinkler systems. Primarily, the major cause of unsatisfactory performance of automatic sprinkler systems has been the result of human action: such action involves the closing of water supply control valves before the fire occurs, or before the fire is completely extinguished. Fire department, security, and industrial personnel have all been involved in the premature closing of water control valves.

Recognizing deficiencies in fire protection and correcting improper firesafety practices are important factors when analyzing large-loss fires. The human element is usually a basic factor in all large-loss fires, since, fundamentally, human beings are responsible for building components that are improperly stored, for structural defects in buildings, and for the installation of inadequate fire protection systems.

## *The Principal Causes of Fires*

There is a simple answer to the question "What are the three principal causes of fires?" That answer is: men, women, and children. The tragic and costly fire loss experienced in this country year after year results from the fact that far too few people are aware of fire prevention or firesafety. All too few members of our society have had the benefit of regularly scheduled fire drills and training. Thus, many members of our population are inadequately prepared to cope with emergency situations. As described in "America Burning," one reason so many people are unaware of fire prevention and firesafety is that it is human nature to avoid thinking about unpleasant subjects:[10]

> ... Though Americans are aroused to issues of safety in consumer products, firesafety is not one of their prime concerns. Few private homes have fire extinguishers ... Too few multiple-family dwellings and institutions have automatic equipment for extinguishing fires ... when fire strikes, ignorance of what to do leads to panic behavior and aggravation of the hazards, rather than to a successful escape.

However, fires can be prevented if people are motivated. About thirty years ago, a large city in western Canada experienced a fire department strike that left the city defenseless. The people in the city were warned of the strike through various media and sources. During the ten days that the fire department went on strike, there were no fires of any consequence in the city. However, when the fire fighters returned to duty, the normal ratio of fires in the city was immediately restored.

Another example of temporary public fire awareness occurred in a large city in the northern United States. A severe blizzard paralyzed the area and streets became impassable. All citizens were warned that the fire department could not respond to their aid in the event of a fire. Fire incidents dropped to practically nothing — but only until the emergency was over.

Every fall most cities and towns in the United States observe Fire Prevention Week. The president and all fifty governors issue proclamations calling for public attention to firesafety; special attention is given to firesafety in the schools, and publicity reminds people to be careful with fire. A study conducted on the effect of Fire Prevention Week in several cities showed that fire losses in these cities began to drop a week or two before Fire Prevention Week, were at a low level during the week and for a few weeks after, but returned to the normal weekly ratio after that time. Fire Prevention Week, and the instances that have previously been cited, prove that people can be motivated to fire prevention — at least for a short time. When motivation occurs, firesafety is usually a major concern. For example, as a result of the large life loss in Boston's Cocoanut Grove nightclub fire, nightclubs all over the world were inspected and laws were passed to provide noncombustible decorations, emergency lighting, safety exits, and control of overcrowding. Our Lady of the Angels School fire in Chicago increased public concern for school firesafety and resulted in better fire protection for schools throughout the country.

However, such examples of public firesafety motivation are only temporary; human actions continue to be a direct cause of fires. Estimated building fires as a direct result of human causes are shown in Table 5.3, and include such human causes as careless smoking habits, misuse of electrical equipment, improper protection of open flames and sparks, poor handling or storage of flammable liquids, and misuse of fireworks and explosives.

***Smoking and Related Fires:*** With millions of people lighting up billions of cigarettes each day, it is not hard to understand that smoking is a leading cause of fires. Cigarettes continue to burn when left unattended or discarded, thus presenting a fire hazard when smokers fall asleep while smoking or when cigarettes are thrown into a container filled with other combustibles. A burning cigarette can ignite clothing, blankets, furniture, draperies, rugs, gasoline vapors, brush and forests, and — under the right conditions — nearly every other kind of combustible solid, liquid, or gas. Fires from careless smoking surpass all other causes of fire in almost all types of occupancies; in one- and two-family dwellings, smoking is the second leading cause of fire.

Smoking-related fires are also caused by the carelessness of smokers when striking or disposing of matches. In addition, the prevalence and availability of matches has led to a large number of deaths and serious burns to children who play with matches. Children of all ages and the misuse of matches are major contributing factors to the causes of many fires.

The development of safety matches that strike only on a box or booklet has helped to decrease smoking-related fires. Although there are available "approved" matches with heads that do not fly off and sticks that are treated for afterglow, "unapproved" safety matches are still in wide use. A recent survey conducted by the Consumer Products Safety Commission reported an estimated 9,500 people were treated in hospitals for match-related injuries such as ignition of clothing and burn injuries to the hand and fingers.

Table 5.3  Estimated Number of Building Fires by
Fire Ignition Sequence, 1970-1974*

|  | 1970 | 1971 | 1972 | 1973 | 1974 |
|---|---|---|---|---|---|
| Heating and Cooking Equipment | | | | | |
| Defective, misused equipment | 79,600 | 87,800 | 89,400 | 97,500 | 93,300 |
| Chimneys, flues | 20,300 | 22,400 | 21,800 | 23,900 | 14,000 |
| Hot ashes, coals | 7,200 | 8,000 | 6,800 | 6,500 | 12,600 |
| Combustibles near heaters, stoves | 35,800 | 39,500 | 37,200 | 37,900 | 40,100 |
| Total: | 142,900 | 157,700 | 155,200 | 165,800 | 160,000 |
| Smoking Related | 107,200 | 118,400 | 109,700 | 115,200 | 121,600 |
| Electrical | | | | | |
| Wiring distribution equipment | 89,500 | 98,800 | 101,600 | 106,700 | 112,200 |
| Motors and appliances | 56,200 | 62,100 | 61,000 | 64,000 | 52,800 |
| Total: | 145,700 | 160,900 | 162,600 | 170,700 | 165,000 |
| Trash Burning | 31,100 | 34,400 | 36,000 | 35,200 | 177,000 |
| Flammable Liquids | 58,800 | 64,900 | 65,200 | 67,300 | 56,100 |
| Open Flames, Sparks | | | | | |
| Sparks, embers | 5,000 | 5,500 | 6,200 | 6,500 | 13,300 |
| Welding, cutting | 8,800 | 9,700 | 8,200 | 9,800 | 11,600 |
| Friction, sparks from machinery | 14,700 | 16,200 | 17,000 | 16,200 | 11,900 |
| Thawing pipes | 5,200 | — | 5,500 | 5,500 | 5,800 |
| Other open flames | 33,500 | — | 35,000 | 32,000 | 34,900 |
| Total: | 67,200 | 74,100 | 71,900 | 70,000 | 77,500 |
| Lightning | 20,100 | 22,200 | 22,700 | 21,600 | 16,600 |
| Children and Fire | 63,800 | 70,400 | 69,200 | 70,800 | 59,600 |
| Exposure | 21,000 | 23,200 | 25,400 | 25,200 | 44,200 |
| Incendiary, Suspicious | 65,300 | 72,100 | 84,200 | 94,300 | 144,400 |
| Spontaneous Ignition | 14,200 | 15,700 | 15,100 | 14,900 | 11,000 |
| Gas Fires, Explosions | 11,400 | 12,600 | 8,700 | 9,600 | 11,900 |
| Fireworks, Explosives | 3,500 | — | 4,200 | 4,300 | 4,200 |
| Miscellaneous Known Causes | 77,800 | 3,800 | 65,900 | 70,500 | 91,700 |
| Unknown Causes | 162,000 | 166,200 | 154,200 | 150,500 | 159,200 |
| Total Building Fires: | 992,000 | 996,600 | 1,050,200 | 1,085,900 | 1,270,000 |

*From *Fire Protection Handbook*, 14th Ed., NFPA, Boston, 1976, p. 1-27.

***Electrical Equipment Fires:*** The principal causes of electrical equipment fires are improper installation, poor maintenance, and careless use of electrical appliances and electrical wiring. Electric cords can become frayed, particularly if they are hidden under rugs. In such instances, the electric cords can smolder for weeks. Overloaded circuits and misused fuses and circuit breakers can cause fires by short circuit faults or by arcs and sparks.

When electrical systems for lighting, power, heating, and other purposes are properly designed, installed, and maintained, they are firesafe. The human element of careless use, improper installation, or poor maintenance, however, has made fires from electrical equipment the leading cause of fires in one- and two-family dwellings. Electrical fires also rank first among fires in hospitals, bowling alleys, motion picture theaters, motor hotels, schools, shopping malls, supermarkets, and many types of industrial operations.

***Open Flames and Sparks:*** If left unattended or improperly shielded from combustibles, open flames in fireplaces, coal- and wood-burning stoves, candles, and burning rubbish piles provide a fire hazard that only requires more fuel. Sparks from welding and cutting operations are a well-known cause of fires, particularly in industries that require these operations. Misuse of equipment used in these operations, or storage of combustibles too near these operations, invites fire.

**Flammable Liquids Fires:** As large-loss fire analysis has shown, careless storage and handling of gasoline, kerosine, and other flammable liquids are primary causes of fire. Vapors from such liquids are primary causes of fire, as they can be easily ignited by an open flame or by a spark from machinery or friction. Spills and leaks in heating or cooking equipment that use flammable liquids can also cause fire.

**Fireworks and Explosives:** The improper handling, storing, and manufacture of fireworks and explosives result in fires and injuries. Gross negligence, misuse, and carelessness by adults and children using fireworks have resulted in fatalities and in minor and serious injuries. (See Table 5.4.) While the form of misuse varies, reports include stories of fireworks being thrown into crowds or at individuals to scare them.

Improper storage or manufacture of fireworks and explosives can also create a serious fire hazard. During the 1976 Fourth of July weekend, fire investigation officials stated that they believed fireworks were a major source of ignition of many other fires, but were unable to develop specific statistics.[11]

## *Analysis of Conflagrations*

In Chapter One, "Fire — The Destroyer," Table 1.3 listed great conflagrations from 1972 to 1977. Although there is no universally accepted definition of a conflagration, the term is usually applied to fires extending over a considerable area and destroying numbers of buildings.

An analysis of conflagrations and the factors contributing to their fire spread is also given in Table 1.3. According to the NFPA *Fire Protection Handbook,* conflagrations have been classified as one or more of the following five general types:[12]

> 1. Fires starting in hazardous occupancies or dilapidated and abandoned buildings ("conflagration breeders") in congested sections which spread in one or more directions before effective resistance is organized to bring them under control. These fires usually spread first to nearby properties lacking exposure protection, cross streets by means of radiated heat, and spread chiefly in the direction in which the wind is blowing. Failure to control such fires is due almost entirely to lack of sufficient water application through heavy stream devices by the fire department, and lack of exposure protection. . . .
> 2. Fires occurring in primarily residential sections which spread beyond control due to closely built combustible construction and wood-shingled roofs. Such conflagrations may occur where such construction practices are

Table 5.4 Selected 1977 Fireworks Incidents*

| Place | Fireworks Device | Incident | Casualty Sex | Age | Injury |
|---|---|---|---|---|---|
| Seattle, WA | Homemade rocket | Victim placed match heads in carbon dioxide-type cartridge, packed it with candle wax, and inserted wick. Device exploded when crimped with a hammer | M | 19 | Fatal |
| Salem, OR | Firecracker, under 1½" long | Firecracker exploded in vent of 55-gallon drum that previously contained a flammable liquid | M | 11 | Second- and third-degree burns to 77 percent of body |
| Raynham, MA | Cherry bomb | Device went off in hand | M | 10 | Tips of two fingers amputated; burns to eyes resulting in loss of sight in one eye |
| Springfield, MA | M-80 | Device exploded in hand; handed to victim by an older child | ** | 6 | Dismemberment of hand |
| Kankakee, IL | Bottle rocket | Device flew into victim | M | 19 | Loss of right eye |
| Allen, TX | Bottle rocket | Victim was struck in eye by fireworks fired by others | M | 15 | Partial loss of sight (possible total loss) |
| Morrill, NB | Public display device | Victim was playing with a public display fireworks device found the day after a display | M | 11 | Serious facial burns; surgery required |
| Moscow Mills, MO | Firecracker; under 1½" long | ** | M | 9 | Second-degree burns to face; unknown damage to eyes |
| Billings, MT | Sparkler | Sparkler ignited clothing | F | 8 | Second- and third-degree burns to 25-to-30 percent of body |
| Topeka, KS | M-80 | ** | ** | 11 | Burns to 20-to-30 percent of face and hands; hospitalization required |
| Buffalo, NY | Unspecified fireworks | ** | M | 20 | Admitted to hospital with severe burns |
| Pasedena, MD | M-80 | Device exploded in victim's hand | M | 46 | Severe lacerations to thumb and fingers; fracture of thumb |

*(Continued)*

*From "1977 Fireworks Incidents," *Fire Journal,* Vol. 71, No. 6, Nov. 1977, pp. 45-47.
**Not reported.

## Table 5.4 Selected 1977 Fireworks Incidents — *Continued*

| Place | Fireworks Device | Incident | Casualty Sex | Age | Injury |
|---|---|---|---|---|---|
| Kansas City, MO | Cherry bomb | Device exploded in glass bottle | M | 12 | Severe cuts and lacerations to eye; lens disruption; hospitalized |
| El Paso, TX | Firecracker | Device exploded near victim's face | M | 28 | Partial loss of sight |
| El Paso, TX | Bottle rocket | Device exploded near victim | M | 23 | Partial loss of sight |
| El Paso, TX | Punk lighter | Victim struck in eye with lighted punk | F | 10 | Partial loss of sight; cornea burned |
| Tacoma Park, MD | Aerial mortar | Device burst near head | F | 18 | Partial loss of sight; second-degree burns to eye and nose |
| Leavenworth, KS | Bottle rocket | ** | M | 18 | Lacerations to eye and eyelid; probable loss of sight |
| Edgewood, TX | Bottle rocket | ** | M | 7 | Probable loss of sight in one eye |
| El Paso, TX | Fireworks | Device struck eye (closed) and exploded | M | 38 | Eye's iris damaged and severe eye hemorrhage; probable loss of sight |
| Minot, ND | Smoke bomb | ** | M | 11 | Damaged eye; probable loss of sight |
| Tulsa, OK | Bottle rocket | Device blew up near face | M | 17 | Laceration of eye |
| Tulsa, OK | Bottle rocket | Device struck eye | M | 15 | Blood in anterior of eye; hospitalized |
| Baltimore, MD | Flare | Victim held device in hand | F | 32 | Third-degree burns to hand |
| Rahway, NJ | Firecracker | Exploded in victim's face | M | 12 | Second-degree burns to face and both hands |
| Topeka, KS | Bottle rocket | ** | ** | 11 | Burns to hands and back; hospitalized |
| Tulsa, OK | Bottle rocket | Parents throwing rockets from car; bottle rocket fell back into car and exploded on child's legs | M | 1 | Lacerations to thigh plus powder burns |
| Garrison, ND | Unspecified fireworks | ** | M | 20 | Burns, lacerations and abrasions to eye, face, hand, and leg |

**Not reported.

allowed, and where fire protection forces are weak and water supplies are inadequate.

3. Conflagrations resulting from extensive forest and brush fires entering a municipality over a wide frontage.

4. Conflagrations due to explosions with resulting fire over a wide area.

5. Conflagrations evolving from multiple fires started in one area or city, caused by earthquake or rioters. Either source [earthquakes or rioters] may hamper or prevent fire fighting.

## *Prevention of Fire Losses*

Once fire losses are analyzed for their cause, fire departments and fire protection organizations can use this information to prevent similar fires from occurring in the future. Increased property inspection or public education programs by fire departments, involvement of fire protection organizations in safeguarding particularly hazardous products or properties, or enactment of state- or national-level standards and codes to eliminate fire hazardous conditions are some of the methods that can be used to prevent fire losses. Chapter Twelve, "Fire Protection Organizations, Information Sources, and Career Opportunities," will discuss the involvement of fire protection organizations in firesafety. Chapter Eleven, "Codes and Standards," will present a discussion of the scope of codes in fire protection control. For the purposes of this chapter, various fire prevention measures that can be enacted by local fire departments will be examined.

***Large-loss Fire Prevention:*** To prevent large-loss fires, fire departments need to examine not only the cause of the fire, but also the performance of fire protection devices. Because of the relationship of structural defects to fire spread, fire department inspection of properties should include an evaluation of fire walls, fire doors, open stairways, and fire cutoffs. Testing and evaluation of existing fire protection systems in a property should be performed to ensure effective operation during a fire. Finally, the storage of combustibles and housekeeping practices should be noted by inspectors, and corrective action for any deficiencies should be advised.

Firesafety education for occupants of potential large-loss fire properties could also contribute to reducing property loss and preventing casualties. If fire extinguishers are available on the property, instructions should be supplied for their use in an emergency. Proper marking of exits and preplanned exit routes should also be known by building occupants for their safe escape.

***Prevention of Fire Ignition:*** Because of the human element present in the fire ignition sequence, public firesafety education can eliminate many fires by alerting people to the potential hazards that surround them. Frequent inspections of residential occupancies can reveal poor housekeeping practices, overloaded or worn electrical equipment, and the lack of proper exits.

Also, firesafety programs prepared for children can warn them of the dangers of playing with matches, fireworks, and explosives and can help to make them more aware of the potential fire hazards in their homes.

***Prevention of Life Loss:*** To eliminate fire deaths, the public should learn how to safely evacuate from a fire. Installation of early warning devices can give building occupants adequate time to escape from deadly fire conditions. In addition, preplanning and training in the use of escape routes in the home, and an awareness of lifesaving methods such as crawling through smoke can decrease the incidence of fire deaths. The potential effectiveness of a firesafety education program in helping to reduce life loss from fires is immeasurable.

***"Group Fire" Prevention:*** The term "group fire" is sometimes used to describe fires within a limit of an industrial plant property or a group of mercantile buildings, particularly when they occupy a single city block. An analysis of the types of group fires that have occurred shows that the lack of exposure protection and the combustible construction of buildings involved in those fires served to increase economic loss. However, many group fires that have occurred have been the result of arson. The following section of this text explores the increase in arson-related fires, and presents detailed information on how arson can be prevented.

## THE GROWING ARSON PROBLEM

Fires that have been deliberately set have become a growing concern to society in general, and to local fire departments and state fire marshals in particular. The deliberate setting of fire, as defined by common law, is called arson. The model law for arson (to be discussed in the section of this chapter titled "Model Arson Law") expands the common law definition to include any property that is burned — including one's own — and designates four degrees of arson: (1) the burning of dwellings, (2) the burning of buildings other than dwellings, (3) the burning of other property, and (4) the attempted burning of buildings or property.

In a September 1976 report of the National Fire Prevention and Control Administration titled "Arson: America's Malignant Crime," arson's growth was estimated as follows:[13]

> Urban fire departments estimate that as much as half of all fire losses in America's cities are from fires that are set on purpose. Far from the cities, about a quarter of all forest fires are similarly set on purpose. Fires of incendiary origin are ripping off America at a rate so high that the metropolitan inner-city areas of some of our major population centers are beginning to resemble London after World War II.
>
> These losses are all around us, and we are beginning to feel their impacts on the American economy which each year may total as much as $10 billion*

through higher insurance premiums, higher prices for what is not burned, lost jobs, and higher taxes. Each year, also, as many as a thousand of us may lose our lives to arson.**

*Based on insurance industry estimates. The Insurance Services Office, an insurance statistical, advisory, and rating organization, estimates that the total actual fire loss due to arson in 1976 could exceed $4 billion. According to the American Insurance Association, incendiary fires currently account for 21 percent of the number of fire insurance claims and 40-50 percent of dollars lost to fire.

**Estimate from International Association of Arson Investigators.

At the second national conference of the National Fire Prevention and Control Administration, Richard L. Best of the NFPA reiterated the property loss and life loss from arson:*

> The increase in the number of incendiary and suspicious fires in the United States in the last ten years has been staggering. In 1966, the number of incendiary and suspicious fires, according to NFPA records, was approximately 37,000. In 1975 that number had soared to 144,000, an increase of approximately 285 percent.
>
> Aside from the dollar loss, other more important costs are the human lives lost in arson and incendiary incidents. That this is all too real is demonstrated by the fact that an estimated 1,000 people, including forty-five fire fighters, die each year in arson fires.

Table 5.5 on the following page further emphasizes the growing number of incendiary and suspicious fires and property losses from these types of fires for a ten-year period.

## Investigation of the Arson Fire

Arson investigation is not always considered a function of fire prevention because of varied responsibility in arson investigation procedures and the definition of the crime of arson in various state laws. Arson is a current, devastating fire problem, and a significant reduction in the number of fires can be accomplished through arson investigation and arson prevention. Before details of the arson investigations are discussed, an understanding of the arsonists, the establishment of an arson fire, and the responsibility for the investigation are needed.

***The Arsonists:*** Firesetters come from all walks of life; neither sex, age, education, nor economic status in any way limits the possibility of becoming a potential arsonist. Professional arsonists are not as prevalent as they were in the past, but they still are in business. Nearly anyone who deliberately sets a

*From an address given by Richard L. Best, fire analysis specialist of the National Fire Protection Association, at the second national conference of the National Fire Prevention and Control Administration, U.S. Department of Commerce, October 18-20, 1976.

Table 5.5  Incendiary and Suspicious Fires and Losses, 1964-1974*

| Year | Number | Property Loss |
|---|---|---|
| 1974 | 114,400 | $563,000,000 |
| 1973 | 94,300 | $320,000,000 |
| 1972 | 84,200 | $285,600,000 |
| 1971 | 72,100 | $232,947,000 |
| 1970 | 65,300 | $206,400,000 |
| 1969 | 56,300 | $179,400,000 |
| 1968 | 49,900 | $131,100,000 |
| 1967 | 44,100 | $141,700,000 |
| 1966 | 37,400 | $94,600,000 |
| 1965 | 33,900 | $74,000,000 |
| 1964 | 30,900 | $68,200,000 |

*From *Fire Protection Handbook,* 14th Ed., NFPA, Boston, 1976, p. 1-28.

fire, however, can usually be classified under one or more of three headings, according to Paul L. Kirk's book, *Fire Investigation:*[14]

> **Arsonists for profit.** This includes perhaps the largest single group of persons who burn their own property for the sake of collecting insurance, or who will burn another person's property for hire. In this context, arson is a calculated act not basically different from burglary or armed robbery, although its perpetrators tend to feel less criminal and more justified by the unfortunate aspects of their individual financial situation, . . . Such persons will certainly meet all the legal qualifications of felons, although to place them formally in this category is often very difficult.
>
> **Arsonists for spite.** Persons in this category are "getting even" or seeking revenge. Someone has wronged them, the wrong being either real or imagined, and the most obvious recourse to these warped minds is to burn the property of their persecutors. Arson, in this context, is undoubtedly more a rural than an urban crime, but no less a crime, and a serious one.
>
> **Arsonists for "kicks."** This includes two different categories of personality, the first being the "firebug" who has a pathological attraction to fires and is happiest when witnessing a fire and its destructive effects. This person is distinctly abnormal . . .
>
> The second category of arsonists for "kicks" is the rather casual but malicious prankster who sets fires merely for the momentary excitement or as a general retaliation against society. This person, generally youthful and delinquent in more ways than one, has no special attraction to fires as such. . . .

***Determining Fire by Arson:*** Intent is the important component of arson fires. Without intent, fire is classified as accidental. Although many fires result from human negligence, no act of arson is involved because the intent to make the fire rage beyond control does not exist.

If there is no clear evidence that the fire was accidental, it should be viewed as if it were the result of arson.

***Responsibility for the Investigation:*** In many states and all Canadian provinces, the primary responsibility for arson investigation rests with the state or provincial fire marshal. In addition to the fire investigation, the state fire marshal may also summon persons to an inquest and may start proceedings to compel testimony under oath.

In the absence of a state fire marshal, police and fire departments investigate the fire and collect evidence to be presented to a prosecuting attorney. Specialists in arson investigation, such as arson squads, operate cooperatively with personnel from police and fire departments. In some states, the head of the fire department or a municipal fire marshal may be given powers equal to those of the state fire marshal.

***Details of the Investigation:*** The NFPA *Fire Protection Handbook* recommends the following points to be essential for the investigation of incendiary fires, for the preparation of evidence from those fires, and for the securing of convictions:[15]

> It must be established that the fire in question actually occurred. This may be done through witnesses, fire records, photographs, sketches, and similar evidence.
> 
> A description of the building in which the fire occurred must be given, together with a statement which gives an accurate picture of the fire.
> 
> It must be indicated that the fire was caused by criminal design. This may be done by positive or circumstantial evidence. Confessions alone are not enough to establish criminal origin; they must be supported by corroborative facts such as the presence of gasoline, kerosine, or other flammables, or explosives, by the presence of more than one fire, by obstructions deliberately placed to impede fire fighting operations, by the removal of valuables just prior to the fire, etc. . . .
> 
> It must be proved that the fire did not occur from accidental causes, except where incendiary origin is so clearly established that further proof is unnecessary. This may be done by eliminating the possible sources of accidental origin.
> 
> Establishment of a motive greatly strengthens other evidence. For example, facts may be presented to show the financial straits of the defendant, or motive may be suggested by proving the existence of a desire to move or to break a lease on the part of the person who had the fire.
> 
> Responsibility for the fire must be directly connected with an individual. This may be done by a confession that is suitably corroborated. Evidence of the circumstances of the fire may also be used to establish the guilt of an individual. Possession of the only available keys to a locked building, for example, constitutes evidence of the sort which may be used to connect a given individual with the fire.

A fire that suggests incendiarism should be reported by the fire department, and the local police or state fire marshal's office should be called in to assist in a further investigation. The district attorney should also be notified for guidance in ensuring a successful arson prosecution.

The fire investigator may be able to note the *modus operandi* of the arsonist in the investigation, which may be valuable in tracing the arsonist and producing information for a trial. An absence of breaking and entering can place suspicion on a person or persons who have access to and familiarity with the building. Wanton destruction that accompanies the fire could point to an individual or individuals seeking revenge on the property owner. This information, when obtained, may be considered a bonus to the entire fire investigation and to the ultimate prosecution of an arsonist.

## Criminal Procedures for Arson

Every fire must be presumed to be accidental until or unless evidence of arson disproves this. However, proof of arson requires more than a proof of burning, which is necessary even if a party has confessed to arson; it also requires that the burning be the result of a criminal action. Arson may be proved by direct or circumstantial evidence; however, because of the absence of eyewitnesses, evidence is often largely circumstantial. Testimony by anyone with special knowledge or experience (for example, the fire investigator, criminologist, or specialists such as chemists and electricians) can be admissible. John Kennedy's book *Fire and Arson Investigation* details three requirements for a conviction in an arson prosecution:[16]

> 1. Proof that a burning occurred. Smoke discoloration is not sufficient. There must be a burning.
> 2. Proof that this burning resulted from a criminal action. This involves the *corpus delicti,* or proof of the incendiary origin of the fire or that the fire was willful and malicious.
> 3. Proof that the person charged caused the criminal burning and . . . is legally responsible.

While convictions for arson are generally difficult to obtain, most of the states have adopted the Model Arson Law (described in the following paragraphs), which has aided many convictions.

**The Model Arson Law:** In 1920 a Model Arson Law was advocated by the Fire Marshal's Association of North America. Later supported by the NFPA, this law is reflected in the legislation of some of the states in the United States. However, there are many differences in its effect; for example, in some of the states the willfull and malicious burning of an automobile is considered arson; in some other states, it is not.

According to the Model Arson Law, first degree arson is committed by "any person who willfully and maliciously sets fire to or burns or causes to be burned. . . any dwelling house . . . whether the property of himself or of another." Arson in the second degree involves the burning of buildings other than dwellings, with the same descriptive statement as first degree arson. Third degree arson involves the burning of other property, with such property "being of the value of twenty-five dollars and the property of another person." Any attempt to burn buildings or property constitutes fourth degree arson, with the following definition of an attempt to burn:[17]

> The placing or distributing of any flammable, explosive, or combustible material or substance, or any device in any building or property . . . or preparation with intent to eventually willfully and maliciously set fire to or burn same, or to procure the setting fire to or burning of same shall, for the purposes of this act, constitute an attempt to burn such building or property.

Burning to defraud the insurer of any building, structure, or personal property is also covered in the Model Arson Law. Such burning constitutes a felony, with the penalty of one to five years imprisonment. Conviction for arson, according to the Model Arson Law, carries the following penalties:

- First degree — two to twenty years imprisonment.
- Second degree — one to ten years imprisonment.
- Third degree — one to three years imprisonment.
- Fourth degree — one to two years imprisonment, or a fine not to exceed one thousand dollars.

*Other Arson-related Laws:* The Statute of Limitations is important to fire investigators because it states that a prosecution for a crime must be started within a certain period of time. Some states have a Statute of Limitations of five years for arson; thus, arson prosecutions in those states must begin within five years of the date of the fire. The law, however, varies from state to state, with some states having no Statute of Limitations for arson. Therefore, the fire investigator should be aware of the Statute of Limitations in the jurisdiction where the fire occurred.

The Federal Fugitive Felon Act is another important arson-related law with which fire investigators should be familiar. The Act prohibits flight from one state to another or to a foreign country to avoid prosecution, custody, or confinement following conviction for arson and certain other crimes. The Federal Fugitive Felon Act also pertains to attempts to commit these crimes, and prohibits interstate flight to avoid giving testimony in any criminal proceeding involving punishment by imprisonment.

## *Arson Prevention*

Despite the Model Arson Law, incendiary fire losses have increased since 1964, as was previously shown in Table 5.5. In addition, the lack of uniformity in state arson laws and the difficulty in apprehending and convicting arsonists have made arson a continuing threat to life and property.

A recent National Fire Prevention and Control Administration report titled "Arson: America's Malignant Crime," was the result of the efforts of professionals in such arson-related fields as fire and police, arson investigation, and insurance. These individuals participated in leadership seminars for developing a coordinated attack on arson. As stated in the report, the seminars revealed nine basic "needs" areas to combat the problem of arson:[18]

1. The need to develop and define responsibilities for arson law enforcement, arson detection, and the victims of arson.

2. The need to reclassify arson in crime reporting systems in order to list it separately to obtain accurate statistics on arson fires.

3. The need to develop and improve public awareness of arson.

4. The need to develop and apply training programs keyed to job-related requirements for arson investigation in the fire service and insurance industry.

5. The need to develop and apply better reporting, data collection, and data analysis to arson cases.

6. The need to promulgate and apply effective laws and regulations for arson.

7. The need to develop and identify sources of funding for arson prevention programs.

8. The need to conduct research into the arson problem and to develop equipment and investigative techniques.

9. The need to develop a consistent, uniform terminology for arson-related laws and data.

## *Summary*

The increased amount of combustibles present in our daily lives has contributed considerably to the national and international fire problem — a problem of such magnitude that only recently has the general public come to realize its seriousness. Accordingly, the scope of fire department responsibility in the area of fire investigation has broadened immensely. With the use of more sophisticated analytical methods, investigations into the fire problem have become more accurate and fire reports and fire reporting systems have been better able to pinpoint particular hazards, although the value of uniform reporting procedures has only recently begun to be appreciated. From the results of fire investigation data, fire departments are better able to understand the basis of local fire problems and are, therefore, better able to cope with the varied types of fire hazards.

With society's increased knowledge of the seriousness of the fire problem has come an awareness of the need to discover new methods for combatting the fire problem. Such new methods will, undoubtedly, result from still further investigation and analysis.

Analysis of fire investigation sequence, fire development, fire casualties, and fire loss can yield information on the fire protection weaknesses of a community, as well as on how to prevent the cause of fires. Investigation of the fire problem is also important in helping to detect and combat the growing problem of arson.

*Investigating the Fire Loss Problem* 139

## Activities

1. Determine: (1) how each of the following materials constitutes a fire hazard, and (2) how the hazard can be eliminated.
   (a) Flammable and combustible liquids.
   (b) Fireworks and explosives.
   (c) Smoking materials.
   (d) Electrical appliances.
   (e) Trash-burning containers and fireplaces.
2. (a) What are three major factors responsible for large-loss fires?
   (b) What are some of the methods that can be used to prevent, or lessen the severity of, large-loss fires?
3. How does each of the following contribute to a fire reporting system?
   (a) Fact finding.
   (b) Fact processing.
   (c) Fact use.
4. Explain the effect that the following laws have on arson investigation and arson prosecution:
   (a) Model Arson Law.
   (b) Statute of Limitations.
   (c) Federal Fugitive Felon Act.
5. How can a knowledge of the former contents of a room in which a fire originated be used to determine the fire ignition sequence and the fire development?
6. Discuss the advantages and objectives of uniform fire reporting systems.
7. How can the findings of a fire investigation be used:
   (a) To develop measures to prevent future fires?
   (b) To ascertain whether criminal activity was involved?
   (c) To provide accurate information for a fire report?
8. Discuss the overall importance of the fire ignition sequence in the scope of fire investigations.
9. How does each of the following contribute to large-loss fire analysis?
   (a) Structural defects.
   (b) Building contents.
   (c) Fire protection weaknesses.
10. What are the major similarities and differences between "group fires" and conflagrations?

## Bibliography

[1] *Fire Protection Handbook,* 14th Ed., NFPA, Boston, 1976, p. 1-19.
[2] _____, 14th Ed., NFPA, Boston, 1976, p. 9-34.
[3] Kennedy, John, *Fire and Arson Investigation,* Investigations Institute, Chicago, 1962, pp. 485-486.

[4]NFPA 902M, *Fire Reporting Field Incident Manual,* NFPA, Boston, 1976, p. 5.

[5]_____, p. 71.

[6]*Fire Protection Handbook,* 14th Ed., NFPA, Boston, 1976, p. 1-22.

[7]NFPA 901, *Uniform Coding for Fire Protection,* NFPA, Boston, 1976, p. 8.

[8]_____, p. 7.

[9]*Fire Protection Handbook,* 14th Ed., NFPA, Boston, 1976, p. 1-24.

[10]"America Burning," May 1973, National Commission on Fire Prevention and Control, Washington, DC, p. 4.

[11]"1977 Fireworks Incidents," *Fire Journal,* Vol. 71, No. 6, Nov. 1977, pp. 42-43.

[12]*Fire Protection Handbook,* 14th Ed., NFPA, Boston, 1976, p. 1-29.

[13]"Arson: America's Malignant Crime," September 1976, National Fire Prevention and Control Administration, Washington, DC, p. 5.

[14]Kirk, Paul L., *Fire Investigation,* John Wiley & Sons, Inc., New York, 1969, p. 159.

[15]*Fire Protection Handbook,* 13th Ed., NFPA, Boston, 1969, pp. 1-34 — 1-35.

[16]Kennedy, John, *Fire and Arson Investigation,* Investigations Institute, Chicago, 1962, pp. 68-70.

[17]"Suggested Model Arson Law of the Fire Marshal's Section of the NFPA," NFPA, Boston, 1934.

[18]"Arson: America's Malignant Crime," September 1976, National Fire Prevention and Control Administration, Washington, DC, pp. 9-28.

*Chapter Six*

# Firesafe Building Design and Construction

*Because fire presents one of the greatest threats to property, building design and construction must take into account a wide range of firesafety features. Not only must the interiors and contents of buildings be protected from the dangers of fire; the building site itself must be planned to ensure accessibility of both fire departments and water supplies.*

## FUNDAMENTALS OF FIRESAFETY DESIGN

Building design and construction practices have changed significantly during the past century. A century ago, design techniques and materials such as structural steel and reinforced concrete were unknown and unimagined.

A hundred years ago, major fires were common occurrences in cities. Because of inferior construction and poor city planning, whole cities were often destroyed by fire. As a result of those disasters, increased attention was given to firesafety in building design. The following excerpt from "America Burning" briefly summarizes the changes in building design:[1]

> Around the turn of the century, in the wake of many conflagrations, so-called "fireproof" buildings began to be constructed. They had thick walls and floors to keep fire from spreading. Like older buildings, they still had windows that could be opened to allow heat and smoke to escape. They had fire escapes or internal fire stairs, and seldom were they too tall for the topmost occupants to escape.
>
> Fires, some of them disastrous, occurred in these buildings, nonetheless. Then, after World War II, a new generation of buildings began to appear: the modern high-rise building. Lighter construction systems and many new materials were used, especially for interiors. Windows were permanently sealed so that central air conditioning would operate efficiently. Walls and floors were left with openings for air conditioning ducts and utility cables. Each of these features compromised the firesafety of these buildings.

## Objectives of Firesafety Design

"America Burning," the report of the National Commission on Fire Prevention and Control, points out a major weakness in building design and construction practice; namely, architects, builders, and building owners all too often think of expense and utility, and do not consciously think of fire protection. Many architects and their clients are content to meet the minimum fire protection standards of the local building code. These same persons often assume, incorrectly, that the code provides adequate measures rather than minimal measures for firesafety.

Before a building designer can make effective decisions relating to firesafety design, the specific needs of the client regarding the function of the building and the general and unique conditions that are to be incorporated into the building must be clearly identified. Decisions regarding the firesafety design and construction of the building should be made in the following areas: (1) life safety, (2) property protection, and (3) continuity of operations.

The importance of firesafety design is described in the NFPA *Fire Protection Handbook:*[2]

> The art of probing to identify firesafety objectives is an important design function. The degree of risk that will be tolerated by the owner and the occupants is a difficult design decision. Consequently, it is often not identified in a clear, concise manner that will enable the designer to provide properly for the realization of the design objectives.

***Life Safety:*** Design considerations for life safety must address two major questions: (1) Who will be using the building? (2) What will the people using the building be doing most of the time? The identification of specific functional patterns, constraints, and handicaps is vital in designing specific fire protection features that recognize occupant conditions and activities. Following are additional design considerations as described in *Operation Skyline:*[3]

> The occupied building provides a great potential for fire because of the presence of large numbers of people, any one of whom could perform a careless or malicious act resulting in fire. Appliances and mechanical or electrical equipment are a potential hazard through misuse, faulty construction, or substandard installation. Accumulations of combustibles, either waiting for disposal or in storage, frequently provide a ready means by which otherwise controllable fires could spread.

***Property Protection:*** One of the most important questions to be asked about the design of buildings with regard to protection of property is: is there any specific high-value content that will need special design protection? The requirements with regard to protection of property within a building are often fairly easy to identify. Materials of high value that are particularly susceptible to fire and/or water damage can usually be identified in advance of building design. For example, vital records that cannot be replaced easily or quickly can be identified in advance as needing special fire protection design considerations.

***Continuity of Operations:*** Continuity of operations, the third major area of building design decision-making, must take into consideration those specific functions conducted in a building that are vital to continued operation and that cannot be transferred to another location. In this regard the owner must identify for the designer the amount of "downtime," or the amount of time an operation can be suspended without completely suspending total operations. The degree of protection required in firesafe building design varies with the number and scope of vital operations that are nontransferable.

## Fire Hazards in Buildings

When the designer and owner either consciously or unconsciously overlook or ignore the possibility of fire in the building to be built, the building and its occupants are endangered. The broad approach to the firesafe design of a building requires a clear understanding of the building's function, the number and kinds of people who will be using it, and the kinds of things they will be doing. In addition, appropriate construction and protection features must be provided for the protection of the contents and, particularly for mercantile and industrial buildings, to ensure the continuity of operations if a fire should occur. Too many fires disastrous to people and to property have occurred, and will continue to occur, because no one has given proper consideration to the threat of potential fire. (See Fig. 6.1.)

***Smoke and Gas:*** Studies of fire deaths in buildings indicate that about 75 percent of these deaths are due to the smoke and toxic gases that evolve as products of the fire. About 25 percent of the deaths result from heat or contact with direct flame. The carbon monoxide developed in many fires, particularly unventilated and smoldering fires, is probably the most common cause of death. Carbon monoxide can be neither seen nor smelled. Exposure to this gas, even in small quantities, can cause impaired mental behavior. When inhaled in large quantities, the smoke given off in most fires can lead to pneumonia and other lung troubles. Smoke obscures visibility and thus can lead to panic situations when occupants cannot see and utilize escape routes.

***Heat and Flames:*** As has already been stated, heat and flames account for 25 percent of fire deaths. Although heat and flame injuries are much fewer than those caused by smoke and toxic gases, the pain and disfigurement caused by burns can also result in serious, long-term complications.

***Building Elements and Contents:*** If the building on fire has combustible interior finish and combustible furniture, flames and toxic gases will spread so

144  Principles of Fire Protection

*Fig. 6.1. This four-story, unsprinklered factory of mill construction in Bellows Falls, VT, was a total loss despite prompt discovery and the efforts of fire fighters from three states. Firesafe building design and the installation of sprinklers could have limited the loss.*

rapidly that occupants may not be able to escape. Poor construction practices, such as failure to protect shafts and other vertical openings, make the work of fire fighters more difficult. Collapse of structural members causes a significant number of deaths and injuries to fire fighters.[4]

> The collapse of structural building elements can be a serious hazard. Although statistically it [the collapse of structural elements] has not resulted in many deaths or injuries to building occupants, it is a particular hazard to fire fighters. A number of deaths and serious injuries to fire fighters occur each year because of structural failure. While some of these failures result from inherent weaknesses, many are the result of renovations to existing buildings that materially, though not obviously, affect the structural integrity of the support elements. A building should not contain surprises of this type for fire fighters.

## Elements of Building Firesafety

The firesafety of a building will depend first on what is done to prevent a fire from starting in the building, and second on what is done through design, construction, and good management to minimize the spread of fire if and when it happens. Good housekeeping is perhaps the major factor in both fire prevention and control. Keeping the fuel load down not only lessens the amount of material that can be ignited but provides less material that can be consumed if a fire breaks out.

Once a fire has started, its spread will depend on the design of the building, the materials used in construction, building contents, methods of ventilation, detection and alarm facilities, and fire suppression systems, if any. Table 6.1 describes the building design and construction features that influence safety. These elements are within the decision-making authority of various members of the design team, based on the assumption that their firesafety objectives are clearly defined by management, the owners, or other responsible parties, both public and private. The design and construction elements are organized in a manner that can give a quick overview of the major aspects that must be considered for firesafety. They show features that include both active and passive design and construction considerations.

The persons responsible for fire prevention are not the same ones responsible for the building design. Table 6.2 describes the elements that comprise firesafety from a prevention consideration. Decisions concerning these elements are predominantly under the control of the building owner or occupant, or both. Table 6.2 includes the elements of emergency preparedness in case of fire that are the responsibility of the owner and/or occupant.

A further note from *Operation Skyline:*[5]

> . . . some owners, operators, and managers [boast] about certain "fireproof" or "fire-resistive" materials in their buildings. Those materials provide a definite advantage, but they should not be confused with "firesafe." They simply indicate that a building with "fireproof" or fire-resistive materials can withstand a burnout of its contents without subsequent structural collapse. Firesafe, on the other hand, indicates that if a fire starts it can be confined and extinguished without jeopardizing life and property elsewhere in the structure.

## *BUILDING AND SITE PLANNING FOR FIRESAFETY*

Two major categories of decisions should be made early in the design process of a building in order to provide effective firesafe design. Early considerations should be given to both the interior building functions and exterior site planning. Building fire defenses, both active and passive, should be designed in such a way that the building itself assists in the manual suppression of fire.

### *Firesafety Planning for Buildings*

Interior layout, circulation patterns, finish material, and building services are all important firesafety considerations in building design. Building design

## Table 6.1  Elements of Building Firesafety*

**Building Design and Construction Features Influencing Firesafety**

1. **Fire Propagation**
   a. Fuel load and distribution
   b. Finish materials and their location
   c. Construction details influencing fire and products of combustion movement
   d. Architectural design features

2. **Smoke and Fire Gas Movement**
   a. Generation
   b. Movement
      — Natural air movement
      — Mechanical air movement
   c. Control
      — Ventilation
      — Heating, ventilating, air conditioning
      — Barriers
      — Pressurization
   d. Occupant Protection
      — Egress
      — Temporary refuge spaces
      — Life support systems

3. **Detection, Alarm, and Communication**
   a. Activation
   b. Signal
   c. Communication systems
      — To and from occupants
      — To and from fire department
      — Type (automatic or manual)
      — Signal (audio or visual)

4. **People Movement**
   a. Occupant
      — Horizontal
      — Vertical
      — Control
      — Life support
   b. Fire Fighters
      — Horizontal
      — Vertical
      — Control

5. **Suppression Systems**
   a. Automatic
   b. Manual (self-help; standpipes)
   c. Special

6. **Fire Fighting Operations**
   a. Access
   b. Rescue operations
   c. Venting
   d. Extinguishment
      — Equipment
      — Spatial design features
   e. Protection from structural collapse

7. **Structural Integrity**
   a. Building structural system (fire endurance)
   b. Compartmentation
   c. Stability

8. **Site Design**
   a. Exposure protection
   b. Fire fighting operations
   c. Personnel safety
   d. Miscellaneous (water supply, traffic, access, etc.)

**Fire Emergency Considerations**

1. **Life Safety**
   a. Toxic gases
   b. Smoke
   c. Surface flame spread

2. **Structural**
   a. Fire propagation
   b. Structural stability

3. **Continuity of Operations**
   a. Structural integrity

*From NFPA *Fire Protection Handbook*.

also has a significant influence on the efficiency of fire department operations. As a result, manual fire suppression activities should be considered during all architectural design phases. Table 6.3 lists pertinent data regarding some recent high-rise fires. It can be assumed that design decisions undoubtedly affected the statistics in this table.

***Fire Fighting Accessibility to Building's Interior:*** One of the more important considerations in building design is access to the fire area. This includes access to the building itself as well as access to the interior of the building.

Table 6.2   Fire Prevention and Emergency Preparedness*

1. **Ignitors**
   a. Equipment and devices
   b. Human accident
   c. Vandalism and arson
2. **Ignitable Materials**
   a. Fuel load
   b. Fuel distribution
   c. Housekeeping
3. **Emergency Preparedness**
   a. Awareness and understanding
   b. Plans for action
      — Evacuation or temporary refuge
      — Handling extinguishers
   c. Equipment
   d. Maintenance — operating manuals available

*From NFPA *Fire Protection Handbook*.

Table 6.3   Partial Listing of Recent High-rise Fires*

| Date of Fire | City | Occupancy | Stories | Time of Fire | Place of Origin | Fatalities | Injuries | Dollar Loss |
|---|---|---|---|---|---|---|---|---|
| Jan. 1970 | Chicago | Hotel | 25 | 6:45 am | Elevator (lobby) | 2 | 36 | $70-100,000 |
| Aug. 1970 | New York | Offices | 50 | 5:45 pm | Office (33rd Floor) | 2 | 30 | $10 Million |
| Dec. 1970 | Tucson | Hotel | 11 | 12:20 am | Fourth Floor | 28 | 71 | $1.5 Million |
| Dec. 1970 | New York | Offices | 47 | 9:50 am | Showroom (5th Floor) | 3 | 20 | $2.5 Million |
| Mar. 1971 | Los Angeles | Offices | 21 | 2:11 am | Restaurant (21st Floor) | 0 | 0 | $378,000 |
| July 1971 | New Orleans | Hotel | 17 | 2:00 am | Room (12th Floor) | 6 | 2 | $150,000 |
| Nov. 1972 | Atlanta | Home for Elderly | 11 | 2:00 am | Apartment (7th Floor) | 10 | 30 | $250,000 |
| Nov. 1972 | New Orleans | Offices | 16 | 1:28 pm | Restaurant (15th Floor) | 6 | 0 | $887,000 |
| Nov. 1972 | Chicago | Apt. & Other | 100 | 4:40 am | Restaurant (96th Floor) | 0 | 1 | $40,000 |
| Dec. 1972 | Ventnor, NJ | Apartments | 19 | 12:45 pm | Apartment (4th Floor) | 1 | 3 | $325,000 |
| Dec. 1972 | Dallas | Apartments | 16 | 7:30 am | Eighth Floor | 0 | 0 | $340,000 |
| June 1973 | Tucson | Offices | 11 | 3:30 pm | Storage (4th Floor) | 0 | 0 | $565,000 |
| Dec. 1973 | Omaha | Home for Elderly | 13 | 5:10 pm | Apartment (13th Floor) | 0 | 4 | $30,000 |
| Sept. 1974 | Virginia Beach | Hotel | 11 | 11:35 am | Room (9th Floor) | 1 | 21 | $145,000 |

*From *Operation Skyline*, NFPA, Boston.

In larger and more complex buildings, serious fires over the years have brought improvements in building design to facilitate fire department operations. The larger the building, the more important access for fire fighting becomes. In some buildings where fire fighters cannot function effectively, the best solution is provision of a complete automatic sprinkler system, as explained in the following excerpt from the NFPA *Fire Protection Handbook:*[6]

Spaces in which adequate fire fighting access and operations are restricted because of architectural, engineering, or functional requirements should be provided with effective protection. A complete automatic sprinkler system with a fire department connection is probably the best solution to this problem. Other methods which may be used in appropriate design situations include access panels in interior walls and floors, fixed nozzles in floors with fire department connections, and roof vents and access openings.

*Ventilation:* Ventilation is of vital importance in removing smoke, gases, and heat so that fire fighters can reach the seat of a blaze. It is difficult, if not impossible, to ventilate a building unless appropriate skylights, roof hatches, emergency escape exits, and similar devices are provided when the building is constructed. The following is a description of the importance of the ventilation factor in building design:[7]

Ventilation of building spaces performs the following important functions:
1. Protection of life by removing or diverting toxic gases and smoke from locations where building occupants must find temporary refuge.
2. Improvement of the environment in the vicinity of the fire by removal of smoke and heat. This enables fire fighters to advance close to the fire to extinguish it with a minimum of time, water, and damage.
3. Control of the spread or direction of fire by setting up air currents that cause the fire to move in a desired direction. In this way occupants or valuable property can be more readily protected.
4. Provision of a release for unburned, combustible gases before they acquire a flammable mixture, thus avoiding a backdraft or smoke explosion.

**Connections for Sprinklers and Standpipes:** Connections for sprinklers and standpipes must be carefully located and clearly marked. The larger and taller the building becomes, the greater the volume and pressure of water that will be needed for a potential fire. Water damage can be very costly unless adequate measures such as floor drains and scuppers have been incorporated into the building design. Confinement of a fire in a high-rise building can only be accomplished by careful design and planning for the whole building. As buildings increase in size and complexity, more dependence on fire detection and suppression systems is necessary. Such systems will be described in detail in subsequent chapters.

## *Firesafety Planning for Sites*

Proper building design for fire protection should include a number of factors outside the building itself. The site on which the building is located will influence the design. Among the more significant features are traffic and transportation conditions, fire department accessibility, and water supply. Inadequate water mains and poor spacing of hydrants have contributed to the loss of many buildings.

*Traffic and Transportation:* Fire department response time is a vital factor in building design considerations. Traffic access routes, traffic congestion at certain times of the day, traffic congestion from highway entrances and exits, and limited access highways have significant effects on fire department response distances and response time, and must be taken into account by building designers in selecting appropriate fire defenses for a building.

*Fire Department Access to the Site:* Building designers must ask the question: Is the building easily accessible to fire apparatus? Ideal accessibility occurs where a building can be approached from all sides by fire department apparatus. However, such ideal accessibility is not always possible. Congested areas, topography, or buildings and structures located appreciable distances away from the street make difficult or prevent effective use of fire apparatus. When apparatus cannot come close enough to the building to be used effectively, equipment such as aerial ladders, elevating platforms, and water tower apparatus can be rendered useless. The importance of the complications resulting from the accessibility factor is further emphasized in the following statement from the NFPA *Fire Protection Handbook:*[8]

> The matter of access to buildings has become far more complicated in recent years. The building designer must consider this important aspect during the planning stages. Inadequate attention to site details can place the building in an unnecessarily vulnerable position. If its fire defenses are compromised by preventing adequate fire department access, the building itself must make up the difference in more complete internal protection.

*Water Supply to the Site:* Another important question that a building designer must ask is: Are the water mains adequate and are the hydrants properly located? The more congested the area where the building is to be located, the more important it is to plan in advance what the fire department may face in its attack if a fire occurs on the property. An adequate water supply delivered with the necessary pressure is required to control a fire properly and adequately. The number, location, and spacing of hydrants and the size of the water mains are vital considerations when the building designer plans fire defenses for a building.

## EXPOSURE PROTECTION

Still another consideration in the design of the building is the possibility of damage from a fire in an adjoining building. The building may be exposed to heat radiated horizontally by flames from the windows of the burning neighboring building. If the exposed building is taller than the burning building, flames coming from the roof of the burning building can attack and damage the exposed building.

150   Principles of Fire Protection

The damage from an exposing fire can be severe. It is dependent upon the amount of heat produced and the time of exposure, the fuel load in the exposing building, and the construction and protection of the walls and roof of the exposed building. Other factors are the distance of separation, wind direction, and accessibility of fire fighters.

Fire severity is a description of the total energy of a fire, and involves both the temperatures developed within the exposing fire and the duration of the burning. NFPA 80A, *Recommended Practice for Protection of Buildings from Exterior Fire Exposures,*[9] describes estimated minimum separation distances under light, moderate, or severe exposures. The severity of the exposure is calculated on the width and height and the percentage of openings in the exposing wall areas and the estimated fire loadings of the buildings involved. Building designers should be aware that the separation distances between the exposing buildings can be reduced by blank walls, closing wall openings, use of automatic deluge water curtains, and use of wired glass instead of ordinary glass. (See Fig. 6.2.)

## INTERIOR FINISH

The way a building fire develops and spreads, and the amount of damage that ensues is greatly influenced by the characteristics of the interior finish in a building. The types of interior finish used in buildings are numerous, varied, and serve many functions. Primarily they are used for aesthetic and/or acoustical purposes. However, insulation and/or protection against wear and abrasion are also considered major functions by building designers. The following statements from the National Commission on Fire Prevention and Control's report titled "America Burning" point out the need for greater concern and attention to the potential fire hazards of interior finishes:[10]

> The modern urban environment imparts to people a false sense of security about fire. Crime may stalk the city streets, but certainly not fire, in most people's view. In part, this sense of security rests on the fact there have been no major conflagrations in American cities in more than half a century. In part, the newness of so many buildings conveys the feeling that they are invulnerable to attack by fire. Those who think only of a building's basic structure (not its contents) are satisfied, mistakenly, that the materials — concrete, steel, glass, aluminum — are indestructible by fire. Further, Americans tend to take for granted that those who design their products, in this case buildings, always do so with adequate attention to their safety. That assumption, too, is incorrect.

### Types of Interior Finish

Interior finish is usually defined as those materials that make up the exposed interior surface of wall, ceiling, and floor constructions. The common interior

Firesafe Building Design and Construction    151

Fig. 6.2. The effectiveness of exposure protection from fire is evidenced in a reinforced concrete fire wall, which protected the dwelling from a fire in an adjacent lumberyard. (From NFPA *Fire Protection Handbook*.)

finish materials are wood, plywood, plaster, wallboards, acoustical tile, insulating and decorative finishes, plastics, and various wall coverings.

While some building codes do not include floor coverings under their definition of interior finishes, the present trend is to include them. Rugs and carpets are now subject to test and regulation under the Flammable Fabrics Act administered by the Department of Commerce. Wool carpets present no particular hazard, but some of the fluffy rugs and carpets made out of synthetics are a factor in fire spread.

Many codes exclude trim and incidental finish from the code requirements for wall and ceiling finish, as explained in the following:[11]

> Interior finishes are not necessarily limited to the walls, ceilings, and floors of rooms, corridors, stairwells, and similar buildings' spaces. Some authorities include the linings or coverings of ducts, utility chases and shafts, or plenum spaces as interior finish as well as batt and blanket insulation, if the back faces a stud space through which fire might spread.

Cellular plastics sprayed on walls for insulation have become popular. Fire retardants can be incorporated in many of these plastics so they can meet building code requirements. However, some plastics containing polyurethane or polystyrene have been involved in serious, rapidly spreading fires. Untreated, highly combustible wallboards have been a major factor in a number of fires causing large loss of life in hotels, hospitals, and nursing homes over recent years. Fire retardant treatments are now required by most codes for this type of material. Without such fire retardant treatments, combustible wallboards not only enable a fire to spread so fast that people may become trapped, but also contribute fuel to the fire and create hazardous concentrations of smoke and toxic gases.

*Wood:* The physical size of wood and its moisture content are important factors that determine whether this material will provide reasonable structural integrity. Wood is the most prevalent material used in the construction of dwellings. If a wood-frame house is subjected to a serious fire, either from burning combustibles inside the house or from an exposure fire, it will not withstand much heat and will have little structural integrity.

Heavy timber construction can resist fire very well. The timbers will char, and the resulting coating of charcoal provides an insulation for the unburned wood. Heavy timber maintains its integrity during a fire for a relatively long time, thus providing an opportunity for extinguishment. Much of the original strength of the members is retained and reconstruction is possible.

*Steel:* The most common building material for larger buildings is structural steel. While steel is noncombustible and contributes no fuel to a fire, it loses its strength when subjected to the high temperatures that are easily reached in a fire. The stress in a steel beam determines its load-carrying capacity. The normal critical temperature of steel is 1,100°F (593°C). At this temperature the yield stress of steel is about 60 percent of its value at room temperature. Buildings built of unprotected steel will collapse relatively quickly when exposed to a contents fire or an exposure fire. The lighter the steel members, the quicker will be the failure.

Another property of steel that influences its behavior in fires is expansion when the steel is heated. Walls can collapse from the movement caused by expansion of steel trusses.

Encasement of the structural steel member has become a very common and effective way of insulating steel to increase its fire resistance. The NFPA *Fire Protection Handbook* describes some of these methods:[12]

> Because unprotected structural steel loses its strength at high temperatures, it must be protected from exposure to the heat produced by building fires. This protection, often referred to as "fireproofing," insulates the steel from the heat. The more common methods of insulating steel are encasement of the member, application of a surface treatment, or installation of a suspended ceiling as part of a floor-ceiling assembly capable of providing fire resistance. In recent years, additional methods, such as sheet steel membrane shields around members and box columns filled with liquid, have been introduced. . . .

In recent years, intumescent paints and coatings have been utilized to increase the fire endurance of structural steel. These coatings intumesce, or swell, when heated, thus forming an insulation around the steel. . . .

Structural steel members can also be protected by sheet steel membrane shields. The sheet steel holds in place inexpensive insulation materials, thus providing a greater fire endurance. In addition, polished sheet steel has been used in recent tests to protect spandrel girders. The shield reflects radiated heat and protects the load-carrying spandrel.

*Concrete:* The resistance of reinforced concrete to fire attack will depend on the type of aggregate used to make the concrete, fire loading, and moisture content. In general, lightweight concrete performs better at elevated temperatures than normal weight concrete.

Usually, reinforced concrete buildings resist fire very well; however, the heat of a fire will cause spalling (chipping and peeling away), some loss of strength of the concrete, and other deleterious effects.

Prestressed concrete is stronger than reinforced concrete and provides better fire resistance. However, prestressed concrete has a greater tendency to spall with the result that the prestressing steel may become exposed. The type of steel used for prestressing is more sensitive to elevated temperatures than the type of steel that is usually used in reinforced concrete construction. In addition, the steel used for this type of reinforced concrete construction does not regain its strength upon cooling.

*Glass:* Glass is a commonly used building material. Modern high-rise buildings, particularly, contain large amounts of glass. Glass is utilized in three primary ways in building construction: (1) for glazing, (2) for fiberglass insulation, and (3) for fiberglass-reinforced plastic building products.

Glass used for windows and doors has little resistance to fire. Wire reinforced glass provides a slightly higher resistance to fire, but no glazing should be relied upon to remain intact in a fire.

Fiberglass insulation is widely used in modern building construction. Fiberglass is popular because it does not burn and is an excellent insulator. However, fiberglass is often coated with a resin binder that is combustible and that can spread flames.

Fiberglass-reinforced plastic building products such as translucent window panels are becoming more common. The fiberglass acts as reinforcement for a thermosetting resin. Usually this resin, which is combustible, comprises about 50 percent or more of the material. Thus, while the fiberglass itself is noncombustible, the product is highly combustible.

*Gypsum:* Gypsum, as reflected in products such as plaster and plasterboard, has excellent fire-resistive qualities. Gypsum is widely used because it has a high proportion of chemically combined water, which makes it an excellent, inexpensive, fire-resistive building material that is far superior to highly combustible fiberboards.

*Masonry:* Masonry (such as brick, tile, and sometimes concrete) provides good resistance to heat, and usually retains its integrity. Because of the prevalence of brick construction in European dwellings, as compared to American wood-frame construction, the dwelling fire record in Europe is much more favorable than the dwelling fire record in America.

*Plastics:* Plastic products are increasing in use by the building industry. Lower cost and aesthetic considerations make the use of plastic building materials desirable. However, all plastics are combustible. (See Chapter 4, "Fire Hazards of Materials.") Presently, there is no known treatment that is able to make plastics noncombustible.

## Test Procedures

Because of the importance of assessing the fire resistance of buildings and building materials, a great deal of work has been done to develop standard test procedures that assess the way building materials and structural assemblies perform under fire conditions. It is possible to estimate the damage that fire can cause to a building by studying: (1) the amount and kind of combustible materials in the building, and (2) the way they are distributed throughout the building. These two factors not only indicate the rate of combustion and the duration of the fire, but also the difficulty that might be encountered when fighting the fire.

The effects of fire on the components of a building (such as the columns, floors, walls, partitions, and ceiling or roof assemblies) are tested against both time and temperature. Results of the tests are recorded in hours or minutes, and indicate the duration of fire resistance.

Test procedures require the loading of the elements being tested to simulate actual conditions under building use. In some cases, in order to qualify for a certain rating, the construction, after the prescribed fire exposure, is subjected to hose streams.

Various criteria are established and used to determine the acceptance of the material or construction being tested. Such criteria include failure to support load, temperature increase on the unexposed surface, passage of heat or flame sufficient to ignite cotton waste, excess temperature on steel members, and structural failure under hose streams.

The standard fire test is designed to define the ability of the test structure to perform its intended function during fire exposure, as well as its subsequent load capacity. The fire test does not measure the suitability of the test structure for future use.

Full-scale room tests with various wall and ceiling finishes demonstrate a phenomenon called "flashover." Thermal radiation from the upper area of the room heats the interior finish and other combustible material in the room to the point where simultaneous ignition (flashover) occurs.

Ratings for flame spread of interior finish materials have been established by the use of the 25-ft (7.6-m) tunnel developed by A. J. Steiner at Underwriters Laboratories Inc. These ratings are used in the NFPA *Life Safety Code* and in other codes to indicate the areas in which finishes of varying flame spread characteristics may be used. The five classifications used in the *Life Safety Code* are:[13]

| Class | Flame Spread Range |
|-------|--------------------|
| A | 0–25 |
| B | 26–75 |
| C | 76–200 |
| D | 201–500 |
| E | over 500 |

The higher the flame spread, the greater the hazard. For example, in a new hospital, Class A materials would be required for most areas, and Class D or E materials would not be permitted at all.

The tunnel tests are measured in a relative scale with cement asbestos board rated 0 and red oak flooring rated 100. Some of the highly combustible wallboards involved in fatal fires have received ratings as high as 1,500.

The principal United States agencies that test assemblies of building construction materials for fire resistance are Underwriters Laboratories Inc. and the National Bureau of Standards. Results of their tests, as well as tests of other laboratories and of building manufacturers, are published and made available to interested parties. One of the best sources of information showing the wide variety of building assemblies and giving the fire-resistance ratings of beams, columns, floors, walls, and partitions is the Underwriters Laboratories Inc. *Fire Resistance Index,* which is published annually.

Because of the various circumstances involving fires and the wide range of materials that might be involved in a fire, testing procedures cannot provide exact statistics applicable to all fire situations. The problem is summarized in the NFPA *Fire Protection Handbook* as follows:[14]

> The nature of materials and the fire environment vary so widely that the development of a fire test becomes a highly complex matter. Three factors must be taken into account in this development process.
> 1. The start and growth of fire in a building is affected by the ignition source, by space geometry, by ventilation, and by nature, amount, and location of other processes and materials.
> 2. The changing conditions during a fire, such as oxygen concentration, rate of heat release, protection systems, etc.
> 3. Variations in form, composition, density, and application of materials present.
>
> With these variables in mind, the difficulty in designing a test that will provide a basis for predicting performance under fire exposure becomes obvious. Equally obvious is the impracticality of designing tests to represent all fire con-

ditions. On the other hand, a test designed to represent a "typical" fire situation or to expose materials to one set of "standard test" conditions may not provide a reliable basis for predicting "real-life" performance of all materials tested. Thus, there is a constant search for improved test methods having a numerical range of results, and for an adequate array of tests to suitably describe the behavior of the various materials available.

Although testing procedures cannot provide exact, "real-life" statistics, they do provide vital firesafety guidelines that should be considered by architects and builders when designing and constructing buildings.

## CONFINEMENT OF SMOKE AND FIRE

To date, little has been done to build dwellings of materials and assemblies that confine a fire to the room or even the floor of origin. However, many other types of buildings are provided with some degree of protection for the occupants. Today, if a building is so constructed that it can neither confine a fire to a given area nor restrict the products of combustion from spreading fire throughout the building, it is likely that proper firesafety safeguards have not been incorporated. Currently it is relatively easy to ensure confinement of a fire through good construction, interior fire protection equipment, and fire-resistive interior materials.

Table 6.4  Estimated Fire Severity for Offices and Light Commercial Occupancies*

Data applying to fire-resistive buildings with combustible furniture and shelving

| Combustible Content<br>Total, including finish, floor, and trim psf | Heat Potential Assumed**<br>Btu/sq ft | Equivalent Fire Severity, approximately equivalent to that of test under standard curve for the following periods |
|---|---|---|
| 5 | 40,000 | 30 min |
| 10 | 80,000 | 1 hr |
| 15 | 120,000 | 1½ hrs |
| 20 | 160,000 | 2 hrs |
| 30 | 240,000 | 3 hrs |
| 40 | 320,000 | 4½ hrs |
| 50 | 380,000 | 7 hrs |
| 60 | 432,000 | 8 hrs |
| 70 | 500,000 | 9 hrs |

*From NFPA *Fire Protection Handbook*.

**Heat of combustion of contents taken at 8,000 Btu/lb up to 40 psf; 7,600 Btu/lb for 50 lbs, and 7,200 Btu for 60 lbs and more to allow for relatively greater proportion of paper. The weights contemplated by the tables are those of ordinary combustible materials, such as wood, paper, or textiles.

## Table 6.5 Fire Severity Expected by Occupancy*

**Temperature Curve A (Slight)**
Well-arranged office, metal furniture, noncombustible building.
Welding areas containing slight combustibles.
Noncombustible power house.
Noncombustible buildings, slight amount of combustible occupancy.

**Temperature Curve B (Moderate)**
Cotton and waste paper storage (baled) and well-arranged, noncombustible building.
Paper-making processes, noncombustible building.
Noncombustible institutional buildings with combustible occupancy.

**Temperature Curve C (Moderately Severe)**
Well-arranged combustible storage, e.g., wooden patterns, noncombustible buildings.
Machine shop having noncombustible floors.

**Temperature Curve D (Severe)**
Manufacturing areas, combustible products, noncombustible building.
Congested combustible storage areas, noncombustible building.

**Temperature Curve E (Standard Fire Exposure — Severe)**
Flammable liquids.
Woodworking areas.
Office, combustible furniture, and buildings.
Paper working, printing, etc.
Furniture manufacturing and finishing.
Machine shop having combustible floors.

*From NFPA *Fire Protection Handbook*.

## Fire Loading

Estimates of the severity and duration of a fire in a building can be made if the fire loading is determined. The fire load is expressed as the weight of combustible material per square foot of fire area, and includes the combustible structural elements and the combustible contents. The fire load also takes into account the kind and quantity of the material or materials involved. In addition to fire load, the way the building and its contents are arranged must be considered in order to determine how rapid the spread of a fire might be. Table 6.4 shows how the severity of a fire increases dramatically as the fire loading becomes heavier.

*The Standard Time-temperature Curve:* The standard time-temperature curve is widely accepted and used by most of the standards and testing agencies. It is based on the maximum indication of the severity of a fire completely burning out an ordinary brick, wood-joisted building loaded with combustible contents. The use of this curve, together with information on the fire loading, is used to estimate the severity of a fire. Table 6.5 indicates the fire severity for typical occupancies.

*Fire Loads by Occupancy:* One of the difficulties faced by fire authorities is change of occupancy. The original occupancy of the designed building may have had only a slight expected fire severity possibility. If new owners or occupants change to another type of business using large amounts of combustible material, the building or area may become unsafe. Changes of occupancy demand the attention of the appropriate building and fire authorities. Building codes define the degree of protection required for walls, ceiling, and floor in accordance with the degree of hazard of the occupancy. If the occupancy is changed, a different degree of protection may be indicated and warranted.

## Fire Doors

Fire doors are the most widely used and accepted means of protecting both vertical and horizontal openings. Suitability of fire doors is determined by nationally recognized testing laboratories; doors not tested cannot be relied upon for effective protection. The doors are tested as they are installed in the field; that is, with the frame, hardware, wired-glass panels, and other accessories necessary to complete the installation.

To protect openings in walls, nearly all building codes use NFPA 80, *Standard for Fire Doors and Windows*.[15] This standard establishes the minimum ratings for the five most commonly encountered types of openings in walls. They are as follows:

1. Class A Openings are in walls separating buildings or dividing a single building into fire areas. Doors for the protection of these openings have a fire protection rating of three hours.
2. Class B Openings are in enclosures of vertical communication through buildings (stairs, elevators, etc.). Doors for the protection of these openings have a fire protection rating of one or one and a half hours.
3. Class C Openings are in corridor and room partitions. Doors for the protection of these openings have a fire protection rating of three quarters of an hour.
4. Class D Openings are in exterior walls which are subject to severe exposure from the outside of the building. Doors and shutters for the protection of these openings have a fire protection rating of one and a half hours.
5. Class E Openings are in exterior walls which are subject to moderate or light fire exposure from outside of the building. Doors, shutters, or windows for the protection of these openings have a fire protection rating of three quarters of an hour.

It is important to note that this classification applies to the various types of openings and not to the fire door itself. A fire door is not a Class A fire door. It is a door that is suitable for a Class A opening.

***Types of Doors:*** Fire doors, shutters, and windows are manufactured in a wide variety of constructions. Some of the more common types are tin-clad doors, composite doors, hollow metal doors, metal clad doors, and rolling steel doors. The type of door and the way it is installed are important to secure the desired degree of protection. For severe exposures, double doors may be called for in Class A openings; tight-fitting doors are needed in hospitals and nursing homes for protection of occupants in rooms and corridors; self-closing doors are indicated for rooms and corridors in hotels and apartments.

The following excerpt from the NFPA *Fire Protection Handbook* is a description of the various types of construction for fire doors:[16]

> *Composite Doors:* These are of the flush design and consist of a manufactured core material with chemically impregnated wood edge banding and untreated wood face veneers, or laminated plastic faces, or surrounded by and encased in steel.
> *Hollow-metal Doors:* These are of formed steel of the flush and paneled designs of No. 20 gage or heavier steel.
> *Metal-clad (Kalamein) Doors:* These are of flush and paneled design consisting of metal covered with steel of 24 gage or lighter.
> *Sheet-metal Doors:* These are of formed No. 22 gage or lighter steel and of the corrugated, flush, and paneled designs.
> *Rolling Steel Doors:* These are of the interlocking steel slat design or plate-steel construction.
> *Tin-clad Doors:* These are of two- or three-ply wood core construction, covered with No. 30 gage galvanized steel or terneplate (maximum size 14 by 20 in.); or No. 24 gage galvanized steel sheets not more than 48 in. wide.
> *Curtain Type Doors:* These consist of interlocking steel blades or a continuous formed spring steel curtain in a steel frame.
> The suitability of a fire door should be judged on the class of opening in which it is to be installed, not on the fire-resistance rating of the wall in which it is to be installed. . . .
> If the opening is in a wall dividing a building into separate fire areas, the door should be suitable for installation in a Class A opening (3-hr fire protection rating). The same door can be used whatever the fire resistance rating of the wall. If a wall encloses a vertical communication, the door should be suitable for a Class B opening. . . .
> The NFPA Fire Doors and Windows Standard* gives recommendations on the installation of suitable approved doors, windows, and shutters, and it also specifies how the opening shall be constructed and how the door or window shall be mounted, equipped, and operated.

## Smoke Control

Control of the smoke that develops in a smoldering fire or in a fire not extinguished in its early stages is important both for reducing the danger of death and injury to people and for the efficiency of fire fighting. Smoke, gaseous products of combustion, and air-borne particulate matter are the leading causes of death in fires.

---

*NFPA 80, *Standard for Fire Doors and Windows,* NFPA, Boston, 1975.

In most dwelling fires and other small building fires, smoke is vented by chopping a hole in the roof. The modern, large-area, single-story factory building is usually provided with roof vents and smoke curtains to vent the smoke. Smoke removal is a difficult problem in high-rise buildings and in windowless and below-ground structures. Because the contents of these buildings cannot be controlled, it is virtually impossible to dictate to a building owner or tenant what can be put into the building. Automatic sprinklers are, at present, the best answer to controlling the products of combustion, despite the contents of the building.

Methods of smoke control generally involve dilution, exhaust, or confinement. If a building is large enough and the smoke is exhausted by normal venting methods, the smoke and products of combustion will be diluted to safe levels. However, dilution can be successful only through infusion of massive quantities of uncontaminated air. People can tolerate 2 or 3 percent of smoke contamination for a short time, but not more than 1 percent for any extended period of time.

It may be feasible to exhaust the smoke in a high-rise building by using the air conditioning and ventilating system and by providing a smoke shaft that runs from the lowest floor to the roof, with dampers at each floor so arranged that the damper will remain closed until a fire occurs at its floor level. The draft needed to force the smoke up the shaft may be provided by the chimney stack effect or by an auxiliary fan.

A third method of smoke control involves confinement. The objective in this case is to provide a barrier to the passage of smoke into specific areas. Sometimes smoke can be kept out of critical areas by the use of dampers, smoke doors, walls, or other physical barriers. The so-called smokeproof towers or stairwells in a tall building can be pressurized to keep smoke out of the area of refuge and exit.

The pressure differential does not have to be very great to make the area safe to breathe in. It is obvious that, in a high-rise building, evacuation to the ground level takes too long, so that removal of the occupants horizontally to safe areas or into safe smokeproof towers and stairways is the practical method for life safety.

## *Summary*

In terms of fire protection, building design and construction practices have improved over the years, but far too many buildings — old and new — are still not firesafe. Many building designers, either through ignorance or for reasons of economy, do not take the necessary precautions to ensure that buildings are firesafe. The magnitude of this problem can be staggering when a high-rise building is involved:[17]

> . . . Because of the capacity of high-rise buildings to accommodate large numbers of occupants, and the impracticality of evacuation to the street or

ground level, fires in high-rise structures can result in staggering death and injury tolls for both occupants and fire fighters. They can also result in heavy physical damage to the buildings, business interruptions, lost tenancies, expensive repairs, years of legal action, and monetary judgments against the owners of the buildings. Rarely are all these losses covered by insurance.

Of course, many fires that have occurred in high-rise buildings have been controlled and extinguished without spread beyond the point of origin and with no loss of life and minimum property damage. The difference between these minimum-loss fires and fires resulting in more serious losses often has been good building design, good fire protection, good maintenance, and a good fire emergency plan.

In terms of quantity, most fires and fire casualties occur in dwellings. Yet little attention has been given to firesafety design and construction of homes. Only recently has the idea of smoke detectors and fire alarms in homes received sufficient response to make the idea attractive to manufacturers and retailers.

Highly combustible contents, even in a well-designed building, can cause severe fire damage. Combustible interior finish has also caused the rapid spread of many fires. Unprotected steel can fail quickly in a fire. A firesafe design and construction consideration would be the careful use of plastics in modern building construction.

Much attention is now being generated to ensure firesafety in buildings. Building codes now use flame spread ratings to reduce fire loss. Tests can be made to determine the ability of the structural elements of a building to resist fire. Estimates of fire loading can determine the severity of a fire. The installation of fire doors is important to contain a fire to a limited area. Smoke and toxic gases from a fire are serious problems, and ventilation measures must be incorporated as a life safety factor.

One of the most critical steps to be taken to ensure life safety is, of course, to educate designers and builders to life safety factors that must be incorporated into buildings:[18]

> Few formal education programs anywhere in the United States for architects and engineers have course requirements in fire protection engineering. (Only the University of Maryland and the Illinois Institute of Technology offer 4-year Bachelor of Science degree programs in fire protection engineering.) While some professional societies have committees concerned with firesafety, few designers take an interest in the committee's work. For lack of training, many designers are unable to understand highly technical reports in firesafety design.
> This absence of training helps to explain the unenthusiastic attention which architects and engineers, when designing buildings, give to firesafety provisions. If the situation were turned around — that is, if architects and engineers were schooled in the principles of firesafety — then undoubtedly they would participate enthusiastically in the search for alternative solutions and better codes consistent with principles of firesafety.

## Activities

1. Explain to your classmates the difference between fire-resistive and firesafe buildings.
2. Describe the effect that modern building design has had on life safety considerations in building design.
3. (a) What three major firesafety factors should a designer take into consideration when designing a building?
    (b) What questions should the designer ask the owner with regard to these three factors?
4. List the major fire hazard factors in buildings, and describe the methods that may be used to reduce these hazards.
5. List the eight major building design and construction features that influence firesafety.
6. (a) Describe the fire defense factors that should be considered when planning a building.
    (b) Describe the fire defense factors that should be considered when selecting a building site.
7. List the major types of interior finish, and describe their effect on firesafety in a building.
8. Describe some of the test procedures used to estimate the firesafety of a building. Then discuss with your classmates the assets and liabilities of these procedures.
9. What steps can be taken to confine smoke and fire in a building?
10. Assume that one of your classmates is a building designer, and convince that classmate of the importance of taking courses in building firesafety design.

## Bibliography

[1] "America Burning," May 1973, The National Commission on Fire Prevention and Control, Washington, DC, p. 58.
[2] *Fire Protection Handbook,* 14th Ed., NFPA, Boston, 1976, p. 6-2.
[3] *Operation Skyline,* NFPA, Boston, 1975, p. 1.
[4] *Fire Protection Handbook,* 14th Ed., NFPA, Boston, 1976, p. 6-3.
[5] *Operation Skyline,* NFPA, Boston, 1975, p. 3.
[6] *Fire Protection Handbook,* 14th Ed., NFPA, Boston, 1976, p. 6-11.
[7] _____, p. 6-11.
[8] _____, p. 6-13.
[9] NFPA 80A, *Recommended Practice for Protection of Buildings from Exterior Fire Exposures,* NFPA, Boston, 1975.
[10] "America Burning," May 1973, The National Commission on Fire Prevention and Control, Washington, DC, p. 58.

[11]*Fire Protection Handbook,* 14th Ed., NFPA, Boston, 1976, p. 6-45.
[12]_____, pp. 6-53, 6-54.
[13]NFPA 101, *Code for Safety to Life from Fire in Buildings and Structures,* NFPA, Boston, 1976.
[14]*Fire Protection Handbook,* 14th Ed., NFPA, Boston, 1976, p. 6-47.
[15]NFPA 80, *Standard for Fire Doors and Windows,* NFPA, Boston, 1975.
[16]*Fire Protection Handbook,* 14th Ed., NFPA, Boston, 1976, p. 6-84.
[17]*Operation Skyline,* NFPA, Boston, 1975, p. 1.
[18]"America Burning," May 1973, The National Commission on Fire Prevention and Control, Washington, DC, p. 77.

*Chapter Seven*

# Fire Protection Systems and Equipment

*Water has long been the most commonly used fire extinguishing agent; however, to better combat the many different types of fires, there have been developed various kinds of extinguishing equipment and systems whose operations are dependent upon a variety of chemical and mechanical actions. Although each works in a specialized manner, all were created to ensure fire protection to life and property.*

## WATER AS AN EXTINGUISHING AGENT

Water has long been the most commonly used fire extinguishing agent. The great majority of fires are extinguished by use of water: from a hose delivering a solid stream or a spray; from a sprinkler system or a water spray system; or from a pump, tank, or bucket. Water, which is usually available at or near the fire scene, has special physical properties particularly suited for fire fighting. The physical characteristics of water that are pertinent to its extinguishing ability and its limitations as an effective extinguishing agent are discussed in the following paragraphs.

### The Physical Properties of Water

Following is a listing that describes the physical properties that contribute to the extinguishing capacity of water. (Besides water, there is no other material normally practical to use as an extinguishant that has the cooling capacity indicated by these characteristics.)

1. At ordinary temperature, water is a heavy, relatively stable liquid.
2. The melting of 1 lb (0.45 kg) of ice into water at 32°F (0°C) absorbs 143.4 Btu, which is the heat of fusion of ice.
3. One Btu is required to raise the temperature of 1 lb of water one degree Fahrenheit, which is the specific heat of water. Therefore, raising the

temperature of 1 lb of water from 32°F (0°C) to 212°F (100°C) requires 180 Btu. (This is explained in more detail in the next section of this chapter.)

4. The heat of vaporization of water (converting 1 lb of water to steam at a constant temperature) is 970.3 Btu/lb at atmospheric pressure.

Another factor that affects the extinguishing action of water concerns the fact that when water is converted from liquid to vapor, its volume at ordinary pressures increases about 1,700 times. This large volume of water (steam) displaces an equal volume of air surrounding a fire, thus reducing the volume of air (oxygen) available to sustain combustion in the fire zone. Applying water in the form of either ice or snow to a fire would, obviously, utilize water's most effective cooling action. However, practical equipment for such application is not currently available.

## *The Extinguishing Properties of Water*

In *Attacking and Extinguishing Interior Fires,* Lloyd Layman explains that the extinguishing action of water involves efficient conversion to steam in the following manner:[1]

> One gallon [3.75 *l*] of fresh water will absorb approximately 1,250 British thermal units (Btu) in the process of raising its temperature from 62°F [16°C] to 212°F [100° C]. This theoretical gallon of water has reached its boiling point (212°F [100°C]) and is ready to start changing from a liquid to a vapor. Absorption of additional heat will not increase the temperature of the water but will reduce the liquid volume by converting some of the water into steam which will escape into the surrounding atmosphere. In the process of vaporization (boiling), this gallon of water will absorb more than six times the volume of heat that is absorbed in the process of raising its temperature from 62°F [16°C] to 212°F [100°C]. When the last drop has been converted into steam, this gallon of water has absorbed a total of approximately 9,330 Btu; 1,250 Btu in the process of raising its temperature from 62°F [16°C] to 212°F [100°C], and 8,080 Btu in the process of vaporization. Based upon these scientific facts, the following truism can be stated: The maximum cooling action of a given volume of water is obtained only when the entire volume has been converted into steam.
>
> Vaporization of water results in the production of steam in ratio of 1 to over 1,600. Water in the volume of 1.05 cu in. [2.66 cu cm] measured at its boiling point will expand into 1,728 cu in. [4,389.12 cu cm] or 1 cu ft [0.30 cu m] of steam. A gallon [3.78 *l*] of water measured at 62°F [16°C] will produce approximately 223 cu ft [67.97 cu m] of steam.

Extinguishment of fire occurs only when the effect of the extinguishing agent is felt at the point at which combustion is occurring. For many decades the principal method used to extinguish fires was to direct a solid stream of water (from a safe distance) into the base of the fire. This same method of fire extinguishment — the application of streams of water through nozzles — continues to be used today to "knock down" those fires that have gone beyond the incipient stage.

A more efficient method is the application of water in spray form, and combination nozzles and/or sprinklers. Other similar devices for this type of application are currently coming into more general use.

***The Cooling Capacity of Water:*** Water has great cooling capacity. This can be demonstrated in several ways. For example, as just mentioned, raising the temperature of 1 lb (0.45 kg) of water from the freezing point (32°F [0°C]) to the boiling point (212°F [100°C]) requires 180 Btu. A Btu, as explained earlier, is the amount of heat required to raise the temperature of one pound of water one degree Fahrenheit. The melting of one pound (0.45 kg) of ice into water at 32°F (0°C) absorbs 143.4 Btu. It takes 970.3 Btu to convert water to steam at a constant temperature and at normal atmospheric pressure.

When water is converted into steam, its volume at ordinary pressures increases about 1,700 times. This large volume of steam displaces an equal volume of air in the fire area, which reduces the oxygen available and necessary to support combustion.

Water in spray form absorbs more heat than water from a solid stream. Spray nozzles that produce water fog are now almost universally used by fire departments for room fires in dwellings.

The four principles that affect the cooling action of water in spray form are as follows:

- The rate of heat transfer is proportional to the exposed surface of the liquid. For a given quantity of water, the surface is greatly increased by conversion to droplets.

- The rate of heat transfer depends on the temperature difference between the water and the surrounding air or burning material.

- The rate of heat transfer also depends on the vapor content of the air, particularly in regard to fire spread.

- The heat-absorbing capacity depends upon the distance traveled and the velocity of the water in the combustion zone. (NOTE: This factor must take into consideration the necessity for projecting a suitable volume of water to the needed location.)

Calculations indicate that the optimum diameter of a water droplet for extinguishing purposes is in the range of 0.3 to 1.0 mm. The best results are accomplished when the droplets are fairly uniform in size. Currently there is no discharge device capable of producing complete uniformity of size, although many discharge devices have been developed that achieve acceptable degrees of uniformity over a fairly wide range of pressures. The droplet must be large enough to have sufficient energy to reach the point of combustion despite air resistance, the opposing force of gravity, or the diverting movement of air caused by thermal updraft and other air currents.

Certain materials decompose chemically when their temperatures are raised. Water can normally be used to cool these materials below the temperature at which decomposition will be self-sustaining, unless the burning material reacts chemically with the water. In a limited number of cases the application of water accelerates combustion; this may be desirable in order to reduce the period of burning.

***The Smothering Capabilities of Water:*** Water can effectively smother fire in a flammable or combustible liquid if the liquid has a flash point of over 100°F (37°C) and a specific gravity heavier than water so that the water can float on the surface. This can be done more efficiently if a foaming agent is added to the water.

Fires in heavy fuel oil and similar viscous flammable liquids can be successfully attacked by using a strong, coarse water spray. A strong, coarse water spray forms a froth on the surface of the burning liquid, and retards the release of the flammable vapors. Sometimes it is also possible to use water to dilute fires in flammable liquids that are water soluble to a point below which the mixture cannot support flame.

Water is most effective on fires in ordinary combustibles such as wood and paper. It can be used effectively on fires in certain flammable liquids. However, because water is a good conductor of electricity, for small fires involving electrical appliances an extinguisher using carbon dioxide or dry chemical is safer and more efficient. In larger fires involving electrical equipment, automatic sprinklers have a good record. Fire fighters may be endangered when playing hose streams on live electrical lines, especially if the voltage is high. Public utilities have adopted rules that indicate that the minimum safe distance for a fire fighter playing a 2.5-in. (6.35-cm) solid hose stream on a high voltage line is 30 ft (9.14 m), and for all hand-held water spray nozzles the minimum safe distance is 10 ft (3.04 m).

***Water Fog:*** Water is converted to steam more readily when the water is in the form of a fog or spray, thus providing the maximum cooling effect. Many people are amazed at the amount of fire that can be controlled by the proper application of fog with a small amount of water. In fog or droplet form, the water surface is greatly increased over the heavier solid water, and this affects the rate of heat transfer favorably. The rate of heat transfer also depends on the cool temperature of the water and the hot air and the hot burning material. It is important that the water fog or spray reach the seat of the fire in order to get the maximum heat-absorbing capacity.

In fires, certain forces (such as air currents, air resistance, and gravity) tend to hamper the effective delivery of the water. In many fires, water playing on the hot surface generates sufficient steam to exert a smothering effect. The steam may also act to carry heat away from the combustion zone. However, most fires are extinguished by the cooling effect of the water.

168    Principles of Fire Protection

It is worth noting that chemicals can be added to water to benefit its penetration. This so-called "wet water" is very useful for special fires, such as the interior of a burning cotton bale or a haystack. Other chemical additives are used to make "thick water." This has been found useful in fighting deep-seated fires in forest undergrowth. Other uses are still under study.

What you have learned in this chapter about the importance of water as an extinguishing agent should help to emphasize the importance of municipal water systems for fire control. This subject area is covered in detail in Chapter Nine, "Municipal Fire Defenses."

## SPRINKLER SYSTEMS

With this country's rapid growth of business and industry and the resultant increase in fire hazards and property values came the need for more adequate protection against fire. The difficulty of reaching a fire with hose streams has often been demonstrated, and such simple fire protection as water pails, standpipes, and hose equipment has proved inadequate unless the fire was discovered in its early stages. Although fire control has been made easier by improved building construction, comparatively little headway was made in reducing fire loss involving delayed detection until the advent of the automatic sprinkler system.

It is safe to say that the development of large industrial and commercial enterprises, the safeguarding of credit, and the ability of insurance companies to assume large risks, were, at least in part, made possible by the installation of automatic sprinkler systems. A typical automatic sprinkler system consists of piping, usually suspended from the ceiling, fed by a water supply, and equipped with sprinkler heads placed at intervals along the pipes. Heat from a fire causes the sprinkler head or heads above the fire to open and discharge water directly onto the blaze.

The great majority of fires in sprinklered buildings are extinguished by the opening of one or two heads over the fire. If obstructions prevent direct discharge onto the fire, or if a rapidly spreading fire opens larger numbers of heads, the sprinklers will usually hold the fire in check until the fire department arrives.

### Development of Sprinkler Protection

In the United States, the practical application of automatic sprinkler protection began about 1878 when the Parmelee sprinkler was first installed. Since the conception of automatic sprinkler systems, many refinements have been made to improve their protection and performance capabilities. The value of automatic sprinkler protection for the safety of people and the protection of property has been repeatedly demonstrated.

## Value of Sprinkler Protection

As previously stated, automatic sprinkler protection aided the development of modern industrial, commercial, and mercantile practices. Large areas, high-rise buildings, hazardous occupancies, large values, or many people in one fire area tend to develop conditions that cannot be tolerated without automatic fixed fire protection.

*Life Safety:* Generally, loss of life in fires in completely sprinklered properties occurs either from the ignition of clothing or bedclothing of persons who are too old, too young, handicapped, or too intoxicated to help themselves, or from the inability of the sprinkler system to operate because a water supply valve has been left closed. Where helpless people live and sleep in such places as hospitals, nursing homes, asylums, orphanages, and the like, complete sprinkler protection is now recognized as the best method of protection.

Objections to sprinklers by uninformed or unthinking people have resulted in slowing down the general public's complete acceptance of sprinklers. Objections range from the fear of excessive water damage to considerations of expense and even aesthetics. None of these objections has proved valid when measured against life safety, and it is felt by many that the disgraceful record of loss of life in ordinary dwelling fires could be substantially reduced if homeowners, architects, and builders could be persuaded to install some sort of modified, inexpensive sprinkler system in all dwellings.

Automatic sprinklers are particularly effective for life safety because they give warning of the existence of fire, and at the same time apply water to the burning area. With sprinklers, there are seldom problems of access to the seat of the fire, or of interference with visibility for fire fighting due to smoke. While the downward force of the water discharged from sprinklers may lower the smoke level in a room where a fire is burning, the sprinklers also serve to cool the smoke and make it possible for persons to remain in the area much longer than if the room were without sprinklers.

Contrary to popular opinion, automatic sprinklers are practicable for dwellings and other small properties. In country areas where water supplies are limited, a pressure tank can be provided with sufficient capacity to control the fire during evacuation.

*Property Protection:* When it comes to the protection of industrial and commercial buildings, the installation of automatic sprinkler systems really pays off. Not only will the system protect the property from a damaging fire, but it will safeguard the business against shut-down, loss of sales, loss of employees, and other devastating business interruptions.

Insurance companies have recognized the value of sprinklers by offering drastic reductions of premiums for sprinklered properties. In most cases, the savings in insurance premiums pays for the complete cost of the sprinkler system installation in a few years' time. When a new building is under con-

*170    Principles of Fire Protection*

struction, the installation of sprinklers permits larger areas and other building code "trade-offs" that lower the construction costs. It should be noted that this discussion concerns complete sprinkler systems. Where partial systems have been installed, the building owner or the installer is gambling on where the fire may start.

Figures available on the fire loss in manufacturing and mercantile properties where sprinklers are installed show a much better loss/value ratio than those properties not so equipped.

***Sprinkler Performance:*** Records of automatic sprinkler performance have been kept by the National Fire Protection Association since the Association was organized more than eighty years ago. These remarkably comprehensive records show that in 95 percent of the some 117,770 fires in sprinklered buildings (where the Association has reliable data), the sprinklers have performed satisfactorily. During the same period there have undoubtedly been numerous unreported small fires extinguished by one or two heads. If these fires could be included in the records, the efficiency of sprinkler performance would be closer to 100 percent. Proof of this has come from Australia. Harry Marryatt, in his book *Fire: Automatic Sprinkler Performance in Australia and New Zealand, 1886-1968,* reports that a record has been kept of all fires involving sprinkler systems since 1886, and that this record shows 99.7 percent satisfactory performance.[2]

The reasons for unsatisfactory performance of sprinkler systems are shown in Figure 7.1. Note that by far the largest reason has resulted from human error — the sprinklers were shut off at the time of the fire. The sprinklers could hardly be expected to function if no water was being delivered to them. As has been previously mentioned, partial sprinkler system protection is an important reason for sprinkler failure. If there are no sprinkler heads over the fire, the fire will probably get out of control.

Rarely do automatic sprinkler systems fail to control fires. Failures are seldom due to the sprinklers themselves, but rather to the lack of water. Even with older types of sprinklers that are no longer approved, the failure of the sprinkler itself has been very infrequent. Under normal conditions failure of the modern types is practically unknown.

## *Sprinkler System Installation*

The terms "sprinkler protection," "sprinkler installations," and "sprinkler systems" usually signify a combination of water discharge devices (sprinklers); one or more sources of water under pressure; water-flow controlling devices (valves); distribution piping to supply the water to the discharge devices; and auxiliary equipment, such as alarms and supervisory devices. Outdoor hydrants, indoor hose standpipes, and hand hose connections are also frequently a part of the protection system.

# Fire Protection Systems and Equipment 171

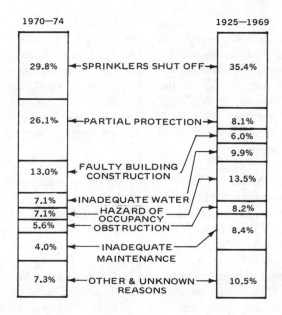

*Fig. 7.1. Reasons for unsatisfactory sprinkler performance — 1925 to 1969, and 1970 to 1974.* (From NFPA *Fire Protection Handbook*)

When considering water supply problems, the performance of sprinklers, dry- or wet-pipe systems, or special arrangements of sprinkler protection, the designation "sprinkler system" applies to the sprinklers controlled by a single water supply valve. Thus, under this particular definition, large buildings could require several sprinkler systems, and a single water system may supply a number of sprinkler systems.

Basically, sprinkler protection involves the principle of the automatic discharge of water, in sufficient density, to control or extinguish a fire in its early stages. In planning for a system that fulfills this objective, many factors must be considered. They can, however, be broadly grouped into four categories: (1) the sprinkler system itself, (2) the features of building construction, (3) the hazards of occupancy, and (4) the water supplies.

NFPA 13, *Standard for the Installation of Sprinkler Systems,*[3] classifies occupancies as light hazard, ordinary hazard, and extra hazard, and provides schedules of pipe sizes, spacing of sprinklers, sprinkler discharge densities, and water supply requirements for each class.

Typical examples of light hazard occupancies are office buildings, schools, churches, hotels, and apartment houses. Ordinary hazard occupancies include warehouses, many manufacturing plants, and piers and wharves. Explosive plants, paint and varnish plants, and oil refineries are examples of extra hazard occupancies.

Automatic sprinkler systems of one type or another have been designed to extinguish or control practically every known type of fire in practically all materials in use today. It is essential, though, that for a given hazard the proper system be used. A sprinkler system designed to control and extinguish fire in an office occupancy with a relatively light amount of combustibles cannot be expected to have the same effectiveness in protecting a hazardous process involving considerable combustible materials, or a storage area where the fire loading is severe. On the other hand, it is not economical to "overprotect" by installing sprinkler equipment capable of controlling and extinguishing a fire of a magnitude beyond any conceivable situation that could arise in the lifetime of a building.

*The NFPA Sprinkler System Standard:* The National Fire Protection Association was organized in 1896 by a small group of men who saw the need for a national set of rules for the installation of automatic sprinkler systems. The Association's first technical committee was formed, and the first standard adopted by the Association was NFPA 13, *Standard for the Installation of Sprinkler Systems,* hereinafter referred to as the NFPA *Sprinkler System Standard.* The standard has been improved and revised many times over the years, and today is followed by manufacturers, insurance companies and bureaus, and by federal, state, and local code and code enforcement agencies.

The NFPA *Sprinkler System Standard* covers the planning and design of sprinkler protection, the type of materials and components used in systems, and the operations necessary for making the installation. The NFPA *Sprinkler System Standard* has had a long and interesting history. When first printed in 1896, it concerned itself principally with sprinkler pipe sizes, sprinkler spacing, and water supplies. Since then, the NFPA *Sprinkler System Standard* has been subject to considerable amplification and refinement, reflecting changes in building construction, materials and techniques, equipment, and types of occupancy conditions that have often posed serious threats to good sprinkler protection. In recent years the standard has been revised to reflect improved efficiency in sprinkler design, and to incorporate modern design and protection methods.

Approval or listing of sprinkler system devices by a recognized testing laboratory is a separate procedure. The use of devices and equipment approved by such a laboratory may be required by an authority having jurisdiction, or the authority itself may approve equipment.

## Types of Sprinkler Systems

The NFPA *Sprinkler System Standard* defines a sprinkler system and its operation as follows:[4]

> A sprinkler system, for fire protection purposes, is an integrated system of underground and overhead piping designed in accordance with fire protection engineering standards. The installation includes a water supply, such as a

gravity tank, fire pump, reservoir or pressure tank and/or connection by underground piping to a city main. The portion of the sprinkler system aboveground is a network of specially sized or hydraulically designed piping installed in a building, structure or area, generally overhead, and to which sprinklers are connected in a systematic pattern. The system includes a controlling valve and a device for actuating an alarm when the system is in operation. The system is usually activated by heat from a fire and discharges water over the fire area.

*Wet-pipe Systems:* The majority of sprinkler systems are called wet-pipe systems. A wet-pipe sprinkler system is under water pressure at all times (in the sprinkler piping water is ready to go to work on the fire when the sprinkler head is actuated) so that water will be discharged immediately when an automatic sprinkler operates.

Water flowing from the wet-pipe-type sprinkler system actuates an alarm valve that gives off a warning signal. Figure 7.2 illustrates the total concept of the wet-pipe automatic sprinkler system.

The essential features of wet-pipe sprinkler systems, which represent about 75 percent of sprinkler installations, include provisions for water supplies, piping, location and spacing of sprinklers, and other pertinent details. This type of system is generally used wherever there is no danger of the water in the pipes freezing, and wherever there are no special conditions requiring one of the other types of systems. Inspection of the wet-pipe sprinkler system at regular intervals is essential, and weekly inspection of all water control valves and alarm control valves is recommended.

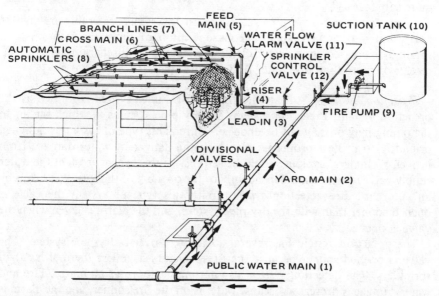

*Fig. 7.2. The total concept of the wet-pipe automatic sprinkler system.* (From Factory Mutual System)

A recent development is an automatic sprinkler system that will turn water off as well as on. With this system the water is shut off when the fire is extinguished, the system reactivates itself if the fire rekindles, and this process repeats as long as necessary.

Where subject to temperatures below freezing, even for short periods, the ordinary wet-pipe system cannot be used since the system contains water under pressure at all times. There are two recognized methods of maintaining automatic sprinkler protection in such locations: (1) through the use of systems where water enters the sprinkler piping only after operation of a control valve (dry-pipe, preaction, etc.), and (2) by the use of antifreeze solution in a portion of the wet-pipe system.

*Regular Dry-pipe Systems:* In locations where there is danger of freezing, it is the usual practice to install a dry-pipe system. In regular dry-pipe systems, the sprinkler piping contains air or nitrogen under pressure instead of water. Admission of the water is controlled by a dry-pipe valve. When a sprinkler is opened by heat from a fire, the pressure is reduced, a "dry-pipe valve" is opened by water pressure, and water flows out of any opened sprinklers.

Most dry-pipe valves are designed so that a moderate air pressure in the dry-pipe system will hold back a much greater water pressure. All sprinkler manufacturers make dry valves, some of the differential type and some of the mechanical type. Both types can be equipped with quick-opening devices that either speed up the air discharge or accelerate the opening of the valve. Particularly for light and ordinary hazard situations, dry-pipe systems have been quite satisfactory.

When two or more dry-pipe valves are used, systems preferably are divided horizontally to prevent simultaneous operation of more than one system, resultant increased time delay in filling systems and discharging water, and receipt of more than one water flow alarm signal.

*Preaction Systems:* Still another type of sprinkler system is known as the preaction system. Preaction systems are dry-pipe systems in which the air in the piping may or may not be under pressure. This type of system is designed primarily to protect properties on which the danger of water damage from broken sprinklers or piping could be serious. In this type of system the water supply valve is actuated independently of the opening of the sprinkler heads by an automatic fire detection system. With this type of system the valve is opened sooner than with the dry-pipe system, and the alarm is given when the valve is opened.

The preaction system has several advantages over a dry-pipe system. The valve is opened sooner because the fire detectors have less thermal lag than sprinklers. The detection system also automatically rings an alarm. Fire and water damage is decreased because water is on the fire sooner, and the alarm is given when the valve is opened. Sprinkler piping is normally dry; thus, preaction systems are nonfreezing and applicable to dry-pipe service.

*Deluge Systems:* For areas of extra hazard occupancies, deluge systems are employed. In these systems all sprinkler heads are open at all times so that when the water comes on, the entire area is flooded. When heat from a fire actuates the fire detecting device, water flows to and is discharged from all sprinklers on the piping system, thus "deluging" the protected areas. Deluge systems are often used in airplane hangars and in areas where flammable liquids are handled or stored.

By using sensitive thermostatic controls operating on the rate-of-rise or fixed temperature principle, or controls designed for individual hazards, it is possible to apply water to a fire more quickly than with systems in which operation depends on opening of sprinklers only as the fire spreads.

*Combined Dry-pipe and Preaction Systems:* These include the essential features of both types of systems. The piping system contains air under pressure. A supplementary heat detecting device opens the water valve and an air exhauster at the end of the feed main. The system then fills with water and operates as a wet-pipe system. If the supplementary heat detecting system should fail, the system will operate as a conventional dry-pipe system.

The intended purpose of a combined dry-pipe and preaction system is to provide an acceptable means of supplying water through two dry-pipe valves connected in parallel to a sprinkler system of larger size than is permitted for a single dry-pipe valve by the NFPA *Sprinkler System Standard.*

Although the NFPA *Sprinkler System Standard* does not restrict the use of combined systems to any particular classes of property, such systems were originally developed for protection of piers where long lines of supply piping could have been subject to freezing if a number of conventional dry-pipe systems had been installed along the length of the pier. Because of the complications of combined dry-pipe and preaction systems and the increased possibility of delayed water discharge, it is general practice to install them only in areas where it is difficult to protect a long supply main from freezing.

The NFPA *Fire Protection Handbook* lists the main features of a combined system as follows:[5]

1. A dry-pipe automatic sprinkler system usually with more than 600 sprinklers and supplied by a long feed main in an unheated area.

2. Two approved dry-pipe valves connected in parallel can be used to supply water to a single large sprinkler system. Two 6-in. [15.24-cm] dry-pipe valves, interconnected with the tripping means for simultaneous operation, are required if a system has more than 600 sprinklers or more than 275 in one fire area. A combination system must have a quick-opening device (QOD) at the dry-pipe valves.

3. A supplemental heat detection system of generally more sensitive characteristics than the automatic sprinklers themselves is installed in the same areas as the sprinklers. Operation of the heat detection system, as from fire, actuates tripping devices which open the dry-pipe valves simultaneously

without loss of air pressure in the system. The heat detection system is also used to give an automatic fire alarm.

4. Approved air exhaust valves, installed at the end of the feed main, are opened by the heat detection system to hasten the filling of the system with water, usually in advance of the opening of sprinklers.

5. Systems with more than 275 sprinklers in one fire area are divided into sections of 275 sprinklers or less by check valves at connections to the feed main. However, not more than 600 sprinklers can be supplied through a single check valve.

6. Provision of means for manual actuation of the heat detection system.

*Limited Water Supply Systems:* Limited water supply systems use automatic sprinklers and follow the standard piping and spacing arrangements. These systems are used where a public water supply or other conventional type of supply, such as a gravity tank or fire pump, is not available for sprinklers with sufficient volume or pressure to satisfy the water supply requirements of the NFPA *Sprinkler System Standard*.

A pressure tank of limited capacity is one source of supply in this type of system which, in other respects, is the same as a conventional system. The minimum sizes of pressure tanks recognized by the NFPA *Sprinkler System Standard* for supplemental supply for limited supply systems contain 2,000 gal (7,570.8 *l*) of water for light hazard occupancies, and 3,000 gal (11,356 *l*) for ordinary hazard occupancies. Approval of plans for all proposed limited supply systems, including the amount of water available for pressure tanks, should be obtained from the appropriate regulating agency.

*Outside Sprinkler Systems:* To protect buildings against exposure fires, outside sprinkler systems are installed that provide a water curtain on the outside wall of the building. Water supply is particularly important here. Large supplies are needed for both the inside and the outside sprinklers.

The use of a water curtain on the outside wall of a building probably antedates automatic sprinklers. In the early years of sprinkler protection, ordinary sprinklers with the struts (structural pieces that were designed to resist pressure lengthwise) removed were used at the peaks of combustible roofs and at the eaves of buildings, particularly wooden buildings. Special types of open sprinklers have since been designed to protect window openings in brick walls. Others, designed to protect combustible cornices, are placed near the top of the window or under the cornice and water is discharged against the glass and frame or cornice, thus providing the desired protection.

## *Sprinkler Heads*

From the time of its conception, the sprinkler head has undergone many changes in design and is now available in a wide variety for all sorts of situations. A major change took place in the mid-1900s when the so-called spray sprinkler was introduced. This is now the standard sprinkler. The new feature of the spray sprinkler that improved the effectiveness of water discharge was the redesign of the deflector. Today's sprinkler heads are designed with temperature ratings ranging from 135°F (57°C) to as high as 500°F (260°C). Ratings of 165°F (74°C) are usual for use in buildings that are maintained at normal, constant temperatures.

In *Automatic Sprinkler & Standpipe Systems,* Dr. John L. Bryan describes the elapsed time of exposure required to operate an automatic sprinkler head as follows:[6]

> The elapsed time of exposure required to operate an automatic sprinkler head when the head is installed in a building and exposed to an accidental fire situation is dependent upon many variables that are related to both the physical environment around the sprinkler head and the design of the sprinkler head. Important physical environmental variables are the distance of the sprinkler head from the ceiling, the surface configuration of the ceiling relative to the heat flow characteristics under the ceiling, the height of the sprinkler heads from the floor, and, most importantly, the characteristics of the fuel array, including its configuration. Sprinkler head design variables include the temperature rating, thermal lag, and the actual condition of the head at the time of the fire occurrence, especially relative to previous high temperature exposures, or any loading of the sprinkler head which might delay or invalidate its operation.

By the turn of the century, the basic design features of the principal components of the automatic sprinkler head had been established. Since that time, any major changes in sprinkler head design have been related to the sprinkler head deflector and the activating element of the head. The link and lever, the triangular strut, the frangible bulb, and the organic or fusible pellet design are the principal activating element arrangements currently utilized in standard sprinkler heads.

## *STANDPIPES*

In order to provide a ready means of manual fire fighting protection, many buildings are equipped with standpipe and hose systems at each floor. There are three classes of standpipe systems. Class 1 systems have 2.5-in. (6.35-cm) hose connections for fire department use to provide heavy fire streams. In nonsprinklered high-rise buildings beyond the reach of fire department ladders, Class I systems can provide water supply for the primary means of fire fighting, *i.e.,* manual attack on the fire.

Class 2 systems have 1.5-in. (3.81-cm) hose, and are for the use of building occupants until the fire department arrives. Lightweight woven jacket rubber lined hose is currently used for this purpose. In the past, unlined linen hose was used; however, the unlined linen hose deteriorated after use when it became wet. The hose is connected to 0.375- or 0.5-in. (0.95- or 1.27-cm) open nozzles or combination spray/straight stream nozzles with shutoff valves. Shutoff or spray nozzles are seldom provided unless the occupancy is one where hand hose would be used frequently. Normally, the hose is kept attached to the shutoff valves at the outlets. Where the occupant-use hose streams can be properly supplied by connections to the risers of wet-pipe automatic sprinkler systems, separate standpipes for these smaller streams are not required.

Class 3 systems have connections for both the 2.5- and 1.5-in. (6.35- and 3.81-cm) hose so that either the fire department or the building occupants can use them. One method for accommodating this multiple use is by means of a 2.5-in. (6.35-cm) hose valve with an easily removable 2.5- x 1.5-in. (6.35- x 3.81-cm) adapter, permanently attached to the standpipe.

Water for standpipes is provided by city water mains, automatic fire pumps, pressure tanks, gravity tanks, or by manually controlled fire pumps operated by remote control devices at each hose station. Two sources of water supply are recommended. For one standpipe there should be sufficient water to provide 500 gpm (gallons per minute) for 30 minutes. The most desirable system maintains water in the standpipe under adequate pressure at all times.

Table 7.1 summarizes the principal specifications for the three classes of standpipes. Fire department connections to standpipe systems are readily accessible on the outsides of buildings so that fire department pumpers can supply water to the standpipe systems. A standpipe system is primarily designed to save time for fire department personnel when placing hose streams in service on the upper floors of buildings.

## *Types of Standpipe Systems*

The four generally recognized standpipe system concepts are:

1. A wet standpipe system, having supply valve open and water pressure maintained at all times. This is the most desirable type of system.

2. A dry standpipe system arranged to admit water to the system through manual operation of approved remote control devices located at each hose station. The water supply control mechanism introduces an inherent reliability factor that must be considered.

3. A dry standpipe system in an unheated building. The system should be arranged to admit water automatically by means of a dry-pipe valve or other approved device. The depletion of system air at the time of use introduces a delay in the application of water to the fire and increases the level of competency required to control the pressurized hose and nozzle assembly during the charging period.

4. A dry standpipe system having no permanent water supply. This type would be used for reducing the time required for fire departments to put hose

Table 7.1  Summary of National Fire Protection Association Standpipe Standards*

| Type | Intended Use | Size Hose and Distribution | Minimum Size Pipe | Minimum Water Supply |
|---|---|---|---|---|
| Class I | Heavy Streams | 2½-in. connections | 4 in. up to 100 ft | 500 gpm 1st standpipe |
| | Fire Department | All portions of each story or section within 30 ft of nozzle with 100 ft of hose | 6 in. above 100 ft | 250 gpm each additional (2,500 gpm maximum) |
| | Trained Personnel | | | |
| | Advanced Stages of Fire | | (275 ft maximum unless pressure regulated.) | 30 minute duration |
| | | | | 65 psi at top outlet with 500 gpm flow |
| Class II | Small streams | 1½-in. connections (Distribution same as Class I) | 2 in. up to 50 ft | 100 gpm per building |
| | Building occupants | | | 30 minute duration |
| | | | 2½ in. above 50 ft | |
| | Incipient Fire | | | 65 psi at top outlet with 100 gpm flowing |
| Class III | Both of above | Same as Class I with added 1½-in. outlets or 1½-in. adapters and 1½-in. hose. | Same as Class I | Same as Class I |

*From NFPA 14, *Standard for the Installation of Standpipe and Hose Systems.*

lines into action on upper floors of tall buildings. This type of system might also be used in buildings during construction, where allowed in lieu of the wet standpipe in unheated areas.

In addition to the preceding four types of systems, systems that are similar to wet standpipe systems are used in some areas. In such systems, the water supply valve that keeps the system full of water involves a limited 1-in. or 0.75-in. connection. These systems are sometimes called "primed" systems, and operating pressure for their use with hose lines is not provided until the public fire department connects to the fire department connection, or until a fire pump is started. Some advantages of the primed system are the reduction in the delivery time of the water as compared with that of the dry standpipe system, and the reduction in corrosive effects on the inside of the standpipe. Dry standpipes may be attached outside of buildings, usually near exterior fire escapes, so that the system will be readily accessible for use by public fire departments.

## *WATER SPRAY FIXED SYSTEMS*

A somewhat specialized use of water is to provide protection for a particular hazard or area by a special fixed-pipe system connected to a reliable water supply, and equipped with spray nozzles for specific water discharge and distribution over the surface or area to be protected. Water spray fixed systems are usually applied to special fire protection problems, and quite often are provided in addition to other forms of protection in order to achieve effective fire control, extinguishment, or exposure protection.

Water spray fixed systems are frequently installed on fractionating or reactor towers to protect them from heat input that could dangerously raise the temperature and pressure of the contained liquids or gases, or could damage the structures. They are also commonly installed for protection of flammable and combustible liquid tanks, gas vessels, oil-filled transformers, oil switches, large electric motors, and for exposed process piping and allied equipment. These are also used for protecting openings through which conveyors or moving stairways pass.

Fixed systems are used in various ways to extinguish a fire, to control the burning, to protect exposures, or even to prevent a fire by diluting or cooling the flammable material involved. These systems are designed to fit the particular situation by determining the appropriate water density required and by locating the nozzles at the right spots. Usually a strong and reliable water supply is required. Ultra-high-speed fixed water spray systems have been designed to protect such hazards as the launching platform for rockets to the moon.

The design of specific systems varies considerably, depending upon the nature of the hazard and the basic purpose of protection. Accordingly, there may be a wide variation in the characteristics of the spray nozzles. As with other automatic systems, no matter how competently they are designed and installed, they are worthless unless they are properly maintained.

### *Use of Water Spray Fixed Systems*

In general, water spray can be used effectively for any one or a combination of the following purposes: (1) extinguishment, (2) controlled burning, (3) exposure protection, and (4) prevention of fire.

***Extinguishment:*** Extinguishment of fire by water spray is accomplished by cooling, smothering from the steam produced, emulsification of some liquids, dilution in some cases, or a combination of these factors.

***Controlled Burning:*** With its consequent limitation of fire spread, controlled burning may be applied if the burning combustible materials are not susceptible to extinguishment by water spray, or if extinguishment is not desirable.

***Exposure Protection:*** Exposure protection is accomplished by applying water spray directly to the exposed structures or equipment to remove or reduce the heat transferred to them from the exposing fire. Water spray curtains mounted at a distance from the exposed surface are less effective than direct application.

***Prevention of Fire:*** Prevention of fire is sometimes possible by the use of water sprays to dissolve, dilute, disperse, or cool flammable materials.

## Application of Water Spray Fixed Systems

As previously stated, water spray fixed systems are most generally used to protect flammable liquid and gas tankage; piping and equipment; electrical equipment such as transformers, oil switches, and rotating electrical machinery; and openings in firewalls and floors through which conveyors pass. The type of water spray required for any particular hazard depends, of course, upon the nature of the hazard and the purpose for which the protection is provided.

NFPA 15, *Standard for Water Spray Fixed Systems for Fire Protection*,[7] calls for piping, valves, pressure gages, and detection systems of an approved type (as does the NFPA *Sprinkler System Standard*). The spray nozzles generally used in these systems are "open," and the pipes, especially those located out-of-doors and subject to freezing weather, are usually dry. Because most spray nozzles have small orifices, strainers capable of removing from the water all solids of sufficient size to obstruct the nozzles must be installed in the water supply lines. These strainers should be checked and cleaned regularly in order to keep the system operable.

In addition to water-flow alarms, a local alarm is usually provided, and this latter device is normally actuated independently of the water flow to indicate operation of the heat-responsive element that triggers the system. All of these devices should be tested periodically; some systems also have trouble alarms to indicate problems in the system (*e.g.,* failure of the power supply to the automatic detection equipment).

## FOAM EXTINGUISHING SYSTEMS

Foam extinguishing systems have been used extensively for many years, especially in the petro-chemical industry, for the extinguishment of flammable liquid fires. The principal kinds of foam are chemical and mechanical (as determined by how they are generated), and these classes are further subdivided. In the past, chemical foam systems for large oil storage tanks commonly consisted of two tanks of stored solutions which, when pumped through a piping system to a mixing chamber, applied foam on top of the burning fuel. Other chemical foam systems functioned through the use of single or double powder generators. The A powder (aluminum sulfate) in water solution with B powder (bicarbonate of soda and foaming agents) generated bubbles containing carbon dioxide gas, which could form a continuous blanket on the surface of a flammable liquid and separate the combustible vapors from the oxygen (air) necessary for combustion.

More recently, mechanical foam, also called air foam, has come into widespread use. It is produced by mixing a liquid foam concentrate in water and mechanically expanding it with air bubbles. These concentrates come in 3 percent and 6 percent compounds for mixing with water in these percentages. The commonly used foam makers produce an expansion in the range of 8- to 16-to-1, and are referred to as conventional foam systems. In ever-increasing use are the high expansion foam systems, which produce volumes of foam (in ratio to the water-concentrate solution) greater than 100-to-1 and up to 1,000-to-1.

Today, special compatible foam concentrates are formulated that result in the generation of a foam that does not break down as readily as ordinary foam when hit with dry chemical. Other special foams are available for application on fires in alcohols, esters, ketones, and ethers (called water-soluble or polar liquids). This type of concentrate produces a foam that does not deteriorate like ordinary foam when in contact with water-miscible solvents.

## Use of Foam Systems

The effective use of foam for fire protection depends upon the general characteristics of foam itself. Foam breaks down and vaporizes its water content under attack by heat and flame. Therefore, it must be applied to a burning surface in sufficient volume and rate to compensate for this loss, and to provide an additional amount to guarantee a residual foam layer over the extinguished portion of the burning liquid. Foam is an unstable "air-water emulsion," and may be easily broken down by physical or mechanical forces. Certain chemical vapors or fluids may also quickly destroy foam. When certain other extinguishing agents are used in conjunction with foam, severe breakdown of the foam may occur. Turbulent air or violently uprising combustion gases from fires may divert light foam from the burning area.

For fires in water-soluble liquids such as alcohol, foam concentrates have been developed which, unlike ordinary air foams, will not break down and lose their effectiveness when used to control this type of fire.

Fire involving aircraft during the fueling operation or after a plane crash can often best be controlled by streams of foam from a crash truck. Protection of an airplane hangar is best accomplished by installation of foam-water sprinkler systems plus portable foam equipment.

Probably the most common method of delivering foam onto a flammable liquid fire is by the fire department use of hoseline foam nozzles. The modern nozzle can provide solid straight streams for range and reach, and also a flat wide "snowstorm" type of foam for application on the burning surface and for protection of the fire fighter.

To protect fuel storage tanks and dip tanks, fixed foam generating equipment can be installed. This is similar to the fixed water spray system described earlier in this chapter. Mobile fire protection vehicles using foam streams are used for crash-rescue trucks at airports and for protecting oil refineries and petro-chemical plants. Many of these are also equipped to discharge dry chemical powder.

## CARBON DIOXIDE SYSTEMS

Carbon dioxide is a noncombustible gas that for many years has been effectively used to extinguish certain types of fires. It acts to reduce the oxygen content in the fire area to a point where it will no longer support combustion. Since carbon dioxide is stored under pressure, it can readily be discharged from its cylinder or extinguisher. Carbon dioxide is inert and will not conduct electricity — two very useful qualities for fire extinguishment. Carbon dioxide can be used safely on energized electrical equipment fires without causing damage to the equipment.

Carbon dioxide in a container is a liquid-vapor mixture. At low temperatures it becomes a solid (dry ice). Carbon dioxide is about 1.5 times as heavy as air, thus helping it to maintain a smothering effect over the burning surface. One pound (0.45 kg) of carbon dioxide in its liquid state will produce about 8 cu ft (2.43 m$^3$) of gas at ordinary pressures.

When the carbon dioxide is released from its cylinder, the heat absorbed during the vaporization cools the liquid to $-110°F$ ($-79°C$), at which temperature it turns into dry ice.

Because carbon dioxide does little or no damage to equipment or materials with which it comes in contact, it is very useful for protection of rooms with contents of high value and contents subject to water damage. Typical of such occupancies are fur vaults, record storage rooms, computer rooms, and rooms housing live electrical equipment. Carbon dioxide is also widely used for extinguishing flammable liquid fires because the carbon dioxide gas rapidly spreads above the surface of the burning liquid and shuts off the oxygen.

Carbon dioxide can be applied for fire control in two ways: (1) it can be discharged onto the surface of the burning material by fixed piping or by hand extinguishers, and (2) the area can be flooded with the gas until the atmosphere in the room becomes inert.

It is important, when using carbon dioxide for fire control and extinguishment, to maintain the supply of gas long enough so that reignition will not occur. Compared to water, carbon dioxide has a low cooling capacity. This makes it difficult to extinguish fires that are deep seated. Although the gas will quickly extinguish the flames, the fire will flare up again if the gas is dispersed over any smoldering fire that might remain.

In using carbon dioxide systems to flood an area, care must be taken to protect people, as a concentration of nine percent is about all most persons can withstand without losing consciousness in a short time. If the area is a small room, an oxygen-deficient atmosphere will be reached quickly. While persons in such an area could get out before the critical concentration is reached, they should be aware of the need to get out quickly.

The cloud of carbon dioxide "snow" in the discharge interferes with visibility and thus makes escape more difficult. There is also a noise factor; carbon dioxide discharges with a roar. Even with small portable extinguishers the noise tends to make an inexperienced operator use the device improperly, with much waste of the extinguishing agent.

## *Extinguishing Properties of Carbon Dioxide*

As has been stated, carbon dioxide is effective as an extinguishing agent primarily because it reduces the oxygen content of the air to a point where it will no longer support combustion. Under suitable conditions of control and application, some cooling effect is also realized.

**Smothering:** Carbon dioxide is stored under pressure as a liquid and, when released, is discharged into the fire area principally as a gas. In general, 1 lb (0.45 kg) of carbon dioxide in its liquid state may be considered as producing about 8 cu ft (2.43 m$^3$) of free gas at atmospheric pressure. When released onto burning materials, it envelops them and dilutes the oxygen to a concentration that cannot support combustion. Table 7.2 shows the minimum carbon dioxide concentration necessary for extinguishing fire.

**Cooling:** When carbon dioxide is released from a storage cylinder, the rapid expansion of liquid to gas produces a refrigerating effect that converts part of the carbon dioxide into snow. This snow, which has a temperature of $-110°F$ ($-79°C$), soon sublimes into gas, absorbing heat from both the burning material and the surrounding atmosphere. Carbon dioxide snow has a latent heat of 246.4 Btu/lb, but since only part of the liquid carbon dioxide is converted to snow, the total cooling effect of the gas and the snow is considerably less than might otherwise be expected. When the liquid is stored at 80°F (26°C), approximately 25 percent is converted to snow upon discharge, giving a total cooling effect of about 120 Btu/lb. When the liquid is stored at 0°F ($-18°C$), approximately 45 percent is converted to snow upon discharge, giving a total cooling effect of about 170 Btu/lb. Water at 32°F (0°C) has a theoretical cooling effect of about 1,150 Btu/lb, assuming that the water is completely evaporated into steam (which rarely occurs).

## *HALOGENATED AGENTS AND SYSTEMS*

About fifty years ago, carbon tetrachloride became popular as a fire extinguishing agent. For the next several decades the small, hand-pump-type of portable fire extinguisher (best known trade name, "Pyrene") was widely used to put out small fires in homes, factories, stores, and automobiles.

Table 7.2  Minimum Carbon Dioxide Concentrations for Extinguishment*

| Material | Theoretical Min. $CO_2$ Concentration (%) |
|---|---|
| Acetylene | 55 |
| Acetone | 26** |
| Benzol, benzene | 31 |
| Butadiene | 34 |
| Butane | 28 |
| Carbon disulfide | 55 |
| Carbon monoxide | 53 |
| Coal gas or natural gas | 31** |
| Cyclopropane | 31 |
| Dowtherm | 38** |
| Ethane | 33 |
| Ethyl ether | 38** |
| Ethyl alcohol | 36 |
| Ethylene | 41 |
| Ethylene dichloride | 21 |
| Ethylene oxide | 44 |
| Gasoline | 28 |
| Hexane | 29 |
| Hydrogen | 62 |
| Isobutane | 30** |
| Kerosine | 28 |
| Methane | 25 |
| Methyl alcohol | 26 |
| Pentane | 29 |
| Propane | 30 |
| Propylene | 30 |
| Quench, lubricating oils | 28 |

*From NFPA *Fire Protection Handbook*.

**The theoretical minimum extinguishing concentrations in air for the above materials were obtained from Bureau of Mines, Bulletin 503.[1] Those marked ** were calculated from accepted residual oxygen values.

Concern about the toxic properties of carbon tetrachloride began to manifest itself about 1960, and in ensuing years the NFPA and the extinguisher testing laboratories discontinued any recommendation of carbon tetrachloride as a fire extinguishing agent.

However, during the past two decades, other halogenated extinguishing agents, commonly referred to as the halons, have become popular. A halon is a hydrocarbon (hydrogen and carbon) in which some of the hydrogen atoms have been replaced by such elements as bromine, chlorine, or fluorine, or by combinations of these in order to create a fire extinguishing gas or liquid. Table 7.3 lists the principal halons and their number designations. The number designations, developed by the Corps of Engineers of the U.S. Army, indicate the chemical composition of the materials and make for easier referral. A

### Table 7.3 Sample Halon Numbers for Various Halogenated Fire Extinguishing Agents*

| Chemical Name | Formula | Halon No. |
|---|---|---|
| Methyl bromide | $CH_3Br$ | 1001 |
| Methyl iodide | $CH_3I$ | 10001 |
| Bromochloromethane | $BrCH_2Cl$ | 1011 |
| Dibromodifluoromethane | $Br_2CF_2$ | 1202 |
| Bromochlorodifluoromethane | $BrCClF_2$ | 1211 |
| Bromotrifluoromethane | $BrCF_3$ | 1301 |
| Carbon tetrachloride | $CCl_4$ | 104 |
| Dibromotetrafluoroethane | $BrF_2CCBrF_2$ | 2402 |

*From NFPA *Fire Protection Handbook*.

number of halons are toxic, thus making them undesirable for general use; however, two of them, Halon 1301 and Halon 1211, have acceptable levels of toxicity and excellent flame extinguishment properties.

Halon 1211 and Halon 1301 are the only two agents recognized by the NFPA Technical Committee on Halogenated Fire Extinguishing Agent Systems. Both Halon 1211 and 1301 are widely used for protection of electrical equipment (both are nonconductors of electricity), airplane engines, and computer rooms. As both of these halons rapidly vaporize in fire, they leave little corrosive or abrasive residue to clean up and do not interfere as much with visibility during fire fighting as foam or carbon dioxide.

These halons can be used for hand extinguishers and also in fixed systems. One of the first uses of a halogenated agent system was for the protection of racing cars at Indianapolis and at Le Mans, France.

### *Halon 1211 (Bromochlorodifluoromethane)*

Halon 1211 (bromochlorodifluoromethane) was first used extensively in Europe for aircraft engine protection systems in the 1960s and, subsequently, in both military and civilian use in a wide range of portable and systems applications. It was not used in the United States until 1973. It is used in portable and wheeled units as its toxicity level is low enough for such use. It is not used for flooding systems because of its toxicity hazard.

The later development of Halon 1211 differs from that of Halon 1301 in that Halon 1211's major areas of use have been established for portable and wheeled units. Although significant use in total flooding systems has been developed in Europe, Halon 1211's toxicity level precludes its use in the total flooding of occupied spaces. However, its toxicity level is low enough for safe use in local application systems and in portable systems in occupied spaces. Small "thermatic" devices using Halon 1211 are available from several European manufacturers, and similar devices are now being developed in the

United States. To satisfy authorities, the development of Halon 1211 in the United States was not started until sufficient toxicity data was collected.

Bromochlorodifluoromethane (Halon 1211) has high effectiveness, and its low toxicity is comparable to that of carbon dioxide. It is a gas at 70°F (21°C), with a vapor pressure of 22 psig and a boiling point of 25°F ($-4$°C). Its relatively high boiling point allows it to be projected as a liquid stream, thus enabling portable extinguishers to have a much greater range than other gaseous materials. The normal overpressurization with nitrogen is from 70 psig at 70°F (21°C), which permits the use of inexpensive hardware.

The range obtained with Halon 1211 is similar to that of dry chemical units and significantly better than that of other liquefied gases. The low pressure makes it easy to handle and also allows for use of lightweight construction materials in the extinguisher. In the latter context, the total weight of a Halon 1211 unit is approximately 40 percent of that of an equivalent carbon dioxide unit. The relatively high boiling point minimizes cold shock to electrical gear and prevents possible skin burns.

## Halon 1301 (Bromotrifluoromethane)

Halon 1301 (bromotrifluoromethane) is the least toxic of the halons. It was first used commercially for aircraft engine protection and is now used for protection of outboard marine engines and computer room facilities. It can be used to flood occupied spaces because of its low toxicity.

In 1954 Halon 1301 was selected by the U.S. Army for use in portable extinguishers, and by the CAA (Civil Aeronautics Administration) and the U.S. Navy for protection of aircraft engine nacelles. Commercial development of Halon 1301 began with the installation of systems for aircraft engine protection on the Lockheed Constellation and Douglas DC-7. Subsequently, it has been used on every commercial aircraft built in the United States. By contrast, Halon 1301 received little attention in Europe until the early 1970s.

Although a 2.5-lb (1.13-kg) portable extinguisher has been available since 1962, high cost has inhibited its development despite its significant technical advantages over carbon dioxide. The most recent work has been in the development of systems for which this agent is ideally suited. Systems range from small 0.5-lb (0.22-kg) units for protection of outboard marine engines to 3.5-ton systems protecting 300,000 cu ft (91.440 m$^3$) oil-processing buildings, the major use being the protection of computer room facilities. The low toxicity of Halon 1301 allows it to be discharged safely from total flooding systems in occupied spaces, an advantage that no other gaseous agent possesses.

## Extinguishing Characteristics of Halons

The mechanism by which the halons extinguish fire is not fully understood. However, there is undoubtedly a chemical reaction as the agents are considerably more effective than heat removal or smothering can account for.

The chemical extinguishants are much more effective because of their ability to interfere with the combustion processes. They act by removing the active chemical species involved in the flame chain reactions (a process known as "chain breaking"). While all the halogens are active in this way, bromine is much more effective than chlorine or fluorine.

## DRY CHEMICAL EXTINGUISHING SYSTEMS

Dry chemical extinguishing agents consist of finely divided powders that effectively snuff out a fire when applied to the fire by portable extinguishers, hose lines, or fixed systems. The original dry powder was sodium bicarbonate (ordinary baking soda). Potassium bicarbonate and other chemical powders, with additives to make the powders free flowing and more moisture resistant, are now in wide use. Dry chemical has been found to be an effective extinguishing agent for fires in flammable liquids and in certain types of ordinary combustibles and electrical equipment, depending upon the type of dry chemical used.

Dry chemical has been used for many years in fire extinguishers, but dry chemical extinguishing systems are of comparatively recent origin. In 1952 the NFPA Committee on Dry Chemical Extinguishing Systems was established, and in 1957 the first edition of NFPA 17, *Standard for Dry Chemical Extinguishing Systems*[8] was adopted. In 1954 the first dry chemical extinguishing system was tested and listed by a nationally recognized testing laboratory.

### Chemical and Physical Properties

Dry chemicals are effective on fires in flammable liquids and on electrical equipment fires. These are usually referred to as "regular dry chemical" or "ordinary dry chemical." Multipurpose dry chemicals can be used on ordinary combustibles such as wood and paper, as well as on flammable liquids and on electrical equipment.

There are now five basic types of dry chemical extinguishing agents. The base chemicals are: (1) sodium bicarbonate, (2) potassium bicarbonate, (3) potassium chloride, (4) urea-potassium bicarbonate, and (5) monoammonium phosphate, which is the base for multipurpose dry chemical. Various additives are mixed with these base materials to improve their storage, flow, and water repellency characteristics. The most commonly used additives are metallic stearates, tricalcium phosphate, or silicones that coat the particles of dry chemical to make them free-flowing and resistant to the caking effects of moisture and vibration. Purple K dry chemical has a potassium bicarbonate base, and Super K dry chemical has a potassium chloride base.

*Stability:* Dry chemical is stable both at low and at normal temperatures. However, since some of the additives may melt and cause sticking at higher temperatures, an upper storage temperature limit of 140°F (60°C) is usually recommended for dry chemical. At fire temperatures, the active ingredients in dry chemical either disassociate or decompose while performing their function in fire extinguishment.

*Toxicity:* The ingredients presently used in dry chemicals are nontoxic. However, particles or dust from the discharge of large quantities may cause temporary breathing difficulty during and immediately after discharge, and may seriously interfere with visibility.

*Particle Size:* Particles of dry chemical range in size from less than 10 microns up to 75 microns. Particle size has a definite effect on extinguishing efficiency, and careful control is necessary to prevent particles from exceeding the upper and lower limits of this performance range. The best results are obtained by a heterogeneous mixture with a "median" particle in the order of 20-25 microns.

## Extinguishing Properties of Dry Chemicals

Dry chemical agents are nontoxic and are stable at low and normal temperatures. While the exact extinguishing action of dry chemical agents is not presently known, the prevailing theory is that the discharge of dry chemical into flames breaks up the combustion reaction in some way. This is believed to be a more important factor than the smothering and cooling effects of the dry chemical.

Dry chemical extinguishing systems are used to protect flammable liquid storage rooms, dip tanks, kitchen range hoods, deep fat fryers, and similar hazardous areas and appliances. Because dry chemical is nonconductive, these systems are useful in the protection of oil-filled transformers and circuit breakers. Dry chemical systems are not recommended for telephone switchboard or computer protection. Dry chemical hose line systems are used in crash trucks and for the protection of aircraft hangars. For many years dry chemicals have been used in portable fire extinguishers, although the more complex and sophisticated systems were not in use until the 1950s.

In fire tests on flammable liquids, potassium bicarbonate-base dry chemical has proved to be more effective than sodium bicarbonate-base dry chemical. Similarly, the monoammonium phosphate-base has been found equal to or better than the sodium bicarbonate base.[9] The effectiveness of potassium chloride is almost equivalent to potassium bicarbonate, and urea-potassium bicarbonate exhibits the greatest effectiveness of all the dry chemicals tested.

Dry chemical, when introduced directly to the fire area, causes the flame to go out almost at once. The exact mechanism and chemistry of the extinguishing action are not definitely known. Smothering, cooling, and radiation shielding contribute to the extinguishing efficiency of dry chemical, but studies suggest that a chain-breaking reaction in the flame may be the principal cause of extinguishment. (See Fig. 7.3.)

**Smothering Action:** For many years, it was a widely held belief that regular dry chemical extinguishing properties relied primarily on the smothering action of the carbon dioxide released when sodium bicarbonate was heated by fire. The carbon dioxide does, undoubtedly, contribute to the effectiveness of dry chemical, as does the like volume of water vapor released when dry chemical is heated. However, tests have generally disproved the belief that these gases are a major factor. Thus, it would appear that dry chemical does not extinguish primarily because of smothering effects. As further evidence to disprove the smothering action theory, it has been shown that certain powdered salts that do not release carbon dioxide, water vapor, or other gases when heated (*e.g.*, sodium carbonate) are effective extinguishing agents. When multipurpose dry chemical is discharged into ordinary burning combustibles, the monoammonium phosphate, decomposed by the heat, leaves a sticky residue on the burning material. This residue seals the glowing material from oxygen, thus extinguishing the fire and preventing reignition.

**Cooling Action:** The cooling action of dry chemical cannot be substantiated as an important reason for its ability to extinguish fires promptly. Studies by C. S. McCamy, H. Shoub, and T. C. Lee,[10] found two extinguishing agents to be equal in extinguishing efficiency. These two agents are dry chemicals containing 95 percent or more sodium bicarbonate (which absorbed 259 cal/g), and borax with 2 percent zinc stearate (which absorbed 463 cal/g). Sodium carbonate, which was only slightly lower in extinguishing efficiency, absorbed an estimated 79 cal/g.

**Radiation Shielding:** Discharge of dry chemical produces a cloud of powder between the flame and the fuel. This cloud shields the fuel from some of the heat radiated by the flame. In reporting the results of their tests to evaluate this factor, McCamy, Shoub, and Lee concluded that the shielding factor is of some significance.

**Chain-breaking Reaction:** The preceding extinguishing actions, each to a certain degree, contribute to the extinguishing action of dry chemical. However, studies reveal that still another factor, which makes an even greater contribution than that of the other factors combined, is present.

The chain-reaction theory of combustion has been advanced by some investigators, such as A. B. Guise[11] and W. M. Haessler,[12] to provide the clue to what this unknown extinguishing factor may be. This theory assumes that free radicals are present in the combustion zone, and that the reactions of these particles with each other are necessary for continued burning. The discharge of

Fig. 7.3. A fire test with a wheeled extinguisher containing a multipurpose dry chemical on a Class A (wood-crib) fire shows the extinguishing action of a dry chemical. (From The Ansul Company)

dry chemical into the flames prevents particles from coming together and continuing the combustion chain reaction. The explanation is referred to as the chain-breaking mechanism of extinguishment.

## COMBUSTIBLE METAL EXTINGUISHING SYSTEMS

There are a number of metals and metal powders found in industrial situations and in transport that will burn, often quite violently. Some metals burn when heated to high temperatures by friction or exposure to external heat. Others burn from contact with moisture or in reaction with other materials. Because accidental fires may occur during the transportation of these materials, it is important to understand the nature of the various fires and hazards involved and the appropriate means for extinguishing such materials. All of these metals and metal powders require special extinguishing agents and special fire fighting techniques. Some develop explosions and very high temperatures, and some react violently with water. Still others give off toxic fumes when burning.

It is apparent, therefore, that to handle a fire involving a combustible metal requires not only the right type of extinguishing agent, but also knowledge of the best way to apply it. Extinguishing agents for combustible metals are designated as dry powders. Some dry powders are used for only one metal, while others can be used to fight fires in several metals.

Where combustible metals are used in industrial processes, both the plant fire brigade and the public fire department should have knowledge of the fire fighting procedures to be employed. In cases in which the metals are in transit, labels and placards that define the hazardous metal should be required so that a local fire department called to the fire scene will have some warning of what it may encounter. Two of the best-known extinguishing agents used to smother fires in combustible metals such as aluminum and magnesium are G-1 Powder and Met-L-X Powder. G-1 Powder is applied by spreading it over the surface of the burning metal. Met-L-X Powder is applied from extinguishers in portable and wheeled units and in fixed pipe systems. There are a number of other dry powders that are commercially available and that can be used to extinguish fires in various metals. Graphite powder, talc, and sand have all been used to smother metal fires.

Some combustible metal extinguishing agents have been in use for years, and their success in handling metal fires has led to the terms "approved extinguishing powder" and "dry powder." These designations have appeared in codes and other publications in which it was not possible to employ the proprietary names of the powders. Such terms have been accepted in describing extinguishing agents for metal fires, and should not be confused with the name "dry chemical," which normally applies to an agent suitable for use on flammable liquid and live electrical equipment fires.

The successful control or extinguishment of metal fires depends to a considerable extent upon both the method of application and the training and experience of the fire fighter. Practice drills should be held on the particular combustible metals on which the agent is expected to be used. Prior knowledge of the capabilities and limitations of agents and associated equipment is useful in emergency situations.

Fire control or extinguishment of metal fires will be difficult if the burning metal is in a place or position where the extinguishing agent cannot be applied in the most effective manner. In locations where there are industrial plants that work with combustible metals, public fire departments and industrial fire brigades have the advantage of being able to conduct fire control drills under the guidance of knowledgeable individuals.

The transportation of combustible metals creates unique problems in that a fire could occur in a location where knowledge and suitable extinguishing agents are not readily available. The Hazardous Materials Bureau of the U.S. Department of Transportation, Office of Hazardous Materials Operations, has anticipated such situations and specifies cargo limitations, labeling, and placarding for the various means of transportation.

## PORTABLE FIRE EXTINGUISHERS

Portable fire extinguishers play an important role in fire protection. A great many fires could be easily extinguished if someone were on hand when the fire started and a suitable portable extinguisher were available. All of the extinguishing agents described in this chapter can be found in approved and reliable portable extinguishers.

Nationally recognized testing laboratories, such as Underwriters Laboratories Inc. and Factory Mutual System, test and label extinguishers clearly to indicate the type of fire on which they can be used.

## Types of Portable Fire Extinguishers

The kind and number of extinguishers needed for particular types of fires are spelled out in NFPA 10, *Standard for Portable Fire Extinguishers*.[13] The following paragraphs describe some of the more common extinguishing agents used in portable fire extinguishers.

*Vaporizing Liquids:* One of the earliest (1908) chemicals employed in portable fire extinguishers was carbon tetrachloride ($CCl_4$). The vapors of $CCl_4$ proved to be quite toxic, and when used on a fire the more toxic decomposition products of hydrogen chloride and phosgene resulted. After World War II a similar but slightly less toxic agent, chlorobromomethane ($CH_2ClBr$), was introduced, and the term "vaporizing liquid" was used to designate extinguishers of this type. In the early 1950s a number of federal agencies banned their use for toxicological reasons. This action, coupled with the availability of more suitable extinguishing agents, resulted in their rapid decline. By the mid-1960s the federal government, many states, cities, and numerous industrial firms no longer permitted the use of any type of vaporizing liquid extinguisher. In the late 1960s listings by most of the leading testing laboratories were discontinued.

*Liquefied Gases:* Although vaporizing liquids have proved to be unacceptable as extinguishing agents, other less toxic halogenated hydrocarbon chemicals, in the form of liquefied gases, are being used increasingly. Bromotrifluoromethane (Halon 1301) was first introduced in 1954 as a high-pressure liquefied gas extinguisher for use on fires in flammable and combustible liquids and on live electrical equipment. Although it bears some resemblance to a small $CO_2$ extinguisher, it is lighter in weight and has a disposable shell which contains 2.75 lbs (1.25 kg) of agent. A low-pressure liquefied gas extinguisher utilizing bromochlorodifluoromethane (Halon 1211) became available in 1973, and by 1974 was being manufactured in a full range of sizes. In 1974 it had an initial UL (Underwriters Laboratories Inc.) listing for flammable liquid and electrical fires, and tests conducted at Underwriters Laboratories Inc. indicated considerable potential for use on fires in ordinary combustibles. In 1974 extensive testing was being conducted on fire extinguishers containing dibromotetrafluoroethane (Halon 2402), which is a liquid at room temperature. The tests indicated a potential for using the extinguishers on all different types of fires (ordinary combustibles, flammable liquids and gases, and electrical equipment).

*Carbon Dioxide:* The first carbon dioxide extinguishers were produced during World War I. During World War II they became the leading extinguishers for flammable liquid fires; however, by 1950 their lead was relinquished to the dry chemical agents.

*Dry Chemicals:* Although the fire extinguishing capabilities of sodium bicarbonate were first recognized in the late 1800s, it was not until 1928 that an effective cartridge-operated dry chemical extinguisher was developed. Considerable developmental work took place during the early 1940s, and an improved, finely granulated agent was introduced in 1943. In 1947 the original extinguisher was replaced with an improved model that utilized the new agent to its best advantage. The rapid increase in the use of flammable liquids and chemicals and their associated process hazards brought about the development of many new dry chemical agents with greater "fire killing" power. The first was potassium bicarbonate-base agent (in 1959), which was about twice as effective as the sodium bicarbonate- ("ordinary") base agent.

*Multipurpose Dry Chemical:* Next (in 1961) came a new type of agent called "multipurpose" dry chemical. In addition to being about 50 percent more effective on flammable liquid and electrical fires, multipurpose dry chemical was also listed as an effective agent on fires in ordinary combustibles. Originally, the less expensive diammonium phosphate was used; however, preference soon shifted to monoammonium phosphate because it had the advantage of being considerably less hygroscopic (*i.e.*, less moisture-retaining). An agent utilizing potassium chloride as a base was first marketed in 1968. It was about 80 per cent more effective when compared to ordinary dry chemical, but also more corrosive and more hygroscopic than potassium bicarbonate. A urea-potassium bicarbonate-base agent was developed in Europe in 1967 and brought to America in 1970. Its comparable effectiveness is judged to be at least 2.5 times better than "ordinary" dry chemical. The first stored-pressure model (rechargeable type) was introduced in 1953; a disposable (nonrechargeable) model first appeared in 1959 and was rapidly adopted for use in dwellings, cars, boats, etc.

*Dry Powder:* The increased use of combustible metals (magnesium, sodium, lithium, etc.) brought about the need for a special agent to extinguish fires in these materials. In 1950 the first dry powder extinguisher using a sodium chloride-base agent was introduced. The designation "dry powder" was specifically adopted to indicate suitability for Class D (combustible metal) fires only, and to clearly establish the differentiation that "dry chemical" would only apply to Class B:C fires or Class A:B:C fires in ordinary combustibles, flammable liquids, and electrical equipment.

## *Application of Portable Fire Extinguishers*

NFPA 10, *Standard for Portable Fire Extinguishers,* classifies fires in four ways, as follows:[13]

> Class A: Fires involving ordinary combustible materials (such as wood, cloth, paper, rubber, and many plastics) requiring the heat absorbing (cooling)

*Fig. 7.4. Markings for extinguishers indicating the classes of fires on which they should be used. Color coding is part of the identification system; the triangle (Class A) is colored green, the square (Class B) red, the circle (Class C) blue, and the five-pointed star (Class D) yellow.* (From NFPA *Fire Protection Handbook*)

effects of water, water solutions, or the coating effects of certain dry chemicals that retard combustion.

Class B: Fires involving flammable or combustible liquids, flammable gases, greases, and similar materials where extinguishment is most readily secured by excluding air (oxygen), inhibiting the release of combustible vapors, or interrupting the combustion chain reaction.

Class C: Fires involving live electrical equipment where safety to the operator requires the use of electrically nonconductive extinguishing agents. *(N.B.:* When electrical equipment is de-energized, the use of Class A or B extinguishers may be indicated.)

Class D: Fires involving certain combustible metals (such as magnesium, titanium, zirconium, sodium, potassium, etc.) requiring a heat-absorbing extinguishing medium not reactive with the burning metals.

The markings for extinguishers indicating the classes of fires on which they should be used are shown in Figure 7.4. Some portable extinguishers are of primary value on only one class of fire; some are suitable for two or three classes; none is suitable for all four classes of fire. The markings on an extinguisher indicate its suitability for the various classes of fires, and also the relative extinguishing effectiveness of the extinguisher.

The tests for establishing the ratings and suitability of portable fire extinguishers are carefully made using reproducible test fires. Ratings are developed as in the following example for foam fire extinguishers:

- The 1.25-gal. (4.73-$l$) foam extinguisher gets a 1A and 2B rating.
- The 2.5-gal. (9.46-$l$) foam extinguisher gets a 2A and 4B rating.
- The 35-gal. (132.49-$l$) wheeled extinguisher gets a 20A and 20B rating.

All other types of portable fire extinguishers are given ratings that show the class or classes of fires on which they can be used and the amount of fire that they can be expected to control.

The selection of the right size and type of extinguisher for the anticipated fire situation is of primary importance. The NFPA *Standard for Portable Fire Extinguishers* defines areas of light hazards, ordinary hazards, and extra hazards to help in determining the probable size of a fire. For example, a small water pump tank extinguisher rated 1A would be suitable for a light hazard occupancy only. A multipurpose dry chemical extinguisher rated 40A would be suitable for an extra hazard occupancy. Sound judgment in selecting the right kinds and numbers of portable fire extinguishers for the area to be protected is essential. There is no one fire extinguisher that is equally suitable and desirable for use on all classes of fire.

Training in the use of extinguishers is another important factor relative to effective fire protection. The same extinguisher in the hands of a trained or untrained person will produce widely different results. Portable fire extinguishers are the first line of defense against fires, and should therefore be available irrespective of other fire control measures.

## FIRE APPARATUS EQUIPMENT

Listings of equipment and appliances needed with fire apparatus are contained in NFPA 1901, *Standard for Automotive Fire Apparatus*.[14] These lists of ancillary equipment for major apparatus categories are arranged in three groupings. The standard lists equipment that the apparatus manufacturer is required to furnish with each type of apparatus unless otherwise specified in special provisions. The Appendix to the standard contains additional equipment lists, including lists of equipment normally carried on each type of apparatus but not necessarily purchased with apparatus and additional equipment needed for each class of service (engine, truck, rescue, etc.), but not necessarily carried on each piece of apparatus of a given type.

### Equipment Carried on Apparatus

Fire apparatus must be equipped with the tools necessary to accomplish fireground operations. Where apparatus is delivered with only the minimum items of equipment, other equipment must be appropriated as needed.

***Forcible Entry Tools:*** Generally, structural fire fighting cannot be successfully carried out without entering the building or the part of the structure where the fire occurs. Fire apparatus carry a variety of tools for gaining access to locked or closed areas. Such equipment permits fire fighters to gain entrance through doorways and windows and to open walls, partitions, ceilings, or roofs to uncover hidden fire, or to perform ventilation procedures. Some fire companies carry patented forcible entry tools that have special features.

Hydraulic or air-powered spreaders are becoming increasingly popular because they allow more leverage than those which can be applied manually. Often a heavy door can be sprung with these tools without permanent damage. Such tools also are useful in freeing persons trapped in elevators and wrecked vehicles, and by machinery.

*Communications Equipment:* All fire department vehicles should be equipped with two-way radios capable of operating on the frequency or frequencies used to dispatch the apparatus, and a fireground communications channel. It is highly desirable to provide radio loudspeakers, both inside and outside the vehicle cab, that can be heard above the usual noise on the fireground so that messages can be received by personnel away from the apparatus. A loudspeaker, and frequently a radio phone, is provided at the pump operator's position, as well as the usual hand set and speaker in the cab. Fire chiefs' vehicles may also be equipped to send and receive hard-copy messages (which may be associated with computerized dispatching and recording), as well as with specialized facilities for receiving visual displays of information, including surveys, maps, and applicable fire hazard data that may be transmitted from the communications center.

*Electric Lights and Generators:* Each piece of fire apparatus should carry a minimum of two hand lights or lanterns. In addition, each officer needs an electric light for work in a smoke-filled building. Such equipment should be carried on the vehicle to which the officer is assigned. The fire service is often faced with the need to provide electric power at fires and emergencies. Power demands may include: (1) power to operate fire apparatus and its appurtenances, (2) power to operate tools and lights on the fireground, (3) emergency power to maintain essential services, such as communications, and (4) emergency power for temporarily replacing essential community services. This demand may be met in several ways: (1) by the provision of generating equipment, transformers, and outlets as part of a vehicle's electric system; (2) by portable generators on fire apparatus; (3) by special "lighting trucks" or mobile generator units; and (4) by standby generators at such locations as fire stations, fire headquarters, and communications centers. Generators may be mounted on the fire apparatus and arranged for automatic starting, but should also be capable of being removed from the apparatus. Some fire apparatus carry two generators so that one can be dropped off and used where needed, while the other provides the required electrical capacity on the unit. In some situations, one portable generator and one generator that is driven by the apparatus engine are provided.

*Portable Pumps:* Portable pumps for fire department service are generally of the centrifugal type. They are grouped in categories based upon the pressure-volume characteristics that make them suitable for various classes of work. Small streams at high pressure are intended mainly for grass and brush fire work. Pumps delivering relatively large volumes at low pressures can serve as a supply pump for fire trucks where the water supply source is beyond the reach of suction hose. They also may be used as a dewatering pump. Fairly large volume flows at higher pressures are considered valuable in hilly areas where ordinary portable pumps do not develop sufficient pressure to overcome elevation of "head."

## Protective Equipment for Fire Fighters

NFPA 1901, *Standard for Automotive Fire Apparatus,* requires that each pumper (engine company) carry at least two pieces of self-contained breathing apparatus approved by the National Institute of Occupational Safety and Health (NIOSH), and that each ladder truck carry six pieces of self-contained breathing apparatus (also NIOSH-approved). The laws of some states require that approved protective breathing apparatus be provided for all fire fighters exposed to smoke and gases from fires and emergencies.

***Breathing Apparatus:*** Protective breathing equipment should be used only after thorough training. Fire fighters wearing masks should not work alone, and should be under the supervision of officers. Masks should not be used by fire fighters who have been subjected to heavy exertion and smoke. The use of masks does not protect an individual against excessive heat, gases, and poisons that attack the body through the skin. Special precautions are necessary when fire fighters use breathing apparatus in pressurized atmospheres, not only because of the increased hazard of fire and explosions, but because the period of protection may be reduced by the higher pressure. Another problem relevant to the use of breathing apparatus exists when fires are difficult to reach and when the time required to return from the fire to the outside air utilizes most of the respirable air of the equipment. Some fire departments have large, wheeled air cylinders arranged for such situations. Many fire departments have scuba equipment (self-contained underwater breathing apparatus) designed for their own purposes. Fire department air or oxygen masks should not be used for underwater work. Appropriate training in the use and care of protective breathing apparatus is an important part of basic fire fighter training. Confidence in the equipment, as well as knowledge of its practical limitations, should increase with use.

***Resuscitators:*** Fire department ladder trucks and rescue vehicles carry at least one resuscitator capable of caring for two patients (plus spare oxygen cylinders). Such equipment should also be carried on engine companies that are assigned to answer rescue calls in their districts. Some states have passed laws requiring that all fire department and police personnel be trained in cardiopulmonary resuscitation (CPR).

***Smoke Ejectors:*** Smoke ejectors are required equipment on ladder trucks and on other apparatus used for performing ventilation service. At serious fires it is not unusual for a number of smoke ejectors to be used at various locations to move the desired quantity of smoke and air. An important use of smoke ejectors is to reduce smoke damage from minor fires. Prompt use helps prevent soot particles from settling, and when smoke ejectors are used in connection with vacuum equipment, most soot may be removed.

*Life Nets:* A rescue item commonly carried on ladder trucks is the life net, or jumping net. Most nets consist of heavy canvas supported by a folding metal frame and springs, and containing a pad to soften impact. A few fire departments also carry rope jumping nets for use in confined places where limited space or obstructions preclude use of the large circular metal-rimmed nets.

*Life Guns:* A life gun designed to shoot a rope line to persons in distress is an item of equipment occasionally carried on ladder or rescue trucks. Primarily, the life gun is used in water rescues (such as for persons stranded on rocks from overturned boats), and to rescue persons from cliffs or canyons.

*Protective Clothing for Fire Fighters:* Fire fighters should not enter burning buildings or buildings that have been subjected to appreciable fire damage unless they are wearing full protective clothing, including fire helmets. At a minimum, protective clothing should include a suitable helmet, a protective coat, boots, and gloves, all of which should be properly sized for the individual wearer. Some fire departments provide a full turnout suit with trousers of material and construction comparable to the protective coat. Protective clothing for fire fighters is designed to protect the wearer from heat and cold, as well as from abrasions. The protective clothing should also be water-repellent, lightweight, and easy to put on. The boots used by fire fighters should have protective insoles. Short "bunker boots" are used with turnout suits. All boots for fire fighters should have straps to facilitate rapid use in donning when a fire alarm is received.

## Summary

Water is the principal fire extinguishing agent because it is readily available, cheap, and has excellent cooling effects. Water discharged through automatic sprinkler systems is a most valuable protection for lives and property in places of public assembly as well as in factories and commercial occupancies. The fire record of completely sprinklered properties is remarkably good, the most common cause of failure being the result of human error: the sprinklers were shut off at the time of the fire. Standpipe and hose systems are used in many buildings to provide fire fighters with water at each floor level. In addition, water fixed spray systems provide protection for special hazards such as flammable liquid tanks and electrical transformers.

Fire fighting foam is used to blanket and smother flammable liquid fires. Mechanical foams, chemical foams, and high expansion foams are used for various purposes and conditions. Carbon dioxide gas is another useful extinguishing agent for flammable liquid and electrical fires. Carbon dioxide is an inert gas and does not conduct electricity. Halons are relatively new extinguishing agents. Halon 1301 and Halon 1211 are the recognized halogenated agents and are used for protecting electrical equipment. Dry chemical extinguishing agents are effective on flammable liquid and electrical fires, and can also be used on ordinary combustibles.

Portable fire extinguishers are valuable as a first line of defense. They are classified and rated for various kinds of fire. The right kind and size of portable fire extinguisher is important for efficient and effective use.

Fire apparatus must be equipped with the tools and appliances necessary to accomplish fireground operations. Where apparatus is delivered with only minimum items of equipment, other equipment must be appropriated as needed.

## *Activities*

1. Describe the physical characteristics of water that are pertinent to its extinguishing ability. Include water's limitations in your description.
2. As a member of the fire service, convince a classmate of the value of automatic sprinkler protection. Your classmate can represent an owner of a building that does not have automatic sprinkler protection.
3. Describe a typical automatic sprinkler installation.
4. Compare wet-pipe and dry-pipe sprinkler systems. Include in your comparison the conditions under which each type would be used.
5. (a) Compare the preaction-type sprinkler system with the dry-pipe sprinkler system.
   (b) Include in your comparison a description of the conditions under which each system would be used.
6. Describe the three classes of standpipe systems and the conditions under which each is used.
7. Explain why, in some types of situations, foam might be required as an extinguishing agent instead of water. Describe the various types of foam in your explanation.
8. Compare the extinguishing properties of carbon dioxide with the extinguishing properties of water.
9. Carbon tetrachloride, a halogen, was once a popular fire extinguishing agent. Today it has been replaced by two major halons. Explain:
   (a) Why the formerly widely used carbon tetrachloride was replaced as an extinguishing agent.
   (b) Why Halon 1211 and Halon 1301 are acceptable as replacements for carbon tetrachloride.
10. Describe five of the common portable fire extinguishing agents, including in your description the various characteristics of each.

## *Bibliography*

[1]Layman, Lloyd A., *Attacking and Extinguishing Interior Fires,* NFPA, Boston, 1955, p. 24.

[2]Marryatt, H. W., *Fire: Sprinkler Performance in Australia and New Zealand, 1886-1968*, Australian Fire Protection Association, Melbourne, Australia, 1971, p. 36.

[3]NFPA 13, *Standard for the Installation of Sprinkler Systems*, NFPA, Boston, 1976.

[4]_____, pp. 5-6.

[5]*Fire Protection Handbook*, 14th Ed., NFPA, Boston, 1976, p. 14-26.

[6]Bryan, John L., *Automatic Sprinkler & Standpipe Systems*, NFPA, Boston, 1976, p. 205.

[7]NFPA 15, *Standard for Water Spray Fixed Systems for Fire Protection*, NFPA, Boston, 1977.

[8]NFPA 17, *Standard for Dry Chemical Extinguishing Systems*, NFPA, Boston, 1975.

[9]Guise, A. B., "Potassium Bicarbonate-Base Dry Chemical," *NFPA Quarterly*, Vol. 56, No. 1, July 1962, pp. 21-27.

[10]McCamy, C. S., Shoub, H., and Lee, T. C., "Fire Extinguishment by Means of Dry Powder," Sixth Symposium on Combustion, The Combustion Institute, Reinhold, NY, 1956, pp. 795-801.

[11]Guise, A. B., "The Chemical Aspects of Fire Extinguishment," *NFPA Quarterly*, Vol. 53, No. 4, Apr. 1960, pp. 330-336.

[12]Haessler, W. M., "Fire and Its Extinguishment," *NFPA Quarterly*, Vol. 56, No. 1, July 1962, pp. 89-96.

[13]NFPA 10, *Standard for Portable Fire Extinguishers*, NFPA, Boston, 1975.

[14]NFPA 1901, *Standard for Automotive Fire Apparatus*, NFPA, Boston, 1975.

Chapter Eight

# Alarm and Detection Systems and Devices

*Communication devices, both manual and automatic, are of great value to fire protection. Different types of signaling systems have been developed that not only alert fire departments to the presence of a fire, but through mechanical detection of smoke, flames, or excessive heat, can warn building occupants that a situation of potential danger is imminent.*

## PUBLIC FIRE SERVICE COMMUNICATIONS

The fire alarm box on a pole on a street corner has been a familiar sight to city dwellers for several generations. The first such municipal fire alarm system was installed in the city of Boston more than a century ago, and most of today's school children have been shown demonstrations, usually by fire fighters, on how to turn in an alarm from a fire alarm box; also, most of these same children have been lectured on the dangers of false alarms.

An effective fire alarm system fulfills two functions: (1) that of receiving alarms from the public through fire alarm boxes located on the street or on private property, and (2) that of transmitting the alarm to the fire companies and personnel who should respond to the emergency.

### Municipal Fire Alarm Systems

All of the earlier municipal fire alarm systems were wired telegraph systems. Today, municipal fire alarm systems are electrically operated and can be classified by their method of operation as either telegraph-type, telephone-type, radio-type, or a combination of the three. As stated above, all types perform the same function of providing a means by which an alarm can be transmitted from a street fire alarm box to the fire department communication center, and from there to the fire companies that are to respond to the alarm.

Notifying the fire companies from the communication center can be done either manually by fire alarm operators or automatically. The larger cities operate their alarm systems with trained operators who retransmit the alarms they receive to the companies to be dispatched to the fire. The installation, maintenance, and use of all municipal fire alarm systems, regardless of the principles of operation, are covered in detail in NFPA 73, *Standard for the Installation, Maintenance, and Use of Public Fire Service Communications.*[1] This standard makes no distinction between a telegraph-type, a telephone-type, or a radio-type alarm system, since, as previously stated, all must perform the same function of providing a means by which an alarm can be transmitted from a street fire alarm box to the communication center, and from the communication center to the fire companies that are to respond to the alarm.

Telegraph-type, series telephone-type, and radio-type fire alarm systems are usually owned by the municipality, while parallel telephone-type systems are usually leased from a public utility.

Generally, a municipal fire alarm system may be used for the transmission of other signals or calls of a public emergency nature, provided such transmission does not interfere with the proper handling of fire alarms. For example, systems employing voice communication between the street fire alarm box and the communication center can be used in transmitting alarms of a nonfire emergency nature. Fire alarm boxes in a radio-type system can be provided with push buttons for calling the police department, the ambulance service, or other emergency services, and signals can be transmitted directly to the service called. In parallel telephone-type systems (each box served by a separate circuit), the fire alarm operator can cross-connect to the proper emergency service, such as the police department, or systems can be arranged so that authorized persons, such as police officers, will be connected directly with their own department. In a series telephone-type system (a number of boxes connected to the same circuit), cross-connecting to other emergency services is usually not practical, but can be done.

Use of the telephone-type box for other than fire alarm purposes should not be extended to corrupt the intended emergency use by extensive calls of a nonemergency nature (such as routine calls to various municipal departments). It should be noted that the parallel telephone-type system may be adapted by proper switching arrangements so that street boxes may also be used as a police-signaling system without interfering with their proper use in a fire alarm system and without involving the fire alarm dispatcher. Various arrangements are available whereby voice communication facilities may be added to a telegraph-type fire alarm system and still ensure the integrity of the telegraph features, even during the use of voice facilities.

Fire alarm systems have been traditionally grouped into two types depending on whether retransmission of alarms is manual or automatic. These two types of systems (Type A and Type B) are briefly described in the following paragraphs, and are more fully detailed in the standards of the National Fire Protection Association.

*Type A (Manual):* The Type A (Manual) system (see Fig. 8.1) is one in which operators are required to check the receipt of alarms and to retransmit them over alarm circuits to fire stations. All fire calls, whatever their origins, are considered to be alarms. One operator is required when more than 600 alarms per year are experienced; two where the number exceeds 1,500.[2] In addition to handling the alarm system, operators may handle telephone calls for fires as well as other telephone business.

With Type A systems there is a sufficiently large number of circuits to use a considerable portion of an operator's time in daily tests of the circuit. One reason for having two operators is that if one suddenly becomes ill in an isolated fire alarm office, the system would not be out of service.

*Type B (Automatic):* The Type B (Automatic) system (see Fig. 8.2) is one in which alarms are retransmitted automatically to fire stations. Type B systems are designed for small communities that have organized fire departments, but that do not have or need a Type A system.

### Fire Alarm Boxes

The effectiveness of a municipal fire alarm system depends on a number of things. For example, to provide proper distribution and protection to the community, fire alarm boxes must be well distributed so they are readily visible and available for public use. As a general rule, boxes should be not more than

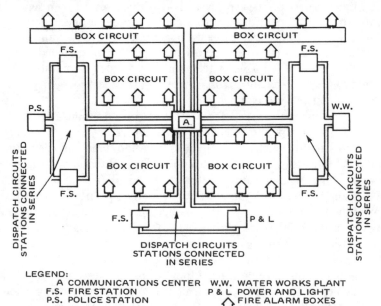

*Fig. 8.1. A schematic diagram of a Type A wired telegraph fire alarm system.* (From NFPA *Fire Protection Handbook*)

# Alarm and Detection Systems and Devices

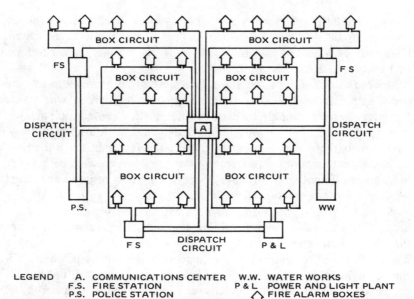

Fig. 8.2. *A schematic diagram of a Type B wired telegraph fire alarm system.* (From NFPA *Fire Protection Handbook*)

500 ft (152.4 m) apart (one block) in congested districts or more than 800 ft (243.8 m) apart (two blocks) in residential areas. There should be a fire alarm box at or near the entrance to every school, hospital, nursing home, and place of public assembly. A fire alarm box should be installed near the entrance to all fire stations, since many persons are more aware of the location of the nearest fire station than of the nearest fire alarm box. Persons have often gone to the fire station only to find that the fire company is away from the station on another alarm and that there is no readily available means of transmitting an alarm to the communication center.

Fire alarm boxes should be located and marked so that they are readily visible, and should be lighted at night with a light of distinctive color. The local fire department should undertake to make the public aware of the function and value of the fire alarm box, the location of the nearest box to their home and place of work, and the fact that false alarms jeopardize life and property protection.

***Telegraph-type:*** The telegraph-type box is actuated by pulling down a lever or turning a handle. This sets off a clockwork mechanism which, in turn, rotates a code wheel that is distinctive for each box. The modern box is noninterfering. Alarms pulled on two or more boxes for the same fire or for another fire will not jumble the signal. Alarm boxes are installed in series on a circuit; thirty boxes to a circuit if the cable is underground; twenty boxes to a circuit if the wiring is overhead.

The wired telegraph-type fire alarm box was developed directly from the one invented by Dr. William F. Channing, a practicing physician in Boston, and University of Maine physics professor Moses G. Farmer, who developed the first municipal alarm system in 1847.

Modern telegraph-type fire alarm boxes are often designed to transmit the coded signal through a ground connection when a circuit wire is broken. However, correspondingly modern communication center equipment is also necessary. Older boxes of interfering design (*i.e.,* code transmission will interfere with that from another box; two interfering boxes will result in a garbled signal) or of noninterfering, yet of the nonsuccession design (*i.e.,* will not transmit if another box is already operating), may result in no alarm being received from one or both of the boxes.

Wired telegraph-type boxes are installed in series on closed circuits that always carry a small current providing continuity supervision (*i.e.,* a break in the wire will interrupt the current) as well as alarm indication (*i.e.,* rhythmic interruption and restoration of the current by the box code wheel). Circuits should be laid out so that disruption of any one circuit will not remove box protection from more than a limited area in case of failure of a circuit. If a circuit utilizes aerial wire in whole or in part, the allowable area is that which would be protected by 20 properly spaced boxes. If the circuit is entirely in underground cable or supported aerial cable, the allowable area is that which would be protected by 30 properly spaced boxes. In situations in which all of the boxes on a circuit and the associated communication center equipment are designed and installed to operate on an emergency ground-return basis, the allowable area may be doubled.

The fire alarm headquarters apparatus of all telegraph-type systems is arranged according to the needs of this class of apparatus, with facilities for testing circuits, sounding devices, controlling supply of current, visual indicators, telegraph keys for sounding signals manually, recording devices, and other equipment. Suitable devices can be provided so that code and voice communication may be established between boxes and fire alarm headquarters, and between fire alarm headquarters and fire stations over box and alarm circuits respectively, without the possibility of interfering with reception or transmission of alarms over the same circuits.

*Telephone-type:* Telephone-type boxes are usually on alarm systems leased to the city by the local telephone company. Each box has its individual pair of wires to the telephone company central office. In large cities the individual circuits can be concentrated to avoid an excessive number of circuits.

In telephone-type boxes the location of the box is definitely established without voice transmission and without interference from other boxes. This type of fire alarm box is essentially a telephone handset installed in specially designed housing. Except in systems where a "concentrator identifier" is installed, each telephone handset is connected to a switchboard at the communication center. The location from which the alarm is transmitted is definitely established without dependence upon voice transmission and without interference from other boxes. Removing the handset from its cradle in the box causes a lamp to light on the switchboard, the number of the box to be visually recorded, and the time to be printed. The circuit is supervised up to and including the handset.

There are two basic arrangements of telephone-type fire alarm systems: These arrangements are: (1) parallel, and (2) series.

1. *Parallel.* In this arrangement, each fire alarm box is served by an individual pair of wires. The circuits to telephone-type boxes are installed in the various cables of the telephone company; they are not segregated, and should be distinctively identified at the points of attachment to the main frames in telephone central office buildings. Care must be exercised in the routing and installation of circuits to avoid conditions that may be accepted in a commercial telephone system, but that would be considered an unacceptable hazard to a fire alarm system. These include such conditions as circuits entering buildings not essential to fire alarm operation, attachment to buildings (particularly of combustible construction), and routing over roofs. While this type of system can be installed for ownership by a municipality, most have been leased from the local telephone company.

2. *Series.* A circuit in a series system serves a number of boxes. The fire alarm box is essentially a telephone handset installed in a specially designed housing. Boxes that also contain a telegraph mechanism are called combination telephone-telegraph-type, and both voice and coded signals can be transmitted to the communication center over the same circuit. The series telephone-type box may also be intermixed with telegraph-type street boxes and master boxes. Master boxes are connected to fire alarm systems that are installed in buildings.

The fire alarm headquarters apparatus provided by the telephone companies is designed to meet the requirements of a Type A system. Street boxes may terminate in a reporting switchboard with both push-button and cord termination, or in various types of standard switchboard equipment, depending on the requirements of the municipality. At fire alarm headquarters, immediate visible and audible signals appear at the switchboard when a street box is activated. Locked-in visual signals for each call box, which can be released only by the dispatcher, indicate the street box and location from which the alarm is received. This assures correct identification of the general location from which the alarm has originated, even though the reporting party may have been too excited to talk, spoke unintelligibly, or replaced the receiver without talking.

***Radio Type:*** Recently, radio-type alarm boxes have come into use. While these types presently transmit coded signals, in the future they may also transmit voice signals. The radio-type boxes may be powered either by batteries or by a prewound mechanism, or may be "user-powered." Battery-powered boxes transmit supervisory or test signals every 24 hours, as well as a warning signal when battery power falls below a predetermined level. Batteries

208   *Principles of Fire Protection*

must be replaced periodically, but solar recharging is permitted. User-powered boxes contain a spring-wound alternator to provide power. The spring is wound when the operating handle is pulled down to uncover the push-button(s). Radio-type boxes are interfering, but signal transmission is so fast that interference is not very probable. Most boxes are designed with random spacing between the required three rounds of signals. When boxes having random spacing are operated at the same time, at least one correct round of the signal will be received from each box. Some battery-powered boxes have an electronic "memory," equivalent to the succession feature of modern telegraph-type boxes. Devices may be added to the box that would indicate to persons using it whether or not the signal had been received at the communication center. While present radio-type fire alarm boxes transmit coded signals, transmission of voice signals can be expected in the near future.

## The Communication Center

A communication center is the location at which alarms are received, and from which appropriate signals are transmitted to the fire department to initiate response of apparatus and personnel. The communication center houses the equipment and personnel to perform these two functions. (See Fig. 8.3.)

A communication center may serve a single municipality, several adjacent municipalities, an entire county, or other large political jurisdiction. The combining of municipal police and fire communications into one communication center is becoming more common.

The communication center should preferably be in a separate building in a park or open space, and should be well protected against fire exposure. If the communication center is located in the fire department headquarters, it must be located so that the probability of interruption from any cause will be minimized.

When an alarm is received at fire alarm headquarters: (1) it should be automatically received and recorded, (2) it should be indicated both visually and audibly, and (3) the time of receipt and the exact location of the alarm should be automatically recorded. Running cards defining the response required for the location from which the alarm has come are filed so that the operator can transmit the necessary information to the fire companies designated on the card.

Even the smallest fire department must have some means of receiving notice of a fire. In small communities, the communication center may consist of a telephone in a private residence or business establishment where someone is always available and where there is a switch to sound a siren or other device to call volunteers. Quite frequently, the local police station is set up to serve as the communication center for a community. As fire departments become larger, a room or area is often provided to house the fire alarm dispatcher and alarm facilities within a fire station.

*Fig. 8.3. Fire dispatchers at the Huntington Beach Fire Department (California) communication center. Dispatchers are seated before cathode-ray tube displays. One display contains information on the incident being reported; the other maintains the status of all incidents.* (From *Fire Command,* NFPA)

In most communities, including those that maintain a complete and reliable municipal fire alarm system, the majority of fires are reported by telephone. Generally, reporting a fire by telephone can be successful, although there have been situations where complete reliance on the telephone to report a fire has been disastrous. Such situations include the following: a person under the stress and excitement of a fire may not give the right information to the fire department; telephones are not always readily accessible; the line may be busy; the fire may have disrupted the telephone circuit; the caller may call the wrong fire department. All such situations, as well as many others, have happened and have resulted in delayed alarms. While everyone should know how to use the telephone to summon the fire department, they should also know where the nearest fire alarm box is and how to use it.

**Fire Department Radio:** Most fire departments now make extensive use of two-way radio equipment for fireground operations. Nearly all of the larger departments have been issued fire radio licenses and frequencies by the Federal Communications Commission (FCC). The rules of the FCC provide that fire radio be used only by authorized personnel of the fire department, and not used for sending nonfire service messages. Most fire departments give training courses in radio procedures, and have specific rules for the operation of their fire radio systems.

Radio is particularly useful for alerting volunteer fire fighters. Many volunteer fire fighters have portable radio receivers in their automobiles, their homes, and their places of business. In fully paid departments, it is now common for off-duty personnel to be furnished with fire radios so they can be alerted when needed for serious fires.

## AUTOMATIC AND MANUAL PROTECTIVE SIGNALING DEVICES

Signaling systems are installed in all sorts of buildings for various types of fire alarm, fire prevention, and fire protection purposes. In general, such systems fulfill the following functions:

1. Notify occupants so they can get out when a fire happens.
2. Summon the appropriate people to fight the fire.
3. Supervise automatic sprinklers or other extinguishing systems to assure their operation when needed.
4. Supervise various industrial operations to warn of situations that develop a fire hazard.
5. Supervise security guards or other personnel to assure performance of their duties.
6. Actuate fire control equipment.

### Classification of Signaling Systems

There are six types of generally recognized protective signaling systems. These systems and the NFPA standards for their installation, maintenance, and use are as follows:

*1. Central Station Systems.* NFPA 71, *Standard for the Installation, Maintenance, and Use of Central Station Signaling Systems.*[3]

*2. Local Systems.* NFPA 72A, *Standard for the Installation, Maintenance, and Use of Local Protective Signaling Systems for Watchman, Fire Alarm, and Supervisory Service.*[4]

*3. Auxiliary Systems.* NFPA 72B, *Standard for the Installation, Maintenance, and Use of Auxiliary Protective Signaling Systems for Fire Alarm Service.*[5]

*4. Remote Station Systems.* NFPA 72C, *Standard for the Installation, Maintenance, and Use of Remote Station Protective Signaling Systems.*[6]

*5. Proprietary Systems.* NFPA 72D, *Standard for the Installation,*

## Alarm and Detection Systems and Devices

*Maintenance, and Use of Proprietary Protective Signaling Systems for Watchman, Fire Alarm, and Supervisory Service.*[7]

6. *Household Fire Warning Systems.* NFPA 74, *Standard for the Installation, Maintenance, and Use of Household Fire Warning Equipment.*[8]

Basic minimum requirements for performance of automatic fire detectors are contained in NFPA 72E, *Standard on Automatic Fire Detectors.*[9] Specific requirements for simple, local alarm units for use with automatic sprinkler systems are contained in NFPA 13, *Standard for the Installation of Sprinkler Systems.*[10] NFPA 101, *Code for Safety to Life from Fire in Buildings and Structures,*[11] suggests provisions for alarm and fire detection systems having particular application to the hazard of life. One provision recommends that fire detectors and the components of fire detection systems be tested and listed by nationally recognized testing laboratories.

**Central Station Systems:** Central station systems are operated by firms whose principal business is the furnishing and maintaining of supervised signaling service. The central station serves various properties subscribing to the service. The alarm and signaling devices on the subscribers' property are connected to the central station where competent operators are on hand to receive the signal and take the appropriate action. Central station operators retransmit alarms to the fire department. Usually a runner is dispatched to the subscribers' premises to check on the situation. Figure 8.4 shows the layout of a typical central station system.

In the central station system, signals register in the office of an independent agency, usually located at a distance from the protected property. The agency has trained and experienced personnel continually on duty to receive signals, to retransmit fire alarms to the fire department, and to take whatever action supervisory signals indicate is necessary.

Central station systems customarily serve a number of properties of different ownership, and are usually operated under contract by an agency that has no direct monetary interest in the protected properties. The ordinary central station furnishes service within a limited geographical area (about 200 square miles [517.9 square kilometers]), is located centrally within that area, and is connected to the plants of its various subscribers by wire facilities that are usually leased telephone lines. To reduce expense, it is customary to connect the plants of several subscribers to a single transmission circuit; each such circuit terminates in a recording instrument, and each subscriber is distinguished by one or more coded signal numbers that are not otherwise repeated on that particular circuit.

In addition to transmission of fire alarms, the central station service includes supervision of sprinkler systems; supervision of other extinguishing systems; supervision of industrial processes, temperature, and humidity; and supervision of guard and "watchman" services. Central station systems have demonstrated a satisfactory performance record over the years.

212   *Principles of Fire Protection*

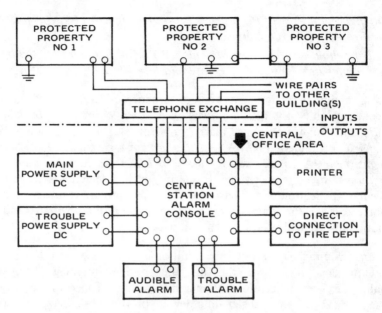

*Fig. 8.4. Diagram of a typical central station system.* (From NFPA *Fire Protection Handbook*)

**Local Systems:** Local protective signaling systems produce a signal at the protective premises for an alarm of fire and for supervisory services required. Such services would include supervision of a security guard's rounds, supervision of sprinkler system waterflow alarm service and of sprinkler systems, smoke alarm service, etc. Local systems are used primarily for the protection of life by indicating the necessity for evacuation of the building, and secondarily for the protection of the property. In cases where the sounding of an alarm might result in panic, the local system can be arranged so that a presignal will alert selected personnel whose responsibility it will then be to sound a general alarm.

In this system, the alarm or supervisory signal registers in the protected premises. A local system is used primarily for the notification of occupants. From the relatively uncomplicated arrangements of wiring and bells on early systems, local signaling systems have been developed that embody a number of more sophisticated techniques directed to assure operability of equipment and to initiate alarm devices.

A first essential of a satisfactory local system is a reliable power supply, which should be one of the following: (1) a dependable commercial power service, (2) an engine-driven generator (where an operator is on duty at all times), (3) storage batteries, or (4) a rectifier. The second essential of a satisfactory local system is the employment of wiring materials and methods suitable for the purpose. The requirements for protective signaling systems contained in NFPA 70, *National Electrical Code,*[12] are applicable.

The third essential of a satisfactory local system is that the equipment shall signal distinctively the presence of trouble, such as a broken or grounded circuit or the failure of a main power source, where such trouble prevents the intended operation of the system. Generally, a local protective signaling system can be expected to operate over troubles commonly encountered, or to signal its inability to operate where such inability exists. Trouble signals for this purpose are distinctive from alarm or supervisory signals and, as a rule, cannot be easily disabled or made inoperative.

*Auxiliary Systems:* The auxiliary type of system connects the appropriate devices in the protected plant with the municipal fire alarm system. Alarms from such a system are received at fire alarm headquarters on the same equipment and by the same alerting methods as alarms transmitted from municipal street boxes. In this system, signals are recorded at a municipal fire department. The connecting facilities between the protected property and the fire department are part of the municipal fire alarm system. The devices in the protected plant are customarily owned and maintained by the property owner. Other equipment, including the equipment that connects the devices to the city's circuits, is owned and maintained by the municipality, or leased by it, as part of the municipal alarm system. The auxiliary type of system is limited to alarm service only.

The three types of auxiliary systems in current use are: (1) local energy type, (2) shunt type, and (3) direct circuit type. The local energy and the shunt type auxiliary systems make use of the receiving equipment and the interconnecting wires of an established municipal fire alarm telegraph system. The direct circuit auxiliary system is one in which the alarms are transmitted over a circuit, usually leased lines, directly connected to the annunciating switchboard at an alarm communication center and terminated at the protected property by an end-of-line resistor or equivalent.

*Remote Station Systems:* Remote station systems are usually used to protect premises on which there is frequently no one present. The alarm or supervisory signal is received at fire alarm headquarters or at the office of a communications agency, usually located at a distance from the protected property. Signals are transmitted and received on privately owned equipment, and the agency receiving the signals may be a municipal fire department or a communications agency capable of receiving the signals and acting upon them. For fire alarms only, the remote station system should be connected with municipal fire alarm headquarters. Of prime importance is the choice of the remote station. If supervisory signals are to be included, the remote station should be located in a commercial signaling agency with personnel trained to provide the proper response to the signals received. Systems of this type are generally leased by the occupant of the protected premises and maintained under contract by the leasing company.

As was previously stated, for purposes of registering fire alarms, the desirable agency is the municipal fire alarm headquarters — if it has personnel constantly in attendance. Supervisory signals, however, require action different from that of the dispatching of fire department equipment. The preferred location for the remote station is in the quarters of a commercial agency that is open continuously, and that has personnel in attendance who are trained to provide proper response to the signals received. Such an agency may also serve in areas where fire department quarters are not continuously staffed for the reception of fire alarm signals. In such cases it is the function of the agency personnel to do whatever is needed to ensure prompt dispatching of fire department equipment.

*Proprietary Systems:* A proprietary protective signaling system is for individual properties where the system is under constant supervision by competent and experienced personnel in a central supervisory station at the property protected. Such systems are almost always found in large industrial plants. In this system, signals are received at a central supervisory station where experienced operators are on duty at all times to take whatever action the signals call for. The central supervisory station is under the control of the owner or occupant of the protected property and is usually on or near that property. The operation of the central supervisory station and the signaling systems connected to it is a function of the personnel employed for this purpose by the owner of the property. The equipment is usually purchased by the user and is subject to no outside control.

Proprietary systems may be considered elaborate local systems to which recording devices at an in-plant central supervising station have been added. The operations of this type of system are under the supervision of operators trained to investigate the situations that the system reports and to take whatever steps are necessary, such as the summoning of the fire department or the calling of in-plant assistance to adjust abnormalities encountered. The distinguishing features of a proprietary system are its ownership by the occupant of the protected plant and the presence of a central supervising station within the confines of that plant. Proprietary systems are almost always used in large industrial establishments where, basically, they are utilized for the supervision of patrolling guards and the reception of emergency signals from them when a fire or other unusual occurrence is discovered. In addition, alarm signals should have precedence over all other signals.

*Household Fire Warning Systems:* Household fire warning systems should provide reasonable protection for the family in ordinary dwellings. Smoke and toxic gases that develop in most dwelling fires are more likely to cause death and injury than exposure to direct heat or flames, except in the case of ignition of clothing and in those cases in which a smoke or heat detector would not be of any help. Many home fires are slow smoldering fires, such as those caused by a discarded cigarette in an overstuffed chair. In many cases, such fires start at night when the family is asleep; thus, a smoke detector located between sleeping areas and the rest of the house offers the most protection. Depending on the particular arrangements of the home, additional smoke and/or heat detectors will be needed unless, of course, there is only one sleeping area in a small home with no basement. When operated, the detector must give a signal that is distinctive and loud enough to wake a sleeping person. The source of power needed to operate the detector can come either from the house electrical supply, or from a battery that will last at least a year and is capable of emitting a distinctive trouble signal to indicate the need for battery replacement.

The two extremes of fire to which household fire warning equipment must respond are the rapidly developing high heat fire and the slow smoldering fire. Either can produce smoke and toxic gases. The detection or alarm systems included in the household fire warning system category are for the sole use of the protected household. If the alarm is extended to any other location, such as a fire department, the system should then be considered one of the aforementioned systems (as applicable), except that the requirements of detector location and spacing, as they apply to home warning systems, would continue to be followed. These home fire warning systems are primarily concerned with life safety — not with protection of property.

## Automatic Fire Extinguishing Systems

In addition to the six types of protective signaling systems described in the preceding paragraphs, there are various automatic fire extinguishing systems that can cause an alarm to be sounded when the system is activated. For example, the fire detecting capability of automatic sprinkers is employed to advantage in protective signaling by using any of the several forms of electrical waterflow alarm switches commonly used in automatic sprinkler systems to provide an alarm during sprinkler system operation. Electrical waterflow alarm switches function when a sprinkler head fuses. Sprinkler systems are frequently equipped with additional devices to signal abnormal conditions that might hamper the effective operation of the system. A supervisory device that will indicate that a sprinkler supply valve has been shut is most useful.

All of the sprinkler manufacturers provide waterflow alarm and other supervisory signaling equipment for wet-pipe and dry-pipe sprinkler systems. Other types of automatic fire extinguishing systems can also serve as fire detection systems, such as those employing foam, foam-water, high expansion foam, halogenated agent, carbon dioxide, dry chemical, water spray, etc.

## Power Supplies for Protective Signaling Systems

Of major concern in the use of protective signaling devices is the dependability of the source of power, or power supply. Usually the electric utility furnishing commercial light and power is the most acceptable primary power supply. An exception to this is that a battery may be the power supply in the case of household fire warning equipment. For the other types of systems, a secondary power supply (such as fully charged storage batteries that are automatically put into service if the commercial power fails) is important to ensure operation.

Experience has demonstrated the need for power supplies of the first quality and with a high reputation for reliability. Distinction is made between primary and secondary supplies. The primary supply furnishes power for transmission and reception, or for sounding of the alarm or supervisory signal. The secondary supply either provides energy to the system for fire alarm signaling in the event that the main supply fails, or provides for trouble signals and other functions that are not essential for alarm transmission but that are associated with the reliability of the system or can provide both emergency signaling as well as trouble signals.

The selection of a primary source of power stresses the importance of continual reliability under most of the conditions that are likely to be encountered on the property. In recent years, commercial light and power supplies have been found sufficiently reliable, in most areas, to permit their use for primary system powering, provided they are used in conjunction with fully charged storage batteries that can be automatically thrown into service should the commercial power fail. Such batteries are expected to have the capacity to operate the system for at least 60 hours. A number of acceptable secondary supplies are suggested by the NFPA standards, including a second phase of commercial power, the first phase of which is used to supply the primary source. Under certain conditions, dry-cell batteries are acceptable.

## FIRE DETECTION MECHANISMS AND DEVICES

This section of this chapter discusses the operating principles of fire detection mechanisms and devices, the spacing of detection devices, fusible links and releases, and the actuation of fire-controlling equipment.

### Heat Detectors

There are a considerable number and variety of heat, smoke, and flame detectors in common use today. Heat detection devices fall into two general categories: (1) those that respond when the detection element reaches a predetermined temperature (fixed-temperature-types), and (2) those that respond to an increase in heat at a rate greater than some predetermined value (rate-of-rise-types). Some devices combine both principles. The same principles apply whether the devices are of the spot-pattern-type, in which the thermally sensitive element is a compact unit of small area, or the line-pattern-type, in which the element is continuous along a line or a circuit.

## Alarm and Detection Systems and Devices 217

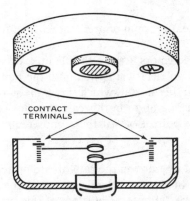

Fig. 8.5. *A bimetallic, snap-action disc, fixed-temperature device.* (From NFPA *Fire Protection Handbook*)

*Fixed-temperature Detectors:* Thermostats are the most widely used fixed-temperature heat detectors in signaling systems. The common form of thermostat is the bimetallic type that utilizes the different coefficients of expansion of two metals under heat to cause a movement resulting in closing of electrical contacts. In their simplest form, bimetallic strip thermostats operate against a fixed contact, and the distance the strip must travel to close against the contact determines the operating point, or temperature rating, of the thermostat. Some thermostats of foreign make take advantage of this property, and by fitting the units with movable stops they are made usable across a fairly wide range of temperatures.

In the United States, however, such variability is not designed into thermostats that are manufactured and used for fire protection purposes, and the usual form encountered is the nonadjustable, prerated unit, responsive to heat at its rated temperature.

Snap-action disc thermostats are similar to the bimetallic strip devices in operation. A metal disc goes from concave to convex when the temperature rating of the thermostat is reached. One special advantage of both these thermostats is that when the temperature goes down, they are restored to their original condition. They are not destroyed or damaged after they have operated. (Fig. 8.5 shows a bimetallic, snap-action, fixed-temperature device.) Disc-type devices develop a greater mechanical force at the point of operation than do ordinary bimetallic-strip-types, and the force is used to close a pair of electrical contacts that are not part of the disc itself. The snap-action disc design eliminates one disadvantageous feature of the bimetallic strip — the relatively low pressure at which the contact of electrically live members is made. As the contacts in a bimetallic-strip-type approach each other, the pressure is so light that false alarms may occur below the rated temperature as the result of jarring or poor contact.

The thermostatic cable is a line-type of thermostat. The cable is made up of two metals separated from each other by a heat-sensitive covering applied directly to the wires. When the rated temperature is reached, the covering melts and the two wires come in contact to initiate an alarm. The section of wire affected must be replaced after operation. The cable can be run around a room or area to provide protection for the entire room or area, rather than for spot locations.

Other forms of fixed-temperature heat detectors are the fusible link, occasionally employed to restrain operation of an electrical switch until the point of fusion is reached, and the quartzoid bulb thermostat, which serves the same function in a slightly different way in that operation in this case depends on removal of the restriction by breaking the bulb. Both of these units require replacement after operation.

*Rate-of-rise Detectors:* Fire detectors that operate on the rate-of-rise principle function when the rate of temperature increases as the detector exceeds a predetermined rate (in number of degrees per minute). Detectors of this type invariably combine two functioning elements, one of which initiates an alarm on a rapid rise of temperature, while the other acts to delay or prevent an alarm on a slow rise in temperature. There are several advantages to rate-of-rise devices: (1) they can be set to operate more rapidly under most conditions of combustion propagation than can fixed-point devices; (2) they are effective across a wide range of ambient temperatures, as, of course, fixed-point devices are not, making them equally useful in low-temperature and high-temperature areas; (3) they recycle rapidly and are usually more readily available for continued service than are fixed-point devices; and (4) they tolerate slow increases in ambient temperature without giving an alarm. The disadvantages of rate-of-rise detectors for some applications are their susceptibility to false alarms where there is a rapidly increasing temperature that is not the result of hostile combustion, and the possibility that they may fail to respond to a fire that propagates very slowly.

Pneumatic tube detectors operate on the rate-of-rise principle. When the temperature increases at a certain rate, the air in the tube expands and causes a diaphragm to move and close a circuit, thus causing an alarm. The device will not cause an alarm if the temperature rise is too slow. These detectors usually operate more rapidly than fixed thermostats, and, because they are effective over a wide range of temperatures, they can be installed in both low-temperature and high-temperature areas without any danger of actuation from temperature extremes. The pneumatic tubing, like the thermostatic cable, can protect an entire area covered by the tubing.

Other types of rate-of-rise detectors include detectors that operate on the thermoelectric principle. In these devices, two sets of thermocouples are arranged so that one set is exposed to convection and radiation of heat while the other is shielded. A difference in temperature between the exposed and shielded thermocouples produces voltage that causes an alarm signal. The thermoelectric principle of fire alarm operation can be used for line-type operation by running a cable assembly of four wires through the exposure area. Two are of metal with high thermal resistivity, and the remaining two are of metal with low thermal resistivity. The variations between the two pairs can be made to cause an alarm.

***Combined Rate-of-rise and Fixed-temperature Detectors:*** Rate-of-rise spot thermostats have been developed to take advantage of the rate-of-rise feature to sense a fast-developing fire; the fixed-temperature part takes care of a fire whose growth is so slow that the rate-of-rise capability might never come into play. The typical form of the rate-of-rise thermostat is a vented air chamber that heats up in a flexible diaphragm carrying electrical contacts. Heat outside the chamber causes air within the chamber to expand. When such expansion exceeds the capacity of the vent to relieve pressure the diaphragm is flexed, thus closing the electrical contacts it controls. Slow changes in ambient temperature near the chamber allow it to "breathe" through its vent, and the diaphragm is not moved sufficiently to cause an alarm. Devices of this type have two ratings, one for the speed of operation and the other for the temperature at which the unit operates. The latter becomes operative only when the former has not functioned.

***Rate Compensation Devices:*** This type of fire detector will give an alarm at some predetermined maximum temperature and will compensate for changes in rates of temperature rise. Even when the rate of temperature rise is low, the device is so built that it will operate to cause an alarm. A typical rate compensation device employs a cylindrical outer shell housing struts in compression on which contacts are mounted. The metal of the shell has a higher coefficient of expansion than the metal of the struts. When the temperature rises, the shell elongates, relieving the compression on the struts and causing the contacts to close. If the rate of temperature rise is low, both the struts and the shell expand; but, because of the difference in coefficients of expansion, the compression on the struts is relieved and the contacts close.

## Smoke Detectors

For the most complete protection of a dwelling, smoke detectors should be installed to protect each separate sleeping area and at the head of each basement stairway; smoke and heat detectors should be installed in all other major areas and rooms of the dwelling. If the detector(s) is installed by the homeowner, the instructions furnished by the manufacturer concerning location, maintenance, and testing should be followed. The value of a household warning system is further enhanced if the family has an escape plan and drill procedure to follow when the alarm is sounded.

Until recently, attempts to persuade homeowners to buy and use any type of alarm or extinguisher for the home have met with appalling indifference. Currently, public interest in home fire warning devices is growing and a number of manufacturers have entered the field. As has been previously stated in this text, statistics show that the majority of fire deaths and injuries from fires occur in ordinary one- and two-family dwellings; thus, the more household fire warning systems installed, the better records will become.

There are five general types of smoke detectors in common use today: (1) photoelectric detectors, (2) beam-type detectors, (3) ionization detectors, (4) resistance-bridge detectors, and (5) sampling detectors. These general types of smoke detection devices are described in the following paragraphs.

*Photoelectric Detectors:* Where the kind of fire anticipated is expected to generate smoke, possibly before a heat detection system comes into play, detectors that operate on a beam of light are used. The smoke either obscures the beam of light directly, or enters a refraction chamber where the smoke reflects the light into the photocell. The change in electric current resulting from partial obscuring of a photoelectric beam by smoke between the receiving element and the light source is measured, and an alarm is sounded when the smoke reaches a sufficient density. Photoelectric detection of smoke, in varying degrees of density, has been employed for several years, particularly where the type of fire anticipated is expected to generate substantial smoke before temperature changes are sufficient to actuate a heat detection system.

One type of photoelectric detector is the spot photoelectric detector, which employs a short beam carried between source and receiver in a ceiling-mounted unit. The change in current resulting from partial obscuring of a photoelectric beam by smoke between a receiving element and a light source is measured, and an alarm is tripped when this obscuration reaches a critical value. Another type of spot photoelectric detector is the refraction-type, which operates on the principle of the reflection of a light source into a photoconductive cell by means of smoke particles. A small chamber, open to the atmosphere, contains a light source and a photoconductive cell so arranged that the beam of light from the light source does not impinge upon the photoconductive cell. When a sufficient quantity of smoke particles from fire enters the chamber, the light is reflected by the smoke particles into the photoconductive cell. This changes the resistance of the cell, and a signal is emitted.

*Beam-type Detectors:* Another type of photoelectric detector employs a beam that is carried between elements at extreme ends or sides of the protected area, and that crosses the area to be protected. The beam is projected into a photosensing cell. Smoke between the light source and the receiving photocell reduces the light that reaches the cell, thus causing actuation of the alarm.

*Ionization Detectors:* A third type of photoelectric detector is the ionization detector. In this type of smoke detector a small amount of radioactive material ionizes the air in the sensing, or ionization chamber, which utilizes a principle wherein air is made electrically conductive (ionized) by bombardment of the nitrogen and oxygen molecules with alpha particles emitted by a minute source of radioactive material. This electrically conductive air permits current to flow through the air between two charged electrodes. When smoke enters the ionization area, the smoke particles decrease the conductance of the air; this causes the alarm.

***Resistance Bridge Detectors:*** Resistance bridge detectors use an electrical bridge grid. Atmospheric changes due to normal environmental conditions are accepted by the grid-bridge circuit, and the bridge is kept in balance. However, increases of smoke particles and moisture present in products of combustion bring about fast impedance changes that upset the balance of the grid-bridge circuit, causing an electronic triggering device to function and to initiate an alarm signal.

***Sampling Detectors:*** Another type of smoke detector consists of tubing distributed from the detector unit to the area(s) to be protected. An air pump draws air from the protected area back to the detector through the air sampling ports and piping. At the detector, the air is analyzed for smoke particles. A cloud chamber smoke detector is a form of sampling detector. The air pump draws a sample of air into a high humidity chamber within the detector. After the air is in the humidity chamber, the pressure is lowered slightly. If smoke particles are present, the moisture in the air condenses on them forming a cloud in the chamber. The density of this cloud is then measured by the photoelectric principle. When the density is greater than a predetermined level, the detector responds to the smoke and the alarm is activated.

## Flame Detectors

Still another important group of detection devices are those that are specifically designed to detect flame and glowing embers. The flames do not have to be visible to the human eye to be detected; thus, radiant energy outside the range of human vision can also be detected. Flame detectors are line-of-sight devices, and must be located so that material in the room or area will not obstruct the view.

There are four basic types of flame detectors: (1) infrared, (2) ultraviolet, (3) photoelectric, and (4) flame flicker.

***Infrared:*** This device has a sensing element responsive to radiant energy outside the range of human vision.

***Ultraviolet:*** As with the infrared device, this device has a sensing element responsive to radiant energy outside the range of human vision.

***Photoelectric:*** This particular device employs a photocell that either changes its electrical conductivity or produces an electrical potential when it is exposed to radiant energy.

***Flame Flicker:*** This device is a photoelectric type that includes means to prevent response to visible light unless the observed light is modulated at a frequency characteristic of the flicker of a flame.

## Gas Detectors

In addition to heat, smoke, and flame detection devices, there are also available various detection devices that will monitor the amount of flammable gases or vapors in an area. Portable gas detectors are used to detect the presence of combustible gas or vapor in basements, sewers, manholes, etc. Other devices will analyze the air samples brought into the device from various points. It is obvious that gas and vapor testing equipment is valuable for preventing fires and explosions in petroleum and chemical plants and in other industries where combustible vapors may be generated. Modern fire departments carry portable gas detectors on their apparatus.

Most instruments in current use utilize the principle of "catalytic combustion." Mixtures of flammable gas or vapor and air cannot be ignited to cause self-sustaining flame unless the concentration of gas or vapor exceeds a minimum value called the lower flammable limit (LFL). Mixtures containing much lower concentrations — approaching zero — can "burn" on the surface of heated platinum, yielding heat in direct proportion to the gas or vapor concentration. This is called "surface" or "catalytic" combustion.

If the heated surface is an electrically heated platinum wire connected in an appropriate circuit, the heat released by catalytic combustion can further increase the temperature of the wire, resulting in a change in electrical resistance and a corresponding deflection of the hand of an electric meter. This is the operating principle of the usual combustible gas indicator of the hot-wire type. Other somewhat similar devices may employ a solid or porous catalytic mass instead of a wire, and may sense the temperature by means other than the increase in electrical resistance.

## Spacing and Location of Detection Equipment

Spacing and location of the previously mentioned heat, smoke, and flame detection devices is shown in appropriate diagrams in the Appendices of NFPA 72E, *Standard on Automatic Fire Detectors* (see Bibliography). Generally, spot-type heat detectors should be located upon the ceiling not less than 6 in. (15.2 cm) from the side wall, or on the side walls between 6 and 12 in. (15.2 and 30.4 cm) from the ceiling. When complete coverage is required, strategically located detection devices should be installed throughout all parts of the building.

Factors to consider when spacing detection devices include ceiling construction, ceiling height, room volume, space subdivisions, the normal room temperature, possible abnormal room temperature conditions due to heat-producing appliances or manufacturing processes, and the draft conditions that may affect the normal operation of the device. Currently, there are devices available that are rated for use with normal ceiling temperatures up to 300°F (149°C) and higher.

NFPA 72E, *Standard on Automatic Fire Detectors,* states that the location of smoke detectors should be based upon an engineering survey of the application of this form of protection to the area under consideration. Some conditions to consider are air velocity, ceiling shape, surfaces and height, configuration of contents, burning characteristics of stored combustibles, the number of detectors required for complete coverage, and location of detectors with respect to ventilating and air conditioning facilities. Typical conditions of occupancy to be evaluated include possible obstruction of photoelectric light beams by the storage and movement of stock, and the presence of dust or vapors that could interfere with the operation of the smoke detection devices.

## *General Utilization of Detection Devices*

Because single station smoke and heat detection units commonly used in the home incorporate the detector, control equipment, and the alarm-sending device in one piece, they may be operated from the house current, by batteries, by a clock mechanism, or by gas in a storage cylinder. However operated, the homeowner must assume the responsibility to see that the power source is working and that the detector is in operating condition. A neglected, inoperative warning device is worse than none.

The various types of heat and smoke detection devices and systems can be and are used to perform a number of helpful fire protection functions. They can: (1) cause fire doors to shut; (2) control dampers in air conditioning systems; (3) open valves to release water to sprinkler systems; (4) operate fixed extinguishing systems employing foam, water spray, carbon dioxide, dry chemical, etc.; (5) open automatic drains; and (6) release dip tank covers. All of these functions are performed by releasing a weight tripped by the system, or by operation of an electrical solenoid. Detection systems can be so arranged that if the fire is small, an alarm is sounded and the system will not turn on the extinguishing agent until the fire gets much larger. In this way, time is given for someone to operate a hand extinguisher or hand hose, thus minimizing the possibility of damage from smoke or fire.

## *Summary*

Fire alarm and detection systems and devices play an important function in saving life and property. Although most alarms of fire to the fire department are received by telephone, street fire alarm boxes are in use in most cities. There are three types of fire alarm boxes: (1) the telegraph-type, (2) the telephone-type, and (3) the radio-type. Fire alarms are received at a communication center from which the fire department response is initiated. A communication center may serve one or more municipalities, an entire county, or other political jurisdictions.

224   Principles of Fire Protection

There are six major types of automatic and manual protective signaling systems. These signaling systems are installed in buildings for fire alarm, fire prevention, and fire protection purposes. Generally, the more complex the property, the more sophisticated the signaling system that is employed.

The utilization of smoke and heat detectors for life safety in dwellings is becoming increasingly more popular, as is the general use of the many types of currently available heat, smoke, and flame detection devices.

## *Activities*

1. An effective fire alarm system fulfills two functions. Describe these two functions.
2. Compare Type A and Type B fire alarm systems.
3. Describe the three types of fire alarm boxes.
4. What are the requirements for distribution and placement of fire alarm boxes in a community?
5. You have been asked to speak before a group of junior-high students. Explain the function and use of fire alarm boxes, and instruct the students regarding the proper way to report a fire by telephone.
6. Describe the procedures to be followed when a fire alarm is received at the communication center.
7. List and describe the six major types of protective signaling systems.
8. List the three classes of fire detectors and give examples of each class.
9. There are a considerable number and variety of heat, smoke, and flame detectors being manufactured for general use today. List and describe at least three types of such detectors.
10. Describe the use and importance of gas detectors.

## *Bibliography*

[1] NFPA 73, *Standard for the Installation, Maintenance, and Use of Public Fire Service Communications,* NFPA, Boston, 1975.

[2] *Municipal Fire Administration,* 7th Ed., International City Managers Association, Chicago, 1967, p. 184.

[3] NFPA 71, *Standard for the Installation, Maintenance, and Use of Central Station Signaling Systems,* NFPA, Boston, 1977.

[4] NFPA 72A, *Standard for the Installation, Maintenance, and Use of Local Protective Signaling Systems for Watchman, Fire Alarm, and Supervisory Service,* NFPA, Boston, 1975.

[5] NFPA 72B, *Standard for the Installation, Maintenance, and Use of Auxiliary Protective Signaling Systems for Fire Alarm Service,* NFPA, Boston, 1975.

[6]NFPA 72C, *Standard for the Installation, Maintenance, and Use of Remote Station Protective Signaling Systems for Fire Alarm and Supervisory Service,* NFPA, Boston, 1975.

[7]NFPA 72D, *Standard for the Installation, Maintenance, and Use of Proprietary Protective Signaling Systems for Watchman, Fire Alarm, and Supervisory Service,* NFPA, Boston, 1975.

[8]NFPA 74, *Standard for the Installation, Maintenance, and Use of Household Fire Warning Equipment,* NFPA, Boston, 1975.

[9]NFPA 72E, *Standard on Automatic Fire Detectors,* NFPA, Boston, 1976.

[10]NFPA 13, *Standard for the Installation of Sprinkler Systems,* NFPA, Boston, 1976.

[11]NFPA 101, *Code for Safety to Life from Fire in Buildings and Structures,* NFPA, Boston, 1976.

[12]NFPA 70, *National Electrical Code,* NFPA, Boston, 1978.

*Chapter Nine*

# *Municipal Fire Defenses*

*The evaluation and planning of fire protection takes into account many factors that influence the strength and effectiveness of a fire department. The ISO Grading Schedule helps determine the level of fire protection for an area. Master planning for fire defenses helps establish good fire practices. The provision of necessary water supply systems ensures effective fire control.*

## PUBLIC FIRE PROTECTION EVALUATION AND PLANNING

The role a public fire department plays in fire protection is often a reflection of community needs, desires, and finances. No matter what the role of a public fire department in a particular jurisdiction, the evaluation of public fire protection capacity and needs is a vital function of any fire department. A number of factors must be taken into account in evaluating and planning public fire protection in a jurisdiction. Such factors as fire hazards, life safety hazards, climate, geography, population distribution, and fire frequency are basic to the evaluation of and planning for public fire protection. As explained in "America Burning":[1]

> Fire protection has been largely a local responsibility, and for good reasons it is destined to remain so. Each community has a set of conditions unique to itself, and a system of fire protection that works well for one community cannot be assumed to work equally well for other communities. To be adequate, the fire protection system must respond to local conditions, especially to *changing* conditions. Planning is the key: without local-level planning, the system of fire protection is apt to be ill-suited to local needs and lag behind the changing needs of the community.

### *Evaluation*

Many factors must be considered when the fire protection needs of any community, large or small, are evaluated. Those factors in which life hazards may be involved are of primary concern. The physical composition of a community or jurisdiction is a major consideration in evaluating the possibility of deaths and injuries from unwanted fires, as well as loss of property from fire. The structural makeup of buildings, the kind and nature of the contents of buildings, the spacing of buildings, the open areas, and the width of streets influence the fire frequency that might be expected and the number of serious fires that might be anticipated.

Once the potentials for life and property loss are evaluated, the amount and kind of public fire protection — i.e., the public fire department, the water supply, building and fire prevention codes and their enforcement — must be weighed and balanced against what the taxpayers of the community are willing and able to provide for their own firesafety.

*Public Fire Defenses:* The resources of a community dictate the level of loss through fire that a community is willing to risk. The costs of fire protection must be balanced with the costs of other important community services. Each community must decide on the level of fire protection it is willing and able to provide. Some costs of fire protection can be transferred to the private sector through requirements for built-in fire protection, such as automatic extinguishing systems.

*Urban Fire Defenses:* Adequate water supply is a major factor in the evaluation of urban fire defenses. In an urban fire situation the relatively large numbers of people in a given area and the likelihood that many buildings are clustered together, thus increasing exposure to the fire problem, can result in inadequate response to initial alarms. It is also likely, as frequently happens, that several simultaneous fires in different parts of an urban area can result in severe personnel and equipment strain, thus producing inadequate fire fighting operations in a fire emergency.

With the exception of the smallest structural fire, effective urban fire fighting procedures require that several operations be carried out simultaneously and frequently. Multiple apparatus must be positioned to carry out these operations. The following is a description of the minimum standard requirements for adequate fire fighting operations in urban areas:[2]

> In its simplest terms, structural fire fighting involves simultaneous operation of three units under a chief officer: (1) a pumper company to make a fast initial fire attack, (2) a pumper to provide adequate water supply for the operation, and (3) a company to handle rescue, ventilation, salvage, and various other truck type services. At large structure fires additional fire fighting personnel are needed to cover the various points of fire attack and, in some cases, various functions can be handled more efficiently by specially trained crews such as rescue companies, salvage companies, etc., operating from specially equipped apparatus.
> 
> In a light hazard residential district, the minimum effective initial alarm response should consist of at least 3 pieces of apparatus, 2 of which should be

equipped to conduct pumping and water supply operations, with the remaining piece equipped for truck type operations. Each piece should carry the necessary emergency tools and appliances to perform the designated operation. The manpower required for reasonably satisfactory operation of this equipment would be 12 fire fighters and a chief officer. A more efficient use of the equipment could be made with a minimum response of 15 fire fighters and officers (12 fire fighters, 2 company officers, and 1 chief officer).

*Personnel:* Effective training and deployment of fire fighting personnel is a major factor in the evaluation of municipal fire defenses. Although sufficient numbers of fire fighting personnel should be available for possible fire emergencies, mere strength in numbers is not, by itself, an indicator that a fire situation will be handled efficiently and well by fire fighting personnel. To be effective and efficient during fire fighting operations, fire fighters must be well-trained and well-coordinated as a team.

The number of fire companies should be limited to those that can be effectively operated by the personnel assigned to them. For example, the effectiveness of pumper companies is measured by their ability to give required hose streams and service quickly and efficiently. A fire company operated by fewer than four fire fighters may be able to apply only half as much water in a given time as a company operated by four or five fire fighters. Understaffed fire companies are limited in their effectiveness in a fire situation, and often must delay operations until additional help arrives.

Consideration should also be given to maintaining additional personnel to provide coverage for multiple alarm fires and minimum coverage for other areas in the jurisdiction in the event of such an emergency. Arrangements should be made in advance for calling back off-shift personnel and for calling for mutual aid from nearby fire departments.

The task force concept, in which units are housed together and trained to follow preplanned tactical procedures, is one of the most efficient fire fighting operations. A minimum task force unit should consist of two pumping engines, a ladder truck, and a chief officer.

*Rural Fire Defenses:* The adequacy of water supply for fire protection needs is a factor that all public fire departments consider and evaluate. However, one of the major distinctions between urban, rural, and suburban fire departments is the depth of consideration given to the water supply problem.

It is not unusual for rural fire departments to be required to relay water from sources up to 2,000 ft (609.6 m), or even greater distances, away from the emergency scene. Because the availability of water for fire fighting is such an important consideration in rural fire fighting, special fire apparatus must be utilized by rural fire departments. Rural fire apparatus must have large water tanks that can be used for initial fire attack while supplementary water supplies are being provided. Following is a description of minimum standard protection that should be available in rural areas:[3]

Minimum standard protection for a rural area would include a pumper with a large water tank and a mobile water supply apparatus on an initial alarm. Properly designed water supply apparatus should be able to transport water at a sustained rate of 100 gpm from a suction source one mile from the fire. If larger hose is required to provide adequate fire protection services, additional tankers must be used or suction sources within reasonable distance of the possible emergency scene must be developed. Rural apparatus should carry 3½ in. or larger lay-in hose for providing adequate water supply at the fire scene. It is always advisable to lay large diameter fire hose from the water supply source as close to the emergency scene as possible in order to avoid extensive friction loss. At the emergency scene large diameter hose is often wyed [made into a "Y" shape] into smaller hand lines. Other pieces of equipment such as rescue and aerial ladder apparatus should be provided as needed to carry out the mission of the fire department. Generally, there is not an extensive need for elevated master streams in rural operations, and normal ladder truck equipment for rescue, forcible entry, ventilation, and salvage type operations must be carried on the other pieces of apparatus.

If the emergency area does not have adequate suction sources, water must be transported to the scene by tanker. The most common tank size is that which has a 1,000 to 1,500 gal (3785.4 to 5678.1 $l$) capacity. As Warren Y. Kimball explains in his book titled *Fire Attack 2:*[4]

Tankers meeting the provisions of NFPA Standard No. 19\* must be able to discharge and reload at a minimum rate of 500 gpm. This is of critical importance in determining the amount of water that can be transported in a given time. A tanker meeting these requirements can unload its contents into the portable suction basin from which the attack pumper drafts, travel one mile, refill, and return in time for the pumper at the fire to sustain a 100 gpm flow through two 1½-inch lines.

As more modern highways and roads are being constructed, fewer rural properties are located beyond the range of rural fire department protection. Areas still beyond the reach of rural fire departments must depend almost entirely on private fire protection and, in some instances, on the agencies connected with the forestry service.

## Insurance Grading Schedules

The *Grading Schedule for Municipal Fire Protection*[5] was originally developed several decades ago by the National Board of Fire Underwriters, an association of stock fire insurance companies, for the purpose of establishing

---

\*This standard was recently revised, and is now referred to as NFPA 1901, *Standard for Automotive Fire Apparatus,* NFPA, Boston, 1975.

230  *Principles of Fire Protection*

basic fire insurance rates. In general, the cities of over 40,000 population were surveyed by teams of engineers from the board, and smaller cities were graded by engineers of the various state insurance rating bureaus. Presently the schedule is published and copyrighted by the Insurance Services Office (ISO), a nationwide insurance group into which many of the state rating bureaus have been consolidated.

The *ISO Grading Schedule* classifies municipalities with reference to the fire defenses and physical conditions, and has influenced their level of fire protection in many municipalities. Reports that result from surveys, while not mandating a particular level of fire protection service, do contain suggested recommendations for correcting any serious deficiencies found during the survey. Even though the ISO makes these surveys for rating purposes and does not attempt to enforce its recommendations in any way, many municipal officials have accepted the insurance surveys as guides for improving municipal fire protection. In addition to improving fire protection services, removal of serious deficiencies can result in a more favorable insurance classification.

The *ISO Grading Schedule* is based on a deficiency point system, as shown in Table 9.1. A community totally unprotected against fire would be represented by 5,000 points of deficiency. The 5,000 deficiency points are divided into ten classes. Every 500 deficiency points that are eliminated move a community into a more favorable category.

The deficiency points on the *ISO Grading Schedule* are allocated to four general subject areas or categories as follows:[6]

> Water supply and fire department each account for a possible 1,950 points, or 39 percent. Fire service communications account for another 450 points, or 9 percent. Firesafety control, including fire prevention and building regulations, counts for 650 points, or 13 percent. Where there is a divergence of more than 500 points between water supply and the fire department, additional deficiency points may be assessed on the grounds that a good water supply requires an adequate fire department to apply it in fire fighting, and a good fire department without an adequate water supply is less effective. If either of these essentials is lacking, up to 900 additional deficiency points may be charged.

*Water Supply:* American communities are said to have the most adequate and reliable water systems in the world. To a large extent, the evaluations and recommendations of underwriter survey teams have effected this.

For fire fighting purposes, the water supply of the community is determined in the *ISO Grading Schedule* by estimating the required fire flow of water. Fire flow is the rate of water flow needed to enable the fire department to confine a major fire to within one block or other relatively small area. Engineers from the ISO make hydrant tests and determine the fire flow required for each section of the community. Fire flows can range from a minimum of 500 gpm (gallons per minute) to a maximum of 12,000 gpm for a single fire. In areas

### Table 9.1  Relative Class as Determined by Points of Deficiency*

| Points of Deficiency | Relative Class of Municipality |
|---|---|
| 0– 500 | First |
| 501–1,000 | Second |
| 1,001–1,500 | Third |
| 1,501–2,000 | Fourth |
| 2,001–2,500 | Fifth |
| 2,501–3,000 | Sixth |
| 3,001–3,500 | Seventh |
| 3,501–4,000 | Eighth |
| 4,001–4,500 | Ninth[1] |
| More than 4,500 | Tenth[2] |

[1] A ninth class municipality is one: (a) receiving 4,001 to 4,500 points of deficiency, or (b) receiving less than 4,001 points but having no recognized water supply.

[2] A tenth class municipality is one (a) receiving more than 4,500 points of deficiency, or (b) without a recognized water supply and having a fire department grading over 1,755 points, or (c) with a water supply and no fire department, or (d) with no fire protection.

*From NFPA *Fire Protection Handbook*.

where there is a possibility of simultaneous fires, an additional 2,000 to 8,000 gpm is required.

The minimum water supply under the *ISO Grading Schedule* must be able to deliver at least 250 gpm for 2 hours or 500 gpm for 1 hour over and above the consumption of water for other purposes at the maximum daily rate. Water supplies providing less than this minimum will be penalized the full 1,950 deficiency points in the *ISO Grading Schedule*.

Where there is a divergence of more than 500 points between fire department and water supply, additional deficiency points may be assessed. A good fire department with a poor water supply will be ineffective, and conversely a good water supply used by a poor fire department will also be ineffective.

Among the items on water supply considered by the *ISO Grading Schedule* and the survey team are adequacy of the water supply, reliability of the source of water supply, reliability of the pumping capacity and power supply, adequacy and reliability of the water mains, and distribution, size, and type of mains and hydrants.

**Fire Department Strength:** Under the *ISO Grading Schedule* a fire department must meet the following minimum requirements:[7]

A fire department must:

- Have a permanent organization under applicable state/local laws
- Be headed by one person responsible for the operation of the department

- Have sufficient membership to provide an alarm response of at least four members
- Conduct training sessions for all active members
- Have at least one piece of suitable fire apparatus with accompanying housing and maintenance
- Provide for 24-hour receipt of alarms and immediate notification of members.

Any fire department that does not meet these minimum requirements will not be graded and will be assigned the full 1,950 deficiency points.

In assessing a fire department, the *ISO Grading Schedule* considers the personnel and officers available for duty; the number and types of fire apparatus; pumper capacity; design, age, and condition of the apparatus; auxiliary equipment such as hose, ladders, special stream devices, masks, and clothing; training provided; response to alarms; operations on the fireground; and all other factors that have a bearing on the efficiency of the department. Location of the companies to provide standard response and required fire flows are also important factors that receive ratings.

It must be noted that the basic purpose of the *ISO Grading Schedule* is to define a fire department and water supply that will prevent conflagrations or sweeping fires and will confine fires to small groups of buildings. Many good fire departments will exceed the requirements of the grading schedule because prefire planning has shown the need for additional measures for life safety and for prompter control of fires in buildings.

In general, structural fire fighting will involve: (1) a pumper company for fast initial fire attack, (2) another pumper to provide the water, and (3) a third company to handle rescue, ventilation, salvage, and other services.

While situations for potential fires will, of course, vary widely, the measure of fire department response needed can be summarized as follows:

For rural areas there should be at least one pumper with a large water tank, one mobile water supply piece, at least six fire fighters, and an officer. Dwelling areas should have at least two pumpers, one ladder truck, twelve fire fighters, and a chief officer. For the average industrial and mercantile area there should be at least three pumpers, a ladder truck, sixteen fire fighters, and a chief officer. High hazard buildings and areas should be equipped with at least four pumpers, two ladder trucks, at least twenty-four fire fighters, and two chief officers. For the last two areas such specialized apparatus as may be needed should also be available:[8]

> Under the *Schedule* the number of engine and ladder companies must be at least equal to the number required for the basic fire flow. Engine and ladder companies must be located so that travel distances for first due, for first alarm companies, and for the maximum number of companies needed to apply required fire flows meet recommended travel distances. Structural conditions and hazards in the municipality may call for more companies than needed to apply basic fire flow. The probability of simultaneous fires, the number and

extent of runs, and the need for placing additional companies in service or for relocating companies during periods of high frequency of alarms are factors considered. Consideration is given to providing protection for all areas during multiple alarms and simultaneous fires.

*Other Important Factors:* The adequacy of fire department protection cannot be based on the size of the community alone. It must be based on the workload that can be anticipated. This depends on many factors such as types of buildings, possibility of several simultaneous fires, availability of apparatus, weather and topography, etc.

*ISO Grading Schedules* are important to fire departments because they provide criteria against which performance can be measured, and encourage fire protection preplanning. Table 9.2 is a checklist for evaluating fire department response capability based on the resources required for various occupancies.

It has been argued that the *ISO Grading Schedule,* with its emphasis on fire department and water supply, tends to overlook the importance of fire prevention measures because only 650 points of deficiency are assigned to fire prevention, building codes, and similar items.

Table 9.2  Evaluation of Fire Department Response Capability*

**High Hazard Occupancies** (Schools, hospitals, nursing homes, explosive plants, refineries, high-rise buildings, and other high life hazard or large fire potential occupancies)

At least 4 pumpers, 2 ladder trucks, 2 chief officers, and other specialized apparatus as may be needed to cope with the combustible involved; not less than 24 fire fighters and 2 chief officers.

**Medium Hazard Occupancies** (Apartments, offices, mercantile and industrial occupancies not normally requiring extensive rescue or fire fighting forces)

At least 3 pumpers, 1 ladder truck, 1 chief officer, and other specialized apparatus as may be needed or available; not less than 16 fire fighters and 1 chief officer.

**Low Hazard Occupancies** (One- , two- or three-family dwellings and scattered small businesses and industrial occupancies)

At least 2 pumpers, 1 ladder truck, 1 chief officer, and other specialized apparatus as may be needed or available; not less than 12 fire fighters and 1 chief officer.

**Rural Operations** (Scattered dwellings, small businesses, and farm buildings)

At least 1 pumper with a large water tank (500 or more gal), one mobile water supply apparatus (1,000 gal or larger), and such other specialized apparatus as may be necessary to perform effective initial fire fighting operations; at least 6 fire fighters and 1 chief officer.

**Additional Alarms**

At least the equivalent of that required for Rural Operations for second alarms; equipment as may be needed according to the type of emergency and capabilities of the fire department. This may involve the immediate use of mutual aid companies until local forces can be supplemented with additional off-duty personnel.

*From NFPA *Fire Protection Handbook.*

***Communications:*** The *ISO Grading Schedule* evaluates the following: the communications center; the communications center equipment and current supply; fire alarm boxes; alarm circuits and facilities, including current supply at fire stations; material, construction, condition, and protection of circuits; fire department radio; fire department telephone service; conditions adversely affecting use and operations of facilities; fire alarm operators; and the handling of alarms. While alarm boxes are not required in residential districts, a credit of up to 20 points is given for such boxes, depending upon coverage.

It is obvious that good fire service communications are essential for the fire department to function effectively. The elements of good municipal fire alarm service were discussed in Chapter 8, "Alarm and Detection Systems and Devices."

## Master Planning

The idea of master planning for all municipal functions, including fire defense, is quite new but is gaining wide acceptance. Master planning for a city or town of any size includes a review of past history, a close study of existing programs, and a projection of the future needs of the community. When this is done, a determination of the methods and costs to fill those needs should be made, and a short- and long-range plan should be adopted.

The National Commission on Fire Prevention and Control, after a two-year study and after soliciting data and advice from thousands of fire departments, recommended that a master plan, designed to bring about an adequate level of fire protection, be devised by the appropriate fire authorities of every community. The guidelines for such a plan were spelled out as follows:

- Identify present and future fire protection needs of the community.
- Identify the best combination of public resources and built-in protection required to meet the problem within acceptable limits.
- Establish goals, programs, and cost estimates to implement the plan.

Following is a description of the goals of master planning as described in "America Burning," the report of The National Commission on Fire Prevention and Control:[9]

> A major section of a community general plan of land use should be a *Master Plan for Fire Protection,* written chiefly by fire department managers. This plan should, first of all, be consistent with and reinforce the goals of the city's overall general plan. For example, it should plan its deployment of manpower and equipment according to the kind of growth, and the specific areas of growth, that the community foresees. It should set goals and priorities for the fire department. Not only is it important to set objectives in terms of lives and property to be saved, but also to decide allocations among fire prevention inspection, fire safety education, and fire suppression as the best way to accomplish the objectives.
>
> Having established goals, the plan should seek to establish "management by objectives" within the fire department. This operates on the principle that

management is most effective when each person is aware of how his [her] tasks fit into the overall goals and has committed himself [herself] to getting specific jobs done in a specified time.

Because fire departments exist in a real world where a variety of purposes must be served with a limited amount of money, it is important that every dollar be invested for maximum payoff. The fire protection master plan should not only seek to provide the maximum cost-benefit ratio for fire protection expenditures, but should also establish a framework for measuring the effectiveness of these expenditures.

Lastly, the plan should clarify the fire protection responsibility for other groups in the community, both governmental and private.

***The ISO Grading Schedule and Master Planning:*** As has been previously described in this chapter, the *ISO Grading Schedule* has, for many years, served as the guide for improvement of deficiencies in community fire protection. Although the *ISO Grading Schedule* was not intended to serve as a guide in making fire department decisions, circumstances invited its use as such. However, it must be remembered that the *ISO Grading Schedule* is a tool that is used to assist in setting fire insurance rates for a community. As such, the *ISO Grading Schedule* is primarily concerned with preventing loss through fire — primarily property loss. Other major aspects of a fire department, such as budget, adapting to a changing community, preplanning, and even fire prevention, receive little or no attention in the *ISO Grading Schedule*. To cope with needs and concerns beyond those in the *ISO Grading Schedule,* local administrators are increasingly turning to the concept of master planning.

***Developing a Master Plan for Fire Protection:*** The following can serve as guidelines to fire department administrators for developing and presenting a master plan for fire protection:[10]

*Phase I*

1. Identify the fire protection problems of the jurisdiction.
2. Identify the best combination of public resources and built-in protection required to manage the fire problem, within acceptable limits:
    (a) Specify current capabilities and future needs of public resources;
    (b) Specify current capabilities and future requirements for built-in protection.
3. Develop alternative methods that will result in trade-offs between benefits and risks.
4. Establish a system of goals, programs, and cost estimates to implement the plan:
    (a) The process of developing department goals and programs should include maximum possible participation of fire department personnel, of all ranks;
    (b) The system should provide goals and objectives for all divisions, supportive of the overall goals of the department;

(c) Management development programs should strive to develop increased acceptance of authority and responsibility by all fire officers, as they strive to accomplish established objectives and programs.

*Phase II*
1. Develop, with the other government agencies, a definition of their roles in the fire protection process.
2. Present the proposed municipal fire protection system to the city administration for review.
3. Present the proposed system for adoption as the fire protection element of the jurisdiction's general plan. The standard process for development of a general plan provides the fire department administrator an opportunity to inform the community leaders of the fire protection goals and system, and to obtain their support.

*Phase III*
In considering the fire protection element, the governing body of the jurisdiction will have to pay special attention to:
1. Short- and long-range goals,
2. Long-range staffing and capital improvement plans,
3. The code revisions required to provide fire loss management.

*Phase IV*
The fire loss management system must be reviewed and updated as budget allocations, capital improvement plans, and code revisions occur. Continuing review of results should concentrate on these areas:
1. Did fires remain within estimated limits?
2. Should limits be changed?
3. Did losses prove to be acceptable?
4. Could resources be decreased or should they be increased?

There are several major reasons that master plans for fire protection are vital for rural communities. The first shopping center or factory in a rural area can represent a huge jump in the demands that could be placed on the fire department's suppression capabilities. It is especially important to plan the location of future fire stations to minimize the distances fire engines must travel and to provide for built-in protection. Since funds, whether tax-based or volunteer, are generally scarce in rural areas, coordinated planning is needed to maximize the payoff in fire protection. Other problems of special concern to rural areas are transportation fires and fires in abandoned buildings. These problems require special consideration in master planning for rural areas.

## WATER FOR FIRE PROTECTION

The amount of water required by a city or town for fighting fires should exceed the maximum anticipated consumption of water for all domestic and industrial purposes.

## Uses of Water Systems

Municipal water systems have two functions: (1) to supply potable (suitable for drinking) water for domestic use, and (2) to supply water for fire protection. Domestic use includes, in addition to drinking water, water used for sanitation, air conditioning, and other similar domestic purposes. Municipal water systems must be capable of supplying adequate water supplies for fire protection and, at the same time, must also be capable of supplying the maximum anticipated amount of water needed for domestic use.

## Rates of Consumption

When a water system is designed, assessment of the rate of water consumption is based on three factors: (1) average daily consumption, (2) maximum daily consumption, which is the maximum total amount of water used in any 24-hour period in 3 years, and (3) peak hourly consumption, which is the maximum amount of water that can be expected to be used in any given hour during an average day.

The maximum daily consumption is normally about one and one-half times the average daily consumption. The peak hourly rate may run two to four times the normal hourly rate. Both maximum daily consumption and peak hourly consumption must be taken into account with regard to a given water system to ensure that adequate water will be available in the event of a fire emergency. To explain further:[11]

> To be considered adequate by the ISO, a water supply must be capable of delivering the required fire flow for a specified number of hours at peak consumption periods. Required fire flow varies according to the type of development in a given district. For example, low-density residential areas need a smaller flow than large industrial parks with buildings that are equipped with sprinkler systems.
>
> A deficiency in available supplies can be caused by an inadequate source of supply, a water utility incapable of delivering the required amount, or a poorly designed distribution system.
>
> 1. Water is usually pumped from wells, rivers, or lakes, and cleansed by filtering and chlorination processes. Then it is either stored in elevated reservoirs or pumped directly into the distribution channels. These sources of water may become inadequate during prolonged spells of hot, dry weather or where development of an area has exceeded the limits of available supplies.
>
> 2. A water company must have sufficient filtering and pumping capacity, and a distribution system adequate for delivering the required amount of water.
>
> 3. A well-designed system would ensure that no large section of the community be dependent upon a single main. Secondary mains would be looped into larger mains, and would be properly spaced. Hydrants would be located according to type of district to be protected, and would be low in friction loss and adequate in size.

## Fire Protection Requirements in Water Systems

The required capacity of a water system is determined by the total amount of water it must provide. The amount needed is the sum of all domestic, industrial, and fire protection requirements for water. In some areas, a heavy industrial demand for water would be a major factor in determining available water supply. In other areas water requirements for air-conditioning systems might place a heavy demand on the water supply. These and all demands on a water system must be evaluated so that the capacity of the water system for fire protection is always adequate.

*Evaluating System Capacity:* The basis of evaluation of a water system is its ability to supply water to meet the maximum daily consumption rate in addition to the water necessary for fire flow. A controlling factor in water system design and evaluation in most large cities is the fact that the peak hourly rate exceeds the maximum daily consumption rate plus fire flow. In smaller communities, the reverse is true; the maximum daily consumption rate plus fire flow becomes the determining factor.

*Pressure Characteristics of Systems:* The requirements of a water system for fire fighting can be estimated on the basis of how many standard hose streams might be needed. A standard hose stream is one with a discharge of 250 gpm at a 40 to 50 pounds per square inch (psi) pressure. A small town with a population of approximately 1,000 people should have a water capacity that will provide two or three hose streams; a city of 40,000 should have water enough for twelve to eighteen streams; a city of 200,000 should have water enough to provide from thirty to fifty hose streams. The exact number required would, of course, depend on the concentration of buildings, their construction, and contents.

The size and arrangement of the supply pipes in a water system is important. For domestic use, small pipes may suffice. However, since for fire fighting purposes the water has to be concentrated, larger mains are necessary. Four-inch or smaller mains are ineffective for fire fighting purposes. Residential areas should have at least 6-in. (15.2-cm) diameter pipes.

Hydrant spacing is also important. For congested areas, 250 ft (76.2 m) between hydrants is acceptable; a spacing of 400 to 500 ft (121.9 to 162.4 m) is acceptable in residential areas.

In many cases the amount of water needed for fire fighting purposes is determined by the engineering surveys of the insurance inspectors working out the *ISO Grading Schedule.* Their reports evaluate every aspect of a municipality's water system. Hydrant tests in various areas are made by the survey engineers to determine the fire flow. Deficiency points are assessed on the water supply depending on the findings of the hydraulic tests.

Water pressure needed for fire fighting is in the range of 65 to 75 psi for most systems. Such pressures will be adequate for ordinary water uses in buildings up to ten stories, and will supply automatic sprinkler systems in buildings of four or five stories. Modern fire department pumpers can develop heavy streams at high pressures from ordinary water systems if an adequate volume of water is provided. A minimum residual pressure of 20 psi should be maintained at hydrants when delivering the required fire flow.

## WATER DISTRIBUTION SYSTEMS

Water for any municipal system may come from lakes, rivers, springs, wells, or from reservoirs that are supplied from any of these sources. If the water does not have to be filtered, it can be pumped directly into the distribution system. If filtration is necessary, the water is pumped into settling reservoirs and filter beds, and the clean water is then transferred to clean water reservoirs and pumped from there into the distribution system.

### Sources of Supply

Sources of water supply fall into two general categories: (1) surface supplies, and (2) ground water supplies.

*Surface Supplies:* Surface water supplies such as lakes, rivers, and impounded supplies are largely dependent on rainfall. However, surface supplies that usually provide an adequate and reliable source of water can be affected by prolonged periods of drought and, in northern areas, by freezing temperatures and resultant ice.

If a lake is large enough, it can provide a reliable supply of water under virtually all conditions. However, ice hazards in some regions can injure an intake structure or crib. If a river is so large that it will not be seriously affected by prolonged drought, it too can be a reliable source of water supply. The major hazards resulting from ponds, lakes, or rivers as sources of water supply are ice hazards and silting.

An impounded water supply is one in which water from a stream or river is dammed or diverted into a natural storage basin. An impounded water supply can be more advantageous than a river water supply that could prove unreliable should a low stream flow occur. In some instances the ocean can be used for drafting water, although such a source of water supply is dependent upon tidal flow.

*Ground Water Supply:* Wells and springs are the sources of ground water supplies. Because ground water supplies are not directly dependent on rainfall, they are less susceptible to radical changes during periods of drought.

*Selection of Supply:* In some cases the selection of a source of water supply is not optional: just one source is available. However, if there is a choice of sources of water supply, the source that should be selected should be the one that provides the most reliability and that gives the desired quantity of water through the use of the least mechanical operation.

## Types of Systems

Gravity flow systems and pumping systems are the two basic types of water distribution systems. Most water systems, however, are a combination of these two types of systems.

**Gravity System:** A gravity system is the most reliable water system because no machinery is involved; thus, no mechanical malfunction problems will occur. In a gravity system, water is impounded at a point sufficiently higher than the distribution system so that gravity allows the water to flow into the distribution system and provide sufficient working pressures.

**Pumping System:** If a gravity system cannot provide sufficient working pressures, pumps must be used to provide sufficient working pressures. Normally, these pumps are located at the source of the water supply.

**Combination Systems:** Pumping systems are often arranged with storage facilities so that the water can be pumped into them during periods when the demand is low. The more water that can be held in storage, the more reliable the system becomes, especially in terms of fire protection requirements.

## Summary

Fire protection is largely a local responsibility responding to the specific considerations and needs of the community the fire department serves. A fire defense system that works well in a rural area may not be appropriate to fire protection needs in an urban area.

For many years, improvements in fire protection service have followed the recommendations of the *ISO Grading Schedule* used to assign community insurance ratings. The municipal *ISO Grading Schedule* used by insurance companies to develop proper base rates for the community has greatly influenced the strength of fire departments and water supplies in our cities, and has been largely responsible for upgrading municipal water supplies.

Every community should have a master plan for fire protection. The concept of master planning for present and future needs involves many additional factors and considerations not addressed by the *ISO Grading Schedule*. Fire departments that have been relying solely on the *ISO Grading Schedule* will find that the master plan involves attention to many more factors and calls for custom-tailoring future priorities to meet local needs.

As communities undertake a basic reassessment of their fire services, they will have to find solutions best suited to their conditions. Some communities are at an early stage of growth where they can consider a number of alternatives to their present system of fire protection. Others have a heavy investment in their present system and can consider only a gradual shift of priorities. For most communities, improving the effectiveness of the fire service calls for gradual changes within the present structure: a shift of priorities toward fire prevention, better deployment systems, and revised and improved management practices. Still other communities will want to consider a major shift from their present system.

Water systems and water supplies needed for fire protection are major considerations to be evaluated when assessing the level of fire protection service available in a community. A community water system must be designed to provide adequate water for fire fighting while at the same time meeting the maximum anticipated consumption needs for all domestic and industrial purposes. An ideal water supply would come from a good watershed into a large impounded lake where it would flow by gravity into the community's distribution system. In many systems, however, pumps must be used in order to provide proper working pressures in the mains.

## *Activities*

1. Discuss why you feel fire protection should or should not be primarily a local responsibility. Defend your opinion with reasons based on what you have learned from this chapter.
2. Describe at least three considerations that are different when planning an urban fire defense system and a rural fire defense system.
3. Describe the *ISO Grading Schedule* and its effect on fire protection services in communities.
4. Describe the criteria for water supply in a community with regard to the *ISO Grading Schedule* rating.
5. List the six minimum requirements for a fire department under the *ISO Grading Schedule*.
6. What are the four major fire protection categories the *ISO Grading Schedule* assesses?
7. Discuss how the *ISO Grading Schedule* can help community fire protection services; then, discuss how you feel it might hurt community fire protection services.
8. What is the difference between "planning" using the *ISO Grading Schedule*, and "master planning?"
9. With a group of your classmates, develop a master plan for fire services for an imaginary community or a community you know.
10. Describe the various water supply systems and how they affect fire protection evaluation.

## Bibliography

[1]"America Burning," May 1973, The National Commission on Fire Prevention and Control, Washington, DC, p. 27.

[2]*Fire Protection Handbook,* 14th Ed., NFPA, Boston, 1976, p. 9-75.

[3]_____.

[4]Kimball, Warren Y., *Fire Attack 2,* NFPA, Boston, 1968, pp. 107-108.

[5]*Grading Schedule for Municipal Fire Protection,* Insurance Services Office, New York, 1974.

[6]*Fire Protection Handbook,* 14th Ed., NFPA, Boston, 1976, p. 9-78.

[7]_____.

[8]_____.

[9]"America Burning," May 1973, The National Commission on Fire Prevention and Control, Washington, DC, pp. 24-25.

[10]*Fire Protection Handbook,* 14th Ed., NFPA, Boston, 1976, p. 9-82.

[11]Didactic Systems, Inc., *Management in the Fire Service,* NFPA, Boston, 1977, p. 225.

*Chapter Ten*

# Fire Department Organization, Administration, and Operation

*Fire departments protect a large number of people and properties, and their organization and operation play a vital role in fire waste control. Fire department administration and management is responsible for maintaining highly trained and efficient operational units for the performance of effective tactical and nontactical operations.*

## FIRE DEPARTMENT ORGANIZATION

There are approximately 25,000 fire departments in the United States and Canada. About 2,500 of these fire departments protect cities and towns having a population of 10,000 or more. About 10 percent of these 2,500 operate with volunteers or with part paid and part volunteer forces. The remaining departments operate in cities and towns with populations less than 10,000, and nearly three-quarters of all these fire departments protect towns with populations less than 2,500. Nearly all of these fire departments operate with volunteer fire fighters.

It is estimated that there are about 185,000 paid fire fighters in our fire departments and over one million volunteer fire fighters. Fire departments comprise a large operation that involves great numbers of people and large expenditures of funds.

The history of fire departments is long and colorful. Up until a hundred years ago fire fighting, even in our larger cities, was accomplished by volunteers. Records indicate that the first big city fire department with fully paid fire fighters was established in Cincinnati in 1853. The apparatus for this department was a horse-drawn steamer. Self-propelled steamers and aerial ladders appeared about 1870, and the automotive fire department pumper dates from 1910. The first NFPA standard on automotive fire apparatus was adopted in 1914.[1]

Fire departments have clearly demonstrated their value in holding down loss of life and property from fires. Perhaps the best way of demonstrating the value and importance of fire departments is from the fire record. Only about 500 out of every million fires involves a loss of $250,000 or more; only about 20 percent of fires involve losses of $500 or more.

## *Fire Department Objectives*

The four traditional objectives commonly accepted by most modern fire departments are:
1. To prevent fires from starting.
2. To prevent loss of life and property when fire starts.
3. To confine the fire to the place where it started.
4. To extinguish fires.

The concept that fire departments should function to prevent fires is comparatively new. The other three objectives have always been recognized as fundamental for any fire department.

It is obvious that, although the objectives of a fire department have always been basically the same, the implementation of these objectives has changed over the years with circumstances and technology. The following statement from *Management in the Fire Service* briefly describes the nature of some of these changes:[2]

> However, the scope of these objectives has changed over the years since individuals realized the necessity for communal and organized fire fighting. In addition, the scope of responsibilities within the fire department has also changed. Highly trained medical professionals in fire departments have broadened responsibilities to include saving lives in disasters other than fires, such as airport crashes or automobile accidents. Today's fire chiefs and officers are no longer only fire scene leaders: they must also know how to train personnel, how to manage both physical and economic resources, and how to effectively manage an entire fire unit. All of these changes in roles and in fire department responsibility have evolved from years of increased technological development in the fire sciences.

To be more meaningful and to meet the demands and requirements that the increased scope and responsibilities have brought to the modern fire department, general fire department objectives should be performance oriented. Each fire department, regardless of its size, should have or develop a set of performance objectives that specify goals to be achieved, results expected, and the period of time required for achievement.

Fire department organizations all have the same general, overall objectives, as described in the following excerpt from *Company Leadership and Operations,* by Anthony R. Granito:[3]

> In the broadest sense the fire department's objective is to save lives and property from uncontrolled fire. Historically the public assumes the fire department has the capabilities to perform effectively if a fire or related emergency occurs.
> In more specific terms the fire department goals can be perceived to include:
> 1. Providing a performance level acceptable to the citizens of the jurisdictional area under consideration.
> 2. Providing a life safety level acceptable to the citizens in the area under consideration.
> 3. Confining fires to a level of incipiency acceptable to the property occupants in the jurisdictional area under consideration.
> 4. Suppressing fires with the least amount of property damage and occupancy interruption possible.
> 5. Providing selected emergency services relating to life safety and property damage to a defined jurisdictional area.
> 6. Meeting the intent of the performance levels established considering economy effectiveness.

As can be noted from Mr. Granito's statements, fire department objectives are geared at being met and adapted to the particular jurisdiction in which the fire department is organized.

## Fire Department Structure

The organization of fire departments takes many forms. The purely volunteer department often operates independently and raises its own funds by fairs, carnivals, and by public subscriptions. Many volunteer fire departments now receive contributions of funds and equipment from the local government of the community. In the larger cities, the fire department is one division of the local government, usually with the fire chief directly responsible to the chief administrative officer of the city. There are a growing number of area fire departments (such as fire districts, fire protection districts, and county fire departments). Fire districts are organized under special provisions of state laws with their own governing bodies, and are supported by a district tax levy. Fire protection districts are set up to contract for fire protection from a nearby fire department, and are supported by taxes paid by property owners in the area.

County fire departments are now found in a number of metropolitan areas. Numerous small communities in the county combine to maintain a large, professionally administered public fire department with countywide communications and fire prevention services. The kind, size, and makeup of the local fire department will of course depend on the financial resources available, the frequency of fires to be expected, the area to be protected, etc.

246  *Principles of Fire Protection*

Fire departments can by staffed by paid, part paid, or purely volunteer personnel.Typical organization charts for small, medium, and large fire departments are shown in Figures 10.1, 10.2, and 10.3. The evolution from a volunteer fire department to a paid fire department follows the growth and congestion of the area. There is no set pattern, however, as to when the changeover takes place. Local circumstances and fire experience of the area will dictate the changeover.

***Principles of Organization:*** Much has been written about fire department organization and organizational principles. The following principles are generally considered to be universal and mandatory.

One of the most basic principles is that work should be assigned to individuals and units based on a careful, well-arranged plan. Another important principle relates to the size of the department. As the department grows as a

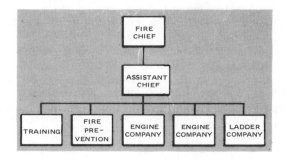

Fig. 10.1. *Typical organizational structure of a small-sized fire department.*

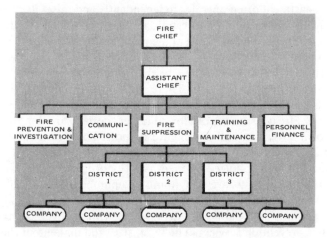

Fig. 10.2. *Typical organizational structure of a medium-sized fire department.*

# Fire Department Organization, Administration, and Operation

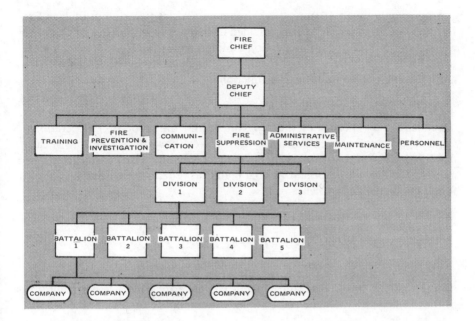

Fig. 10.3.  *Typical organizational structure of a large-sized fire department.*

result of more complex community needs, careful consideration must be given to coordination of the various departmental and divisional functions.

Another well-known principle involves lines of authority. Individuals should be aware of their relationship to the total organization. In the same manner, each operational unit should also be aware of its place and function in the total organization. Additionally, it is important that when responsibility is given to an individual or to a unit, the authority to carry out that responsibility must also be clearly given. In this way when authority is necessary to complete and carry out a given responsibility, it is there.

Unity and clarity of command and command levels are important principles of fire department organization. When an individual receives conflicting orders from several superiors, or when a superior has too many individuals and functions to coordinate and supervise, poor organization is reflected in the general confusion that usually results.

*Line Funtions:* Line functions in fire departments normally refer to the activities directly involved with fire suppression operations. Officers directing fire suppression are primarily considered line officers. This does not mean, however, that these officers do not have other functions. As these officers are promoted to higher levels within the department, their line responsibilities may be equally divided with staff responsibilities. At the highest officer levels within the department, line responsibilities diminish, while staff responsibilities increase.

*Staff Functions:* Staff functions are those activities that do not involve fire fighting. When the department is divided into divisions or bureaus (such as fire prevention, training, communications, maintenance, and personnel), a staff officer will usually be assigned to supervise each such division. Staff officers are normally not involved in line functions.

*Organizational Plans:* The manner in which fire departments are organized is dependent upon the size of the department and the scope of its operations. Organizational plans are designed to illustrate or show the relationship of each operating division to the total organization. It is essential that each fire department have available an organizational plan that reflects the current status and functions of the department.

## ADMINISTRATION AND MANAGEMENT

The management of a paid fire department should be much the same as that of any well-run business establishment. The governing body is normally that of the municipality with the power to levy taxes for the support of the fire service, to own property such as fire stations, and to pay the personnel. The fire chief is the manager of the department and is usually appointed by the mayor, city manager, or chief administrative officer of the city or town.

### Function of Management

Fire department management is responsible for maintaining highly trained and efficient operational units to perform assigned tasks both in the prevention and suppression of fires. Fire department management is generally the responsibility of the local government that supports the service. As with any governmental or business operation, fire department management involves three major areas of responsibility: (1) fiscal management, (2) personnel management, and (3) productivity.

Generally the fire chief determines the number and distribution of fire companies. The total equipment and personnel to be made available is determined by the governing body. The fire chief decides which records are to be kept and which reports of department activities are to be made, and is also responsible for any necessary planning and research. The fire chief usually has charge of accounting and budgeting, and also functions as the personnel officer of the department.

Other important administrative functions of the fire chief include the following: department public relations, management of water for fire protection, fire prevention inspection procedures, fire department regulations, investigation of fires, training procedures, communication facilities, and maintenance of buildings and apparatus.

## Personnel

The proper selection of fire fighters is important to the success of any fire department. Personnel should be selected to meet appropriate standards. NFPA 1201, *Organization for Fire Services,*[4] recommends that the fire chief serve as the personnel officer once personnel standards and procedures have been established at the management level. The fire chief may delegate personnel matters to an assistant fire chief who would then function as the department personnel officer.

The scope of personnel activities can be limited by outside factors, however, such as those described in *Management in the Fire Service:*[5]

> . . . The specific personnel activities that might be conducted by a fire department, however, may depend on the scope of personnel services provided by municipal or state personnel or civil service agencies. State, provincial, and civil service commission legislation may set specific standards of pay, hours, working conditions, working schedules, and other features of personnel policy that would limit the authority of the municipality and the fire department.

**Recruitment:** Paid fire fighters are not, as a rule, recruited directly by the fire department but by a municipal personnel or civil service agency. Recruitment is subject to federal, state, and local employment standards. Local regulations with regard to age, physical fitness, and education can vary widely.

In many states the Civil Service Commission maintains a roster for individuals who wish to join a paid fire department. To become eligible, a candidate usually must pass both a written examination and a physical examination. Those who qualify are usually entered on the roster in order of their test results. Several names from the top of the list are submitted to a fire department for specific selection. Either a personnel officer, the company officer, or a small selection team must then choose from these candidates the one who is best qualified to fill the position.

Attempts are being made to establish national professional fire fighter qualifications standards, sponsored by the Joint Council of National Fire Service Organizations, using the NFPA standards-making system. The Joint Council has appointed a National Professional Qualifications Board. NFPA 1001, *Standard for Fire Fighter Professional Qualifications,*[6] was adopted in 1974. This standard covers entrance requirements, including medical examination, and three grades based upon nondiscriminatory and performance criteria that result from evaluation of performance of required tasks after appropriate instruction has been given by qualified fire department instructors. The standard requires a high school diploma or equivalent, age of at least 18 years, sound mental and physical fitness, and an evaluation of the candidate's general character. The first level of progression in the department is Fire Fighter I. The requirements for Fire Fighter I are described in detail in the standard which covers such subjects as forcible entry, protective breathing apparatus, first aid, ropes, salvage, fire hose, nozzles and appliances, fire streams, ladders, ventilation, inspection, rescue, sprinklers, fire alarm and communications, personal safety, and the chemistry of fire.

Similar and progressively more detailed knowledge in all these areas is indicated for Fire Fighter II and Fire Fighter III. Fire Fighter I works under direct supervision of a company officer; Fire Fighter II works under minimum direct supervision; Fire Fighter III works under minimum supervision but under orders.

Other standards covering fire service officers, fire inspectors, fire investigators, fire service instructors, and fire service driver-operators have been adopted or are in the process of being adopted. The success of these or any other methods of selecting and developing personnel depends on the training program used by the department to prepare its personnel to meet specified levels of performance.

*Promotion Practices:* Promotion practices and procedures in most paid departments are also administered by municipal personnel departments or by state or local civil service authorities. A good outline of what is involved in the duties of company officers and of chief officers is found in NFPA 1201, *Organization for Fire Services* (see Bibliography). This standard suggests that the following steps be used by the chief in a promotion program in a department. These steps are as follows:

- Prepare lists of members for promotion.
- Arrange assignments so that promotion candidates may have a variety of duties and experience in various staff work.
- Require supervisors to report on the aptitudes and attitudes of candidates to aid in evaluating qualifications for promotion.
- Require candidates to successfully complete an in-service training program.
- Arrange assignments for education at accredited schools.

*Staffing Practices:* With the cost of staffing each position on fire apparatus around-the-clock ranging from $50,000 to $90,000 annually, today's fire departments can no longer afford haphazard staffing procedures. Working hours per week, number of shifts, duty tours, vacations, and sick and injury leave have to be taken into consideration. Staffing procedures covering hours and wages, provisions for overtime, off-shift response, and mutual aid agreements are also covered by federal, state, and local regulations. The laws and regulations regarding employment are so varied that each fire department, large or small, has to develop its own plans and procedures for adequate staffing of fire apparatus. Many fire departments are not adequately staffed because of the heavy costs involved. The tendency now is to have fewer companies adequately staffed, rather than more companies with too few fire fighters to operate the apparatus efficiently.

Formerly, staffing practices were relatively haphazard. Companies were generally furnished with from four to seven fire fighters on each shift, depending on the location and fire frequency in the area, and little thought was given to days off for vacations, sick leave, injuries, etc. In many places fire fighters were poorly paid and worked long hours. Seldom were they given proper training in standard fireground tactics.

Times have changed and with higher pay and shorter hours, the staffing of fire companies is necessarily more carefully planned. Depending on the tours of duty involved, staffing is based on the essential positions that must be covered 24 hours a day for 365 days a year. Additional staffing must be provided for vacations, sickness, injuries, and various other compensated absences from duty. Usually 20 percent additional personnel is provided to the fire department to cover such absences.

Work weeks for fire fighters now average from 40 to 56 hours. It is estimated that about half of the paid fire fighters work 54 hours or less per week. A 42-hour work week for fire fighters is becoming common.

The cost of staffing fire departments with paid personnel has risen so rapidly that the contribution that volunteers have made in protecting the small communities of America has been highlighted. It has been estimated that to replace this free community labor with minimum staffing by fully salaried personnel would involve an added national fire protection cost of approximately three billion dollars per year.

## *Intergovernmental Relations*

Fire departments are but one agency of local government, and although their success depends upon their working relationships with other state, local, and federal agencies, their most important working relationship is with local or municipal government.

The fire chief, as the administrator of the department, has to work closely and effectively with other municipal departments. For example, cooperation with the police department is an obvious necessity. Regular police response to fire alarms is necessary to control traffic and spectators. Police help is also needed in fire investigations and arson control.

The city water department and the city building department provide other important city services essential to the operations of the fire department. Also, in many cities, the personnel department recruits fire fighters. In addition, there may be a city purchasing department that affects the specifications for purchase of fire apparatus and equipment as well as a city data processing department into which fire department records can be fed.

**Building Department:** Federal, state, and local laws are giving increased attention to the proper construction and arrangement of buildings as vital factors in fire protection programs. Before building permits can be issued, many laws and ordinances now require written approval by the head of the fire department with regard to specified fire protection features. Chapter Eleven, "Codes and Standards," describes various building and fire codes and standards and their effect on fire protection considerations in buildings. The fire chief handles code enforcement responsibilities in smaller communities. In larger communities, this most important responsibility is generally delegated in its entirety to the fire prevention bureau.

*Police:* Cooperation between the fire and police departments is essential. Because the police are often the first at the scene of a fire or emergency, they should have basic training in what to do and what not to do at fires. This includes notifying fire departments of traffic accidents where fire department service is required to assist in rescues or to control fire hazards. Police cooperation is also needed in many fire investigations, particularly where there may be criminal law violations; in some jurisdictions, members of the fire prevention bureau staff have special police powers.

*Water Department:* The water department or water division has the responsibility of ensuring adequate water supplies for fire protection. However, because the water authorities have the most up-to-date knowledge of fire flow requirements, it is recommended that a fire officer maintain a liaison with the water authority. This liaison should include ensuring that fire hydrants are properly located and set. In many fire departments, liaison with the water authority is one of the responsibilities of the training division. Chapter Nine, "Municipal Fire Defenses," explains in detail the water supplies necessary for adequate fire protection.

*Personnel Department:* Members of fire departments are public employees, and as such, in some jurisdictions, recruitment and promotion often involve cooperation with the personnel agency that is responsible for conducting entrance and promotional examinations. Usually the fire department executive officer, executive secretary, or chief training officer is assigned to maintain liaison with the personnel department.

*Finance Department:* The fiscal business of the fire department is usually conducted by the finance department of a municipal or local governmental agency. Generally fire chiefs or their executive officers are responsible for coordination between the fire department and the finance department.

*Purchasing:* If a purchasing department is part of the fire department's intergovernmental relations, close liaison is necessary to assure that specifications are properly drawn to meet fire department needs, and that when bids are opened, any proposals that deviate from specifications are rejected. This may be handled directly by the fire chief, by the chief's executive officer, or by the chief of apparatus and equipment.

***Data Processing:*** Fire departments are increasingly utilizing electronic data processing for keeping fire and payroll records, and for statistical analysis. Ideally, each fire department should have persons knowledgeable in the use of data processing. In large fire departments, it may be desirable to have one officer appointed as coordinator of this work.

## Personnel Utilization

The principal resource of a fire department is its highly trained personnel. The most effective utilization of fire fighting personnel must be made by fire department management. Thus, the work schedules of the fire fighting personnel must be carefully planned for most effective utilization. Time must be provided for training, inspections, apparatus and equipment maintenance, as well as for other essential activities, in order to get the most for the dollar expended.

In a modern paid fire department, there are many staff functions that can be handled by secretaries and clerks rather than by paid fire fighters. Also, janitorial duties can be performed by employees other than fire fighters. Today, the trend is to turn over much of the paper work to office staffs rather than to expect fire officers to handle such chores. The following statement from "America Burning" emphasizes the need to consider better ways of utilizing the capabilities of fire fighters:[7]

> The nature of the job of most fire fighters requires much standby time which is not devoted to reducing fire losses [fighting fires]. Most leaders in the fire services agree that the productive time of fire fighters ought to be increased. And most agree that whatever additional services fire fighters are called upon to render, the services ought to utilize fire fighters' special capabilities. Painting signs and registering bicycles are useful activities, but they don't meet this criterion.

## FIRE DEPARTMENT OPERATIONS

Most of the resources of fire departments are committed to fire suppression efforts. However, as has been previously stated in this chapter, much more attention needs to be given to utilizing these resources for fire prevention as well. This concept is reinforced in "America Burning":[8]

> Activities which ought to receive topmost priority in extending fire fighters' productivity lie in the area of fire prevention. A recurring theme of this report is that a much heavier investment of time and resources in fire prevention is the most expeditious route to reduce life and property losses from fire. While many departments recognize responsibilities in fire prevention, too few are doing all they should or could.

For example, the majority of personnel within the department are assigned to fire suppression duties at locations throughout the community where they can also perform an invaluable service by aiding in the fire prevention effort through in-service inspection activities, prefire planning, training, and other related assignments.

## Organizations for Fire Suppression

The basic tactical unit of fire suppression is the company. Each company is normally composed of one or more pieces of apparatus and a complement of personnel under the supervision of a company officer. Typical fire companies are engine, ladder, and rescue. Two other types of tactical companies are manpower squads and task forces; although not widely accepted, these two types of tactical companies have proved successful in communities that use them.

*Engine Company:* The engine company is the basic unit for fire fighting. Many small volunteer fire departments operate with just one engine company. Larger departments add pumper companies as needed. The basic pumper is usually a triple combination unit that consists of a pump, hose, and water tank. Some departments prefer a two-piece engine company comprised of a triple combination pumper and a separate hose wagon specifically designed to respond with the pumper. In small departments with one or two pumpers only, all aspects of fire fighting, including ladder work, are carried on by the pumper unit. In larger departments, engine companies usually function in initial fire attack, interior and exterior fire fighting, and rescue operations.

*Ladder Company:* The basic unit of apparatus for ladder companies may be either the aerial ladder or the aerial platform. Ladder companies perform rescue operations and ventilation, and are also used to carry on forcible entry, salvage, and overhaul work. In some fire departments that do not have rescue companies, or divisions, the ladder companies are assigned to all of the nonfire-related rescue operations.

*Rescue Company:* Some fire departments also utilize rescue companies for both fire- and nonfire-related rescue and other activities. Personnel trained in all phases of rescue are assigned to a special vehicle equipped to perform most rescue tasks.

*Tactical Control Units:* Some large fire departments utilize special tactical control units to supplement the regular company forces at working fires or to augment assigned companies during peak periods of activity.

*Task Forces:* In very large departments, special task forces may be created. These special task forces consist of several pieces of apparatus with appropriate personnel under a commander who can direct the force using the best combination of personnel and apparatus for the particular fire to be handled. A basic task force may consist of two engines and one aerial ladder, plus personnel, including an officer in command. The principal advantage of the task force concept is that, with several pieces of apparatus and a complement of personnel, task force leaders have the flexibility to use the best combinations of personnel and equipment to accomplish assignments.

## *Tactical Operations*

When fighting a fire, there are certain tactical operations that are employed. Tactical operations are described in the NFPA *Fire Protection Handbook* as follows:[9]

> Tactical operations are the basic means employed by fire suppression forces to cope with fire incidents. There are several tactical operations that may or may not be employed at each fire incident and at times there may be several tactical operations being carried out simultaneously during multicompany operations at fires. In some instances the situation may dictate only one tactical operation. Regardless of the number of operations needed, it is essential that each individual, depending upon company assignment, be trained in carrying out these operations.

*Size-up:* Size-up is a continuous mental evaluation of the situation and all related factors that may determine the success or failure of the fireground operation. This mental evaluation should begin as soon as companies are alerted; it should be continuous throughout the incident. The officers and fire fighters involved should constantly size up the situation to be encountered, based on knowledge of the building and contents, the climatic conditions encountered, and other factors having a bearing on the exact circumstances at the fire scene.

*Rescue:* Rescue of people is the first and most important consideration at any fire incident. Rescue may entail personal risk to the fire fighters, and may preclude any attempts to extinguish the fire. Effectiveness of rescue operations will depend on many factors (such as time of day, prompt or delayed alarm, kind of occupancy, height, area, and construction of the building, and the number of fire fighters available at the time).

In formulating rescue strategies, the officer in charge needs to consider the degree of danger to which occupants and fire fighters will be exposed, the resources available (such as rescue apparatus, first-aid materials, and personnel with emergency medical training), and how to assign them.

*Exposures:* Preventing a fire from spreading into adjacent structures is a vital tactical operation. It will depend on the construction and occupancy of the building on fire and also of adjacent structures, access to the fire and nearby area, and the amount of personnel and equipment on hand. A warning from the NFPA *Fire Protection Handbook:*[10]

> The failure to adequately protect exposed structures often leads to large loss fires that extend beyond the building of fire origin. The problem of exposure protection may be compounded by closely spaced buildings, combustible construction, the type of occupancy, the lack of fire department access, and the lack of fire department resources. Exposure protection is a vital and necessary tactical operation, and should be anticipated. Exposure protection should commence as soon as possible to prevent the extension of fire to exposed structures.

*Confinement:* The confinement of fire to its area of origin is often a complex operation because it first requires locating the fire. The fire must be located and surrounded (over, under, and around). Ways in which the fire can spread must be determined and controlled. All of the factors (such as type of fuel involved, construction, built-in fire protection such as sprinklers, smoke conditions, etc.) must be evaluated.

*Extinguishment:* The problems associated with confinement (such as the type of fuel involved, the location of the fire, and the degree of involvement) are generally applicable to extinguishment operations. Getting the water to the seat of the fire, the number of hose lines needed, the use of special extinguishing agents such as foam or dry chemical, and the degree of involvement at the time of arrival are just some of the factors that affect the proper measures for putting out the fire.

*Ventilation:* Systematic removal of heat, smoke, and fire gases from the fire is needed for protection of people, for locating the seat of the fire, and for the visibility and comfort of the fire fighters. Sometimes ventilation is necessary before rescue and fire fighting operations can be commenced.

*Salvage:* An important part of minimizing damage to property is taking measures to protect the building and its contents from heat and flame, smoke, and water. Good salvage work will drastically reduce the amount of loss. Salvage is an integral part of tactical operations and should commence as soon as possible to prevent additional damage to both structure and contents. As explained in *Management in the Fire Service:*[11]

> The term salvage operations includes all operations required to protect a property from unnecessary damage caused by excessive water or other extinguishing materials. Salvage operations include covering objects with salvage cloths and removing water from the property so that it doesn't seep through floors and cause damage to the contents of lower floors. In this way a building can be restored to a reasonable condition before the fire fighters leave.

*Overhaul:* Once the fire is under control, overhaul operations are commenced to make sure that the fire is completely extinguished, to see the building is restored to a safe condition, and to determine how and why the fire started. It is important that extensive overhaul not be commenced prior to a thorough investigation to determine the cause of the fire. Once the investigation is complete, overhaul should continue in order to ensure that the premises are left as safe as possible, and that all fires are extinguished.

## Nontactical Operations

There are additional tasks performed by fire suppression personnel that are not tactical operations, but are equally important. These are prefire planning, fire prevention activities, and training.

*Prefire Planning:* An efficient fire department will not only carry out tactical operations at the fire scene but will do a good deal of planning for each area and building under its jurisdiction long before fire strikes. Prefire planning involves knowledge of each building and its contents, including construction, occupancy, special hazards, water mains and hydrants, exposures, built-in fire protection, and other features that will affect the situation should a fire occur. This is especially important for properties such as hospitals, nursing homes, homes for the aged, asylums, prisons, places of public assembly, and dangerous manufacturing processes. The lives of fire fighters may well depend on what has been done to plan for the emergency before it occurs. The NFPA *Fire Protection Handbook* recommends the following steps in prefire planning:[12]

1. Information Gathering — Collecting pertinent information at the selected site that might affect fire fighting operations, such as building construction features, occupancy, exposures, utility disconnects, fire hydrant locations, water main sizes, and anything else that would be pertinent if a fire should occur.
2. Information Analysis — The information gathered must be analyzed in terms of what is pertinent and vital to fire suppression operations, a plan formulated, and put into a usable format that can be used on the fireground.
3. Information Dissemination — All companies that might respond to each prefire planned location should receive copies of the plan so that they become familiar with both the plan and the pertinent factors relating to it.
4. Class Review and Drill — Each company that might be involved at the preplanned location should review the plan on a regular schedule. Periodic drills with all companies should be scheduled on the property if possible.

*Training:* Training opportunities and procedures have evolved steadily over the years until today almost all fire department personnel have access to good training either in their own department or through state training programs.

Most of the larger fire departments have their own drill tower and training officer. In-service training at company quarters is conducted by the company officer on a daily basis. Refresher courses and special training are conducted at the drill school. Officer training may be given at the drill school or at the state training center. State training schools for basic fire fighting instruction are available to volunteer fire fighters. Some states send training instructors to various departments for periods of training. Many of the state training programs provide special courses for officer training, fire administration, fire and arson investigation, communications, apparatus maintenance, fire prevention inspections, and public education in fire prevention.

The goal of any fire department training program should be the provision of the best possible training for each person within the department, regardless of rank, so that each person will perform at acceptable performance levels. Ideally, training program courses should have their own instructional objectives, a list of enabling objectives showing how the instructional objectives can be reached, and stated methods that explain how anticipated or desired behavioral changes can be measured.

## *Transportation Incidents*

A growing number of serious fires endangering fire fighters and the public and causing substantial losses are those that involve the transportation of flammable liquids, gases, and chemicals by truck, train, plane, or ship.

It is obviously impossible to predict where a transportation fire will occur, but a certain amount of preplanning for such fires can be done by identifying locations where particularly dangerous situations could develop, carrying on special drill evolutions for fighting transportation fires, and by study of the special hazards of fires in which gasoline, oil, or important hazardous chemicals may be involved. Special techniques for handling tank truck fires, airplane crashes, railroad derailments and crashes, and fires in ship's holds must be understood and planned for to safeguard the lives of fire fighters and the public and to avoid heavy property loss from such fires.

Transportation incidents differ from structural fires or fires at industrial plants in that they may occur in any part of a fire department's area of protection. It is, therefore, difficult to preplan action on a precise basis; nevertheless, a plan of action should be developed that can be applied to any possible location where the incident might occur.

## *Mutual Aid*

Many fire departments have mutual aid plans with neighboring fire departments. The plans are to be employed when serious fires or disasters strike. Following are the principal functions of a sound mutual aid plan: (1) immediate joint response of several departments to alarms of fire from high-risk properties, (2) response to alarms adjacent to the boundaries between municipalities, (3) covering in at fire stations to take care of other alarms when the existing forces are at working fires, (4) supplementing the local department forces at major fires, and (5) supplying specialized apparatus and equipment not available to the department involved with the fire.

Mutual aid plans need to be carefully drawn to provide for the legal requirements governing the operation of the department when it is outside its normal jurisdiction. Mutual aid procedures have proved their value over the years to all fire departments, large and small, paid and volunteer.

## Fire Prevention

Chapter Eleven, "Codes and Standards," covers the various building and fire laws, standards, and codes that are commonly employed by federal, state, and local fire prevention agencies. A brief description of the fire prevention agencies in each of the fifty states and of the laws and regulations that they enforce can be found in the appendix of the fourteenth edition of the NFPA *Fire Protection Handbook*. In many cases the state authority will designate local fire prevention officers to act as agents for the state in certain areas of inspection, enforcement, and investigation. The state agencies will function in those areas that are beyond the jurisdiction of municipal, county, or fire district organizations.

Fire prevention activities in municipal paid fire departments are of comparatively recent origin. Fifty years ago only a handful of the larger cities had established fire prevention bureaus. Today most paid departments have at least one officer designated as fire marshal or chief of the fire prevention bureau, and a number of the progressive volunteer departments now devote some time and attention to fire prevention inspections and public education. Unfortunately, the concept that it is just as much the duty of a fire department to prevent fires as it is to extinguish them is still not fully recognized.

In its report titled "America Burning," the National Commission on Fire Prevention and Control, in discussing the nation's fire problem, made the following recommendation:[13]

> Response to important social changes is a key to improving the nation's record in fire protection. A consideration of equal importance is the need to change priorities in the field of fire protection. Currently, about ninety-five cents of every dollar spent on the fire service is used to extinguish fires; only about five cents is spent on efforts — mostly fire prevention inspections and public education programs — to prevent fires from starting. Much more energy and funds need to be devoted to fire prevention, which could yield huge payoffs in lives and property saved.

Effective fire prevention work by a fire department involves many factors,

including the following:
- Inspection of all properties in the community to uncover and correct fire hazards.
- Education of the citizens of the community concerning the dangers of fire and the ways to prevent fires.
- Enforcement of the fire prevention code so that built-in safeguards (such as automatic sprinklers) are provided where needed.
- Investigation of the cause and origin of fires to control arson and to learn how to prevent future similar fires.

The model ordinance establishing a Bureau of Fire Prevention in the fire department defines the duties of the bureau as follows.

To enforce all laws and ordinances covering the following:
1. The prevention of fires.
2. The storage and use of explosives and flammables.
3. The installation and maintenance of automatic and other fire alarm systems and fire extinguishing equipment.
4. The maintenance and regulation of fire escapes.
5. The means and adequacy of exit in case of fire from factories, schools, hotels, lodging houses, asylums, hospitals, churches, halls, theaters, and all other places in which numbers of persons work, live, or congregate from time to time for any purpose.
6. The investigation of the cause, origin, and circumstances of fires.
7. The maintenance of fire cause and loss records.

***Organizations for Fire Prevention:*** Most states have a state fire marshal charged by law with the responsibility for the administration and enforcement of state laws relating to safety to life and property from fire. All of the Canadian provinces have a provincial fire marshal or fire commissioner who performs similar functions.

All of the larger cities and many smaller ones have a fire prevention bureau in the fire department engaged primarily in the prevention of fires through property inspections and enforcement of a fire prevention code or other fire laws and regulations. These bureaus also carry on public education programs in fire prevention and investigate the cause and origin of fires. Some counties and fire protection districts also maintain similar functions.

Following is a description from the NFPA *Fire Protection Handbook* concerning the various fire prevention departments or offices:[14]

> *State Fire Marshal:* The makeup of state fire marshal offices differs from state to state. Most receive their authority from the state legislature and are answerable to the governor, a high state officer, or a commission created for that purpose. In some states the fire marshal's office may be a division of the state insurance department, state police, state building department, state commerce division, or other state agency. Few are organized as separate agencies.

State or provincial agencies normally function in those areas that go beyond the scope of the municipal, county, or fire district organizations. Local fire protection organizations are sometimes granted the authority to act as agents for the state in stipulated areas of inspection, enforcement, and investigation.

*Chief of Fire Prevention or Local Fire Marshal:* The laws of the county, municipal, or fire districts delegate the responsibility and authority of fire prevention to the fire chief or fire department head. Provision is then made for him to delegate this authority to an individual or division, depending on the size of the department. The individual or head of the division should be a high ranking chief officer and should also function as a staff officer to the fire chief. This division of the fire service is normally called the fire prevention bureau, and its top officer is chief of fire prevention or local fire marshal. Where size permits, the bureau is divided into subdepartments of inspections, investigations, and public education. These subdepartments are then headed by subordinate chiefs.

*Fire Inspector or Fire Prevention Officer:* The term fire inspector or fire prevention officer has different meaning depending on department classification. Sometimes the two titles are the same and denote the position responsible for conducting fire inspections assigned to the fire prevention bureau. In bureaus not large enough for three subdepartments, the fire inspector is also responsible for fire investigations and public education duties.

**Inspections:** Inspections by the fire department are extremely important if fires are to be prevented. Inspections should be made by the fire companies of all properties in their district, not only to uncover and correct fire hazards but also to familiarize the fire fighters with the buildings in their district before a fire starts in any of them. Any special hazards disclosed by company inspections are referred to the Bureau of Fire Prevention for action.

The Bureau of Fire Prevention will make regular inspections of target risks — *e.g.,* hospitals, nursing homes, homes for the aged, hotels, theaters, halls, and other places where the public may congregate, and of all important manufacturing and business establishments — to make sure no code violations exist in these occupancies.

A key provision of the ordinance establishing a Bureau of Fire Prevention gives the fire marshal or chief of the fire prevention bureau the following power: whenever any inspector shall find in any building or upon any premises or other places combustible or explosive matter, or dangerous accumulation of rubbish, or unnecessary accumulation of waste paper, boxes, shavings or any highly flammable materials especially liable to fire, which is so situated as to endanger property; or shall find obstructions to or on fire escapes, stairs, passageways, doors or windows liable to interfere with the operations of the fire department or egress of occupants in case of fire, he shall order the same to be removed or remedied, and such order shall forthwith be complied with by the owner or occupant of such premises or buildings.

1. *Home Inspections.* Home inspections are extra work and are not required by law, but are of immense value to the public and to the fire department. The National Commission on Fire Prevention and Control in its report

issued in 1973 strongly recommended that home inspections be undertaken by every fire department in the nation and suggested that any federal financial assistance to fire jurisdictions should be contingent upon their implementation of effective home inspection programs.

Dwelling inspections are an extra service by the fire department to its citizens. If carefully planned and publicized in advance, such inspections are welcomed by the great majority of families and not only eliminate many hazards, but engender goodwill toward the fire department.

The purpose of home inspections by fire departments is to point out such common fire hazards as needless accumulations of rubbish, overloaded electric circuits, improper storage of paints and flammable liquids, space heaters too close to flammable material, matches kept within reach of children, etc.

2. *Business Inspections.* Most of our larger manufacturing and mercantile establishments recognize the disruption that a fire would cause to their business, and not only spend money on fire protection but conduct their own inspections and organize private fire brigades. These and smaller business properties are also subject to inspections by public fire forces, insurance carriers, and the Occupational Safety and Health Administration. Good inspection procedures from whatever source are the best way to keep fire losses down. The policy and attitude of the management of the business toward fire prevention and protection is of prime importance. As a rule, plants with obvious fire hazards (such as oil refineries, chemical and explosive manufacturing, and plants manufacturing plastics) usually have a good fire record because of their acute awareness of the fire hazards of their processes and products. A large number of small plants and stores generally depend on inspections by their insurance carriers or by the fire department for calling potential fire hazards to their attention.

***Public Education:*** The fire prevention arm of the fire department is usually given the responsibility for a continuing program of public education. There are many ways to motivate people to do the things that will save them and their families from the threat of fire. Certainly the fire department should persuade the school authorities that all school children should be taught at least the basics that they need to know about what to do when fire strikes and how to eliminate the common fire hazards in their own homes. In addition, the fire department should send personnel to the schools to talk about the danger of false alarms, conduct demonstrations of fire hazards, etc. Many departments invite children and the public to visit the fire station to see how the department works and to learn how they may cooperate to reduce fire losses. Another excellent procedure is the distribution of home inspection blanks to pupils. Many parents have been alerted to fire hazards in the home by having to sign their child's home inspection blank.

The fire prevention division should also inform the public through television, radio, and the press with regard to special information about hazardous toys, garments, or home appliances; special fire department problems in times of large fires, winter storms, floods, and other disasters; and special programs directed at particular fire hazards.

*Seasonal Activities:* The fire department sponsors or participates in Fire Prevention Week, Spring Clean Up Week, Christmas holiday warnings, Fourth of July safety from fireworks, and similar seasonal campaigns. Many fire departments sponsor junior fire department programs, such as Sparky's Junior Fire Department, and encourage fire prevention activities by Boy Scouts, Girl Scouts, etc. One excellent activity, endorsed by the Fire Marshals Association of North America, is Operation EDITH (Exit Drills In The Home). Homeowners are urged to conduct a fire drill in their homes at a certain day and time during Fire Prevention Week. For most effective results, public education in fire prevention has to be constantly updated and upgraded to maintain public interest and support.

*Enforcement of Codes:* Most communities have adopted a fire prevention code or at least some ordinances covering storage and handling of flammable liquids, electrical wiring, heating appliances, etc. It is the duty of the fire marshal or fire prevention bureau to see that such codes and ordinances are enforced. The many codes and standards available for public use are described in Chapter Eleven, "Codes and Standards."

A good local fire prevention code will include not only the common hazards found in any community, but any special hazards that the fire department may face in industrial or mercantile operations of that particular community. The code should also cover the interior fire protection required in various occupancies (such as automatic sprinklers, fire detection and alarm systems, portable fire extinguishers, and standpipes). Many of the national standards referred to in building and fire prevention codes are primarily for purposes of preventing fires. Typical of these are the many storage standards covering indoor and outdoor storage of combustibles, rack storage, tire storage, records, explosives and flammable liquids, pyroxylin plastics, and forest products.

Other well-known fire prevention standards are on oil burning equipment,[15] fuel gas,[16] cutting and welding processes,[17] dust explosions,[18] and the safe installation of chimneys and fireplaces.[19]

## Investigation of Fires

Another important function of the fire marshal or fire prevention bureau is the investigation of fires to determine cause and origin. (See Chapter Five, "Investigating the Fire Loss Problem.") At first glance it may seem that fire investigation is not related to fire prevention. However, careful investigation to discover all the factors that influence the start and spread of a fire will provide the data required to develop the codes and inspection procedures needed to prevent future fires. The fire prevention bureau should maintain records to show not only the number of fires by location and occupancy, but fire cause, time of day, room or floor in which the fire occurred, and other data basic to an understanding of the problems that the department faces.

The importance of fire investigation is recognized by the courts. Fire marshals are usually given broad powers to investigate fires, and to subpoena persons and records. Arson can often be discouraged if the potential firesetter knows that fires are carefully investigated to determine any suspicious circumstances. In general, the steps to follow in a good investigation of a fire, particularly a suspicious fire, are:

1. A review of the fire scene covering structure and contents, the way the fire was fought, and the time sequence.
2. Reconstructing as many of the events prior to and during the fire, as completely as possible.
3. Determining the path of heat travel and the point of origin.
4. Establishing approximate burning time and temperatures.
5. Evaluating the combustion characteristics of the materials involved.
6. Comparing the fire with fires involving similar materials and situations.
7. Fitting the facts to all the possibilities of the situation.
8. Corroborating the information developed from occupants and witnesses.

## *Summary*

North America is protected by approximately 25,000 fire departments. The majority of these departments are small volunteer fire departments. About 10 percent of the paid fire departments in the United States protect large communities. Although fire departments have evolved to fit local needs, much can be done to improve performance in the areas of both fire protection and fire prevention. Through the development of specific fire department goals and objectives that can be measured and evaluated by performance standards, the organization and management of fire departments can be made increasingly more efficient and effective.

In addition to more effective utilization of the resources of the fire department itself, consideration must be given to intergovernmental relations with regard to effective coordination between other public agencies and the fire department, so that the needs and requirements of the fire department will be clearly communicated and implemented. In addition, fire department effectiveness is further ensured by careful prefire planning and training.

The importance of fire prevention activities, as well as fire protection, with regard to fire department responsibilities, is fast becoming an area of national concern. Most fires start from such common causes as poorly maintained heaters and furnaces, careless smoking, frayed electric cords, and improper storage of gasoline and paint. Every school should teach elementary precautions about fire. Most states have a state fire marshal, and most cities have a fire marshal or a fire prevention bureau in the fire department. A fire prevention code covering the principal fire hazards is now law in many states and cities, with the result that inspections for fire hazards should be regularly carried out by personnel in industries and commercial establishments and by state and city fire marshals. Such inspections are particularly important in places such as hospitals, homes for the aged, nursing homes, hotels, theaters, halls, and other places of public assembly. Home inspections by fire departments are of great value and should be universal. Public education should also be a responsibility of fire departments.

## Activities

1. Discuss with your classmates how the development of specific performance objectives would result in more efficient fire department practices and procedures.
2. Write five specific performance objectives for your fire department or for a fire department with which you are familiar.
3. Give five reasons why effective management of a fire department results in better fire prevention and fire protection service to a community.
4. Describe the governmental agencies in your community with which the fire department must interact, and suggest ways in which this interaction can be improved.
5. Describe the organization of your fire department or of a fire department with which you are familiar.
6. List and describe the eight major tactical operations that are the responsibility of a fire department.
7. Choose a classmate to play the role of a fire chief who does not believe in prefire planning. Convince this "fire chief" of the importance of prefire planning.
8. Describe your community and the various transportation incidents for which the fire department should be prepared.
9. With your classmates acting as the "local board," convince this board of the need to allocate additional funds for fire prevention activities.
10. List five reasons why investigation of fires is a vital fire department responsibility.

## Bibliography

[1]NFPA 1901, *Standard for Automotive Fire Apparatus,* NFPA, Boston, 1975.

[2]Didactic Systems, Inc., *Management in the Fire Service,* NFPA, Boston, 1977, p. 147.

[3]Granito, Anthony R., *Company Leadership and Operations,* NFPA, Boston, 1975, pp. 6-7.

⁴NFPA 1201, *Organization for Fire Services,* NFPA, Boston, 1977.

⁵Didactic Systems, Inc., *Management in the Fire Service,* NFPA, Boston, 1977, p. 157.

⁶NFPA 1001, *Standard for Fire Fighter Professional Qualifications,* NFPA, Boston, 1974.

⁷"America Burning," May 1973, The National Commission on Fire Prevention and Control, Washington, DC, pp. 37-38.

⁸_____, p. 38.

⁹*Fire Prevention Handbook,* 14th Ed., NFPA, Boston, 1976, p. 9-26.

¹⁰_____, p. 9-26.

¹¹Didactic Systems, Inc., *Management in the Fire Service,* NFPA, Boston, 1977, p. 253.

¹²*Fire Protection Handbook,* 14th Ed., NFPA, Boston, 1976, p. 9-26.

¹³"America Burning," May 1973, The National Commission on Fire Prevention and Control, Washington, DC, p. 7.

¹⁴*Fire Protection Handbook,* 14th Ed., NFPA, Boston, 1976, p. 9-30.

¹⁵NFPA 31, *Standard for the Installation of Oil Burning Equipment,* NFPA, Boston, 1974.

¹⁶NFPA 54, *National Fuel Gas Code,* NFPA, Boston, 1974.

¹⁷NFPA 51B, *Standard for Fire Prevention in the Use of Cutting and Welding Processes,* NFPA, Boston, 1977.

¹⁸NFPA 63, *Standard for the Prevention of Dust Explosions in Industrial Plants,* NFPA, Boston, 1975.

¹⁹NFPA 211, *Standard for Chimneys, Fireplaces, and Vents,* NFPA, Boston, 1977.

*Chapter Eleven*

# Codes and Standards

*Throughout history there have been building regulations for preventing fire and restricting its spread. Over the years these regulations have evolved into the codes and standards developed by committees concerned with fire protection. In many cases, a particular code dealing with a hazard of paramount importance may be enacted into law.*

## THE HISTORY AND DEVELOPMENT OF FIRE PROTECTION REGULATIONS

King Hammurabi, the famous law-making Babylonian ruler who reigned from approximately 1955 to 1913 B.C., is probably best remembered for the *Code of Hammurabi,* a statute primarily based on retaliation. The following decree is from the *Code of Hammurabi:*

> In the case of collapse of a defective building, the architect is to be put to death if the owner is killed by accident; and the architect's son if the son of the owner loses his life.

Today, we no longer endorse Hammurabi's ancient law of retaliation but seek, rather, to prevent accidents and loss of life and property. From these objectives have evolved the rules and regulations that represent today's building codes and our current standards for fire prevention, fire protection, and fire suppression.

### Early Building and Fire Laws

The earliest recorded building laws were apparently concerned with the prevention of collapse. During the rapid growth of the Roman Empire under the reigns of Julius and Augustus Caesar, the city of Rome became inundated with hastily constructed apartment buildings — many of which were erected to considerable heights. Building collapse due to structural failure was frequent, so laws were passed that limited the heights of buildings first to 70 ft (21.3 m) and then to 60 ft (18.2 m).

Later in history there evolved many regulations for preventing fire and restricting its spread. In London during the 14th century an ordinance was issued requiring that chimneys be built of tile, stone, or plaster; the ordinance prohibited the use of wood for this purpose. Among the first building ordinances of New York was a similar provision, and among the first legislative acts of the town of Boston was one requiring that dwellings be constructed of brick or stone and roofed with slate or tile (rather than being built of wood and having thatched roofs with wooden chimneys covered with mud and clay similar to those the early settlers had been accustomed to in Europe). Obviously, the intention of these building ordinances was to restrict the spread of fire. As an inducement for helping to prevent fires, a fine of ten shillings was imposed on any householders who had chimney fires. This encouraged the citizenry to keep its chimneys free from soot and creosote. Thus, the first fire code in America can be said to have been established and enforced.

In colonial America, the need for laws that offered protection from the ravages of fire developed simultaneously with the growth of the colonies. The laws outlined the fire protection responsibilities of both the homeowners and the authorities. Some of these new laws were planned to punish people who exposed themselves and others to fire risk. For example, in the town of Boston no person was allowed to build a fire within "three rods" of any building, or in ships that were docked in Boston Harbor. It was illegal to carry "burning brands" for lighting fires except in covered containers, and arson was punishable by death. Regardless of such precautions, in Boston and in the other municipalities that were springing up, fires were everyday occurrences; it therefore became necessary to enact more laws with which to govern building construction and to make further provisions for public fire protection. Thus emerged a growing body of rules and regulations concerning fire prevention, protection, and control. From these small beginnings various codes and types of codes have evolved in this country, ranging from the most meager of ordinances to comprehensive handbooks and volumes of codes and standards on building construction and firesafety.

## Development of Building and Fire Codes

The rapid growth of early American cities inspired much speculative building, and the structures in our early communities were usually built close to one another. Construction was often started before adequate building codes had been enacted. For example, the year before the great Chicago fire of 1871, Lloyd's of London stopped writing policies in Chicago because of the haphazard manner in which construction was proceeding, and other insurance companies had difficulty selling policies at the high rates they had to charge. Despite these excessively high rates, many insurance companies suffered great losses when fires spread out of control.

The National Board of Fire Underwriters, organized in 1866, realized that the adjustment and standardization of rates were merely temporary solutions to a serious technical problem and began to emphasize safe building construction, control of fire hazards, and improvements in both water supplies and fire departments.* As a result, the new tall buildings that were constructed of concrete and steel conformed to specifications that helped limit the risk of fire. These buildings were called Class A buildings. In 1905 the National Board of Fire Underwriters published the first edition of its *Recommended Building Code*[1] (now titled the *National Building Code*[2]).** This was a first and very useful attempt to show the way to uniformity.

In San Francisco in 1906, although there were some new Class A concrete and steel buildings in the downtown section, most of the city was made up of fire-prone wood shacks. Concerned with such conditions, the National Board of Fire Underwriters wrote that "San Francisco has violated all underwriting traditions and precedents by not burning up."

On April 18 of that same year, the city of San Francisco did burn up in a conflagration (started by an earthquake) that destroyed some 28,000 buildings. (See Table 1.1. on page 13.) Although the contents of many of the new Class A buildings were destroyed in the San Francisco fire, most of the walls, frames, and floors remained intact and could be renovated. (See Fig. 11.1.)

Following analysis of the fire damage caused by the San Francisco disaster and other major fires, the National Board of Fire Underwriters became convinced of the need for more comprehensive standards and codes relating to the construction, design, and maintenance of buildings. With this increasing recognition of the importance of fire protection came more knowledge about the subject. Engineers started to accumulate information about fire hazards in building construction and in manufacturing processes, and much of this information became the basis for many of our original standards and codes. Obviously, regulations based on such detailed codes could help prevent many fires and reduce fire loss in those that do occur. In recent years, many attempts have been made to standardize such codes, keep them up to date, and eliminate obsolete and obstructive requirements.

The American Insurance Association publishes its recommended fire code, and several building code organizations have written, or are in the process of writing, fire codes for use in conjunction with their own building codes. The major objective of any successful fire code is, of course, to provide a reasonable degree of safety to life and property from fire.

---

*In September 1964, the National Board of Fire Underwriters merged with the Association of Casualty and Surety Companies and the former American Insurance Association (founded in 1953) to become the new American Insurance Association (AIA).

**Recommended Building Code,* 1st Ed., National Board of Fire Underwriters, New York City, 1905. Since that time, many revisions have been made in order to include the most recent developments in the building industry and to present the most current thinking of experts in the field of firesafety and fire spread control. Now titled the *National Building Code,* it has been adopted in its entirety by many cities and towns and used by others in the framing of, or revision of, existing building regulations.

## Relationships Between Building and Fire Codes

It is often difficult to differentiate between the items that should go into a fire prevention code and those that should be included in a building code. Generally, whatever requirements deal specifically with construction should be part of a building code and should be administered by the building department. A fire prevention code, on the other hand, should include information on equipment incidental to a process or hazard in a building, and should be regulated by the fire department. Usually, requirements for exits and fire extinguishing equipment are found in building codes, while the maintenance of such items is covered in fire prevention codes. The following general outline from the NFPA *Fire Protection Handbook* presents the areas of concern usually covered by building codes, and those that are usually covered by fire prevention codes.[3]

Fig. 11.1. The great earthquake and the ensuing conflagration devastated San Francisco in 1906. As many as 452 people were killed, and 28,000 buildings were damaged or destroyed. Total financial loss was $350,000,000.

Most municipal building codes cover, in general, the following items: (1) administration, which spells out the powers and duties of the building official; (2) classification of buildings by occupancy; (3) establishment of fire limits or fire zones; (4) establishment of height and area limits; (5) establishment of restrictions as to type of construction and as to use of buildings; (6) special occupancy provisions which stipulate special construction requirements for various occupancies such as theaters, piers and wharves, garages, etc.; (7) requirements for light and ventilation; (8) exit requirements; (9) materials, loads, and stresses; (10) construction requirements; (11) precautions during building construction; (12) requirements for fire resistance, including materials, protection of structural members, fire walls, partitions, enclosure of stairs and shafts, roof structures, and roof coverings; (13) chimneys and heating appliances; (14) elevators; (15) plumbing; (16) electrical installations; (17) gas piping and appliances; (18) signs and billboards; and (19) fire extinguishing equipment.

The principal provisions usually found in a fire prevention code cover the following: (1) administration, which includes the organization of the bureau of fire prevention and defines its power and duties; (2) explosives, ammunition, and blasting agents; (3) flammable and combustible liquids; (4) liquefied petroleum gases and compressed gases; (5) lumberyards and woodworking plants; (6) dry cleaning establishments; (7) garages; (8) application of flammable finishes; (9) cellulose nitrate plastics (pyroxylin); (10) cellulose nitrate motion picture film; (11) combustible metals; (12) fireworks; (13) fumigation and thermal insecticidal fogging; (14) fruit-ripening processes; (15) combustible fibers; (16) hazardous chemicals; (17) hazardous occupancies; (18) maintenance of fire equipment; (19) maintenance of exit ways; (20) oil burning equipment; (21) welding and cutting; (22) dust explosion prevention; (23) bowling establishments; (24) auto tire rebuilding plants; (25) auto wrecking yards and junk yards; (26) manufacture of organic coatings; (27) ovens and furnaces; (28) tents; and (29) general precautions against fire.

Although there might seem to be an overlap in the administration of these requirements, closer scrutiny will show that the installation of the original fire prevention item (*e.g.,* duct, vent, exit, or sprinkler system) should be supervised by the building department, but that its maintenance or the determination of its continuing adequacy is the responsibility of the fire department.

## *Responsibility and Enforcement of Safety Provisions*

While there are standards for both fire prevention and building codes, confusion and rivalry sometimes exist between the departments of building and fire prevention as to who is responsible for which safety provisions, and for the enforcement of same. Where problems exist, they can usually be resolved by means such as open communications and continuous meetings between high-level officials from both departments. Cooperation between the two departments is essential for the public safety. One way to ensure cooperation is the practice of having a knowledgeable fire officer review and approve all plans submitted to the building department for new buildings or for changes in construction or occupancy of the building. In some European countries many problems have already been eliminated by making fire department officials responsible for both building and fire codes. Similarly, some local governments in the United States have elected to assign fire fighters to inspect for building code violations at the same time they are conducting their fire code violation inspections.

## FORMATION OF CODES AND STANDARDS

It is difficult to envision the staggering proliferation of fire laws, fire codes, and fire protection standards that have been developed over the last 50 to 75 years. Some idea can be provided by comparing the fact that in 1900 the NFPA, one of the principal sources of consensus fire protection standards and codes, had published five standards totalling approximately 50 printed pages of text, and by 1978 the NFPA's *National Fire Codes* were printed as a sixteen-volume compilation that contained more than 11,000 pages.*

### The Standards-making Process

Incorporated in many fire protection laws are standards developed and processed by private voluntary concensus standards-making organizations such as the NFPA. The example of the standards-making process presented in the following paragraphs is based on the NFPA's procedures. (See Fig. 11.2.)

The development of basic firesafety standards for processes, materials, and operations involving a degree of fire hazard is one of the NFPA's most important functions. While these standards are often adopted and incorporated into state and local ordinances, the NFPA considers its status to be only advisory. From the time of its organization in 1896, the NFPA has developed standards to cope with fire hazards in specific industries such as lumber, metalworking, and electronic computer/data processing; in specific occupancies, from hospitals to mobile homes; in transportation vehicles and systems on land, at sea, and in the air; and in a variety of situations involving flammable liquids and gases. NFPA 70, *National Electrical Code,*[4] is especially influential and carries the force of law throughout the United States. Two of the most widely used sets of safety requirements in the world are the NFPA's *National Electrical Code* and NFPA 101, *Code for Safety to Life from Fire in Buildings and Structures.*[5] The NFPA has also developed NFPA 1, *Fire Prevention Code.*[6] The *National Fire Codes* of the NFPA are revised periodically.

---

*The 1978 edition of the *National Fire Codes* includes 238 codes, standards, guides, manuals, recommended practices, and model laws.

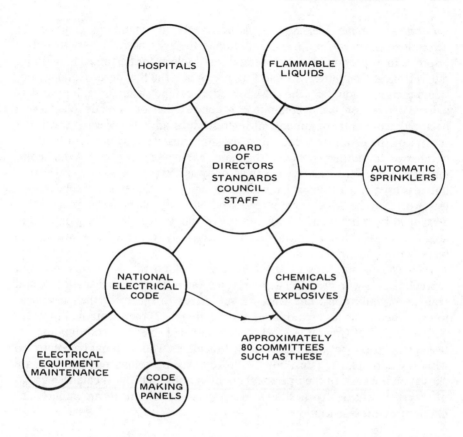

*Fig. 11.2. The standards-making process of the National Fire Protection Association. (From* The Voluntary Standards System of the United States of America, *reprinted by permission of the American Society for Testing and Materials.)*

***Publicizing Intent to Initiate a New Standard:*** The first step in the NFPA's standards-making process is the publication of its intent to initiate a new standard when it recognizes, or is alerted to, the need for a standard on a subject within its scope of activity. The Association researches material and viewpoints on the proposed project, and solicits information relating to the existence and extent of the topic's coverage by any other organization. Furthermore, the Association seeks an indication of the general interest to participate in the development of an original NFPA standard on the subject or project in question. Next, the Standards Council, which is appointed by the Association's Board of Directors and which consists of nine persons (including a chairperson to provide administration of the Association's standards-development activities), determines whether or not the proposed project will be undertaken after it has completed this preliminary investigation and has gaged the extent of public response.

***Formation of the Technical Committee:*** Once the Standards Council has determined the need for a new standard and the existence of favorable public response to its proposal for such a standard, it selects the personnel to make up the Technical Committee. In selecting personnel, the Standards Council considers each individual's technical knowledge, competence, and ability to participate actively on such a committee. A concerted effort is made to achieve a balanced representation among all interests substantially concerned with the subject of the proposed standard; a simultaneous effort is made to limit the number of committee members to a manageable working group. While committee sizes vary from as few as 8 to as many as 250 persons, the average committee is usually made up of from 20 to 40 members. Often, small subcommittees work within larger committees. The NFPA's codes and standards are developed by more than 150 committees of the Association. The committees are made up of voluntary members, not necessarily from the Association, who serve without compensation.

Technical Committees of the NFPA are composed of: (1) organization representatives who are affiliated with interested international, national, and regional organizations, and who are authorized to speak for their organizations in the particular area of committee activity; (2) representatives of other NFPA Technical Committees, who are designated to ensure coordination between related projects; and (3) qualified members who need not be connected with any particular organization or already existing committee, and who are chosen on the basis of their personal qualifications. When this balanced group of experts is selected by the Standards Council, appointments are publicly announced by the Association.

***Determining Scope and Content of Standards:*** Once the first basic step in the standards-making process — the organization of the Technical Committee — has been accomplished, the scope and technical content of the projected standard must be determined by the committee. This is accomplished by utilization of all the relevant research data and field experience available. This second step is of great importance, because it is here that the committee must apply its expertise in revising and amending existing standards as well as in developing new ones. Normally, the re-evaluation of existing standards comprises the greatest bulk of the committee's work.

***Submission of Standard as a Draft:*** The next step is the preparation of a draft, or report. When in final form, this report is submitted to all members of the Technical Committee for study and vote. If a report is to be submitted for Association action, it must be accepted by at least two-thirds of the committee members eligible to vote. Results of the balloting, with inclusion of the reasons for any negative votes, are made known to all committee members.

***Publication and Review of Report:*** The NFPA publishes the committee's report in *Technical Committee Reports,* which is distributed to members of the

Association and others upon request. A 60-day period is reserved for the public to review the report and comment on it before further action is taken. At the same time, the proposed standard may be submitted to the American National Standards Institute (ANSI) for review and approval as an American National Standard. After this period of public review, the amendments or changes that have been suggested are carefully considered by the Technical Committee, which formally documents its disposition toward each proposal. This written disposition of the amendments proposed to each standard is published in a compilation entitled *Technical Committee Documentation,* which is distributed upon request to Association members and others in advance of the Association's Annual Meeting or Fall Meeting.

*Adoption by the NFPA:* At either the Annual Meeting or the Fall Meeting, the standard, in its final form, is presented for vote. The meeting can adopt the standard as published, return a report to the Technical Committee, or adopt an amended report. Following its adoption by the NFPA, each standard is published in pamphlet form and made available for voluntary adoption by any organization or jurisdiction having enforcement powers; each standard is also included in the appropriate volume of the next revised edition of the *National Fire Codes.*

## Enforcement of Codes and Standards

Today the life and property of every citizen is safeguarded to at least some extent by fire legislation enacted by the Congress of the United States, state legislatures, city councils, town meetings, and all other jurisdictions and levels of government. The implementation and enforcement of this tremendous mass of legislation is in the hands of administrative agencies of government such as federal departments and agencies, state fire marshal offices and other appropriate state agencies, and local fire deparments, building departments, electrical inspectors, and others.

In this country's earlier days, the protection of citizens from fire was solely the concern of the local community. Fire fighting is now, and always has been, carried on by local fire departments. The idea that the fire department should assume some responsibility to prevent fires did not become accepted in most cities and towns until about fifty years ago. While most communities have had some type of building code since the beginning of the 20th century, they did not have fire prevention codes until many years later.

With the need for more detailed, comprehensive standards and codes relating to the construction, design, and maintenance of buildings came the knowledge that regulations based on such codes could certainly prevent most fires and reduce losses in the fires that did occur. Codes are, in themselves, only standards; in order to achieve the purpose for which they were created, they are dependent upon regulations that cover their enforcement.

Regulations relating to firesafety are determined and enforced by different levels of government. While some functions overlap, federal and state laws generally govern those areas that cannot be regulated at the local level.

*Federal-level Authority:* There is a substantial amount of federal regulation with respect to firesafety. Under the Constitution, the Congress has the power to regulate interstate commerce; this power has been interpreted to permit Congress to pass laws authorizing various federal departments and agencies to adopt and enforce regulations to protect the public from fire hazards.

Any federal department or agency can promulgate firesafety regulations only if authority to do so is granted by a specific act of Congress. These regulations have the force of law, and violations can result in legal action.

In general, such federal laws can be enacted to provide: (1) that all state laws on the same subject are superseded by the federal law, (2) that state laws not conflicting with the federal law remain valid, or (3) that any state law will control if it is more stringent than the federal law.

All federal fire regulations affecting the general public must be published in the *Federal Register* before they become law so that interested citizens can make comments (favorable or unfavorable) to the proper department or agency prior to the adoption of such regulations into the law. After adoption, all fire regulations are published in the *Code of Federal Regulations.* The *Code of Federal Regulations* is revised and updated annually.

Probably the most extensive set of fire regulations ever adopted by a department of the federal government are the regulations promulgated by the Occupational Safety and Health Administration (OSHA) of the Department of Labor, governing the health and safety of workers in industry and commerce.

The Consumer Product Safety Commission, an independent agency, has broad regulating authority over the firesafety aspects of products sold to consumers. It can prescribe and enforce mandatory product standards.

Firesafety standards for hospitals, nursing homes, and other health care facilities are issued by the Public Health Service of the Department of Health, Education, and Welfare. These standards must be met as a condition for grants and loans for building and improving public and private health care facilities. The Social Security Administration, also a division of the Department of Health, Education, and Welfare, requires that hospitals and nursing homes participating in the Medicare and Medicaid programs comply with the NFPA's *Life Safety Code.*

*State-level Authority:* The Tenth Amendment to the Federal Constitution reserves for the states certain police powers affecting public health and safety. State and local laws concerned with public firesafety are based on this exercise of powers.

Although all of the states have adopted some fire legislation, the amount and effectiveness of such legislation varies widely. For example, one state has adopted the latest edition of the NFPA *National Fire Codes,* while another state leaves fire legislation largely to local fire officials in its cities and towns. Most of the states have laws covering such areas as the storage, use, and sale of combustibles and explosives; installation and maintenance of automatic and other fire alarm systems and fire extinguishing equipment; construction, maintenance, and regulation of fire escapes; means and adequacy of fire exits in factories, asylums, hospitals, churches, schools, halls, theaters, nursing homes, and all other places in which numbers of persons live, work, or congregate from time to time for any purpose; suppression of arson; and the investigation of the cause, origin, and circumstances of fires.

In all but four of this country's fifty states, the principal fire official is the state fire marshal. The four states with no state fire marshal are Colorado, Idaho, New Jersey, and New York.

For the most part, the state fire marshal is the statutory official who is charged by law with the responsibility for the administration and enforcement of state laws relating to safety to life and property from fire. Usually the state fire marshal also has the power to investigate fires and to suppress arson. In some states other state agencies have been delegated the responsibility for some fire regulations, such as enforcement of the *National Electrical Code* or the *LP-Gas Regulations.*[7]

In eight states, the office of the state fire marshal is an independent agency of the state government. In fifteen states, the office of the state fire marshal is part of the Department of Public Safety or of the State Police. In eleven states, the office of the state fire marshal is in the office of the State Insurance Commissioner: in the remaining states, it is under various state departments such as the Attorney General's Office or the State Department of Commerce.

In most states, the fire marshal has the legal power to draft rules and regulations covering various fire hazards; such regulations have the effect of law. In most cases, the office of the state fire marshal serves as a central agency for sponsoring and promoting all kinds of fire prevention activities. Some state fire marshals are also responsible for reviewing construction or remodeling plans of state buildings, schools, nursing homes, and hospitals. All of the Canadian provinces have a provincial fire marshal with, generally, the same powers and duties of the state fire marshal.

The responsibilities of the office of state fire marshal are in the following general areas:

1. Prevention of fires.
2. Storage, sale, and use of combustibles and explosives.
3. Installation and maintenance of automatic alarms and sprinkler systems.
4. Construction, maintenance, and regulation of fire escapes.
5. Means and adequacy of exits in case of fire in public places or buildings in which a number of persons live, work, or congregate (such as schools, hospitals, and large industrial complexes).
6. Suppression of arson, and the investigation of the cause, origin, and circumstances of fire.

While the state fire marshal's office has legal authority for fire prevention, much of this power is delegated to the local fire departments and to local government. Fire departments may carry out inspection of private properties to determine if there are fire hazards or code violations on the premises, and local authorities are given the power — through "enabling acts" — to adopt their own regulations relating to fire prevention.

***Local-level Authority:*** Local fire legislation is for the most part embodied in a fire prevention code administered by the fire department and a building code administered by the building department. In many cities and towns the *National Electrical Code* is adopted; in these cities and towns it is enforced by a local electrical inspector or electrical inspection department. Certain fire matters also may be handled by the police or some other similar local departments. Most localities use nationally recognized standards and codes as the basis for their codes, sometimes modifying them to fit local needs.

Special "fire laws" may be enacted locally to handle life and firesafety problems in the community such as piers and wharves, oil refineries, grain elevators, chemical plants, etc. In all too many cases a local fire prevention code was adopted after a major fire disaster in the community, with heavy emphasis on preventing another similar disaster. Thus, while local codes and ordinances might seem to be of concern only at the local level, in them are reflected and incorporated many of the standards and codes set up by both state and private organizations. Some states have adopted uniform codes in special areas, such as in the area of building construction. Such uniform codes naturally supersede existing local ordinances.

The local government has the power to enforce state regulations that support fire codes where they exist and to enact its own ordinances. In turn, the fire department has the responsibility for the enforcement of these regulations and ordinances. Also, should there be changes within the district that make existing fire codes inadequate, it is the fire department's responsibility to suggest the need for changes and to help develop new codes and regulations where they are needed.

To be most effective, regulations should be supported with inspections. Thus, buildings in which large numbers of people work, live, or meet must be inspected to ensure that they are free from any known hazards and that they conform to the standards and codes specified in existing regulations and ordinances. The identification of fire hazards by means of inspections does not always result in compliance with established regulations. For example, a building owner can refuse to remove the fire hazards or renovate a building so that it conforms with the standards. To eliminate such possibilities, it is necessary that firesafety ordinances (regulations built around a model fire code) not only specify their inspection procedures, but are enforceable and carry penalties for violators. Violators may be fined, certificates of occupancy withheld, or permits for specific businesses or manufacturing processes withheld until compliance with the codes is obtained.

As was previously stated, laws for local firesafety generally fall into two categories: (1) those relating to buildings, and (2) those relating to hazardous materials, processes, and machinery that may be used in buildings. Often, local planners disagree as to what should go into a building code and what should go into a fire code. Generally, requirements relating to construction go into the building code and are enforced by the building inspector and the building inspector's department, and requirements relating to the safe operation of machinery or equipment are the responsibility of the fire department.

## Adoption of Standards into Law

Consensus safety standards are developed so that they can be adopted by jurisdictions that have the power to enforce them. In most cases these jurisdictions are public authorities. However, certain private organizations, such as trade associations that have the means to "enforce" in one way or another, cannot be ignored.

The legal procedure for adopting a standard into law may vary from one enforcing jurisdiction to another. Usually, the simplest and best way is to adopt by reference. This method, which is applicable to both public authorities and private entities, requires that the text of the law or rule cite the standard by its title and give adequate publishing information to permit exact identification of the standard. The standard itself is not reprinted in the law. All deletions, additions, or changes made by the adopting authority are noted separately in the text of the law. Adoption of a current edition of a standard obviates outdated editions maintained as law until a new law referencing a new edition of a standard is adopted.

Where local laws do not permit adoption by reference, a standard can be adopted by transcription. This requires that the text of the adopted standard be transcribed into the law. Existing material can be deleted and new material added, only if such material does not change the meaning or intent of the existing or remaining material. Under adoption by transcription, the standard cannot be rewritten, although changes can be made for administrative provisions. Because the text of the standard is transcribed into the law, due notice of the copyright of the standard's developer is required. As a result, most developers copyright their standards to prevent misuse and unlawful use.

## *TYPES OF CODES AND STANDARDS*

The standardization of firesafety requirements and safety testing in the United States coincides roughly with the widespread introduction of electricity at the end of the 19th century. Insurance industry concerns, the need for governmental regulations, and the prospect of substantial variations in requirements that may be set independently by regulatory authorities suggested the need for reasonable, fair, and objective safety standards that would be

280   Principles of Fire Protection

uniform nationally and applied by any jurisdiction. The private sector met this need with the development of safety standards that are based on a balanced consensus of all relevant interests.

## "Model" Fire Prevention Codes

Currently, there are five "model" fire prevention codes available in the United States. Each of these five "model" codes utilizes NFPA standards as the basis for the technical details of its fire prevention and fire control measures. These five codes are: (1) the NFPA's *Fire Prevention Code,* which was developed by a committee representing state and city fire marshals and other concerned interests, and which was adopted after going through the usual NFPA standards-making process and submission procedure at an annual meeting of the association; (2) the American Insurance Association's *Fire Prevention Code,*[8] which was the first model fire prevention code issued, and which has been adopted by many communities; (3) the *Uniform Fire Code,*[9] published by the International Conference of Building Officials in cooperation with the Western Fire Chiefs Association; (4) the *BOCA Basic Fire Prevention Code,*[10] published by the Building Officials and Code Administrators International, Inc.; and (5) the *Southern Standard Fire Prevention Code* of the Southern Building Code Congress.[11] (The general content of these model fire prevention codes was outlined earlier in this chapter in the section subtitled "Relationships Between Building and Fire Codes.")

Each community, when adopting its fire prevention code, will select the hazards to be covered for its local needs and add the ordinance or other legal document needed to put the code into effect. It should be obvious that while no fire prevention code can be specific enough to cover *all* fire hazards in *all* situations, the fire codes can serve as general guides for suggesting the proper measures to be followed.

## Model Building Codes

As illustrated by the excerpt at the beginning of this chapter, building codes have been in existence since King Hammurabi's ancient law of retaliation for poor building construction. Today, in almost all local communities, a local building code was adopted long before a local fire prevention code.

Today's building codes set forth minimum requirements for design and construction. These minimum standards are established to protect the health and safety of the general public, and generally represent a compromise between optimum safety and economic feasibility. Although builders and building owners often establish their own requirements, the minimum standards must be met. The construction features covered include structural integrity, fire protection, life safety of occupants, mechanical and utility systems, and interior finish.

Building codes have had a long evolution. While the earliest codes were designed to protect against building collapse, modern building codes incorporate fire protection features. Today, about 80 percent of a building code is concerned with life and property safety from fire.

A building code sets forth minimum requirements for design and construction of the building to protect the health and safety of the people occupying the building.

There are two types of building codes. Specification codes detail what materials can be used, how large (or small) the components are, and how the components should be assembled. Performance codes detail the objective to be met and establish criteria for determining if the objective has been met. The designer and builder are, thus, allowed freedom in selecting construction methods and materials as long as it can be shown that the performance criteria can be met. Performance-oriented building codes still embody a fair amount of specification-type requirements, but permit substitution of alternate methods and materials if they can be shown to be adequate.

Building codes usually establish fire limits in certain areas of the municipality. Only certain types of construction are allowed within the fire limits, thus restricting the conflagration potential of the more densely populated areas. Outside the fire limits, the restriction of certain construction types is relaxed due to decreased population density, increased spacing between buildings, etc. Obviously, the establishment of fire limits approximates occupancy and utilization zoning.

Another example of the impact of building codes on fire protection and prevention is the establishment of height and area criteria. These criteria establish how big and how high a particular building can be, based on what the building will be used for. Unfortunately, these requirements vary considerably from area to area, and there is no nationally recognized standard for setting height and area limitations. NFPA 206M, *Guide on Building Areas and Heights*,[12] is a basis for establishing such criteria. The types of building construction are important factors in establishing height and area limitations.

Some other building code requirements that directly affect fire protection include: enclosure of vertical openings, such as stair shafts, elevator shafts, and pipe chases; provision of exits for evacuation of occupants; requirements for flame spread of interior finish; and provisions for automatic fire suppression systems. Exit requirements found in most building codes are based on those found in the NFPA *Life Safety Code*.

Today, when a code commission or building official is revising an outdated building code, there are four model codes that are usually used as guides. These four model codes are: (1) the *National Building Code*, (2) the *Uniform Building Code*,[13] (3) the *Standard Building Code*,[14] and (4) the *BOCA Basic Building Code*.[15] (See Bibliography.)

**National Building Code:** Today's building codes are more uniform than the building codes formulated earlier in the century. The publishing of the National Board of Fire Underwriters' *National Building Code* in 1905 was an early attempt at uniformity. This Code, which is revised and republished periodically, has been adopted by many cities and towns.

The purpose of the *National Building Code* is to provide for safety, health, and public welfare through structural strength and stability, means of egress, adequate light and ventilation, and the protection of life and property from fire and hazards incident to the design, construction, alteration, removal, or demolition of buildings and structures. As stated in the *National Building Code,* the provisions of the Code are as follows:[16]

> The provisions of this Code apply to: (1) the construction, alteration, equipment, use, occupancy, location, and maintenance of buildings and structures; (2) additions to buildings and structures; and (3) appurtenances to buildings and structures; whether hereafter erected or, where expressly stated, existing; and whether on land, over water, or on water, permanently moored to land, and substantially a land structure. The provisions of this Code also apply to buildings and structures, and equipment for the operation thereof, hereafter moved or demolished.
>
> The provisions of this Code based on occupancy apply to conversions from one existing occupancy classification to another, to conversions to or from an occupancy which is controlled by the provisions of Article III of this Code, to conversions of any occupancy within a building having mixed occupancy, and to conversions of any occupancy to an occupancy used for the manufacture, sale, or storage of combustible products or merchandise.

**Uniform Building Code:** In 1927, the International Conference of Building Officials produced its *Uniform Building Code.* Originally, this Code was widely adopted on the west coast; more recently, its usage has spread to other areas of the country. The purpose and scope of the *Uniform Building Code* are as follows:[17]

> *Sec. 102.* The purpose of this Code is to provide minimum standards to safeguard life or limb, health, property, and public welfare by regulating and controlling the design, construction, quality of materials, use and occupancy, location and maintenance of all buildings and structures within the city and certain equipment specifically regulated herein.
>
> *Sec. 103.* The provisions of this Code shall apply to the construction, alteration, moving, demolition, repair, and use of any building or structure within the city, except work located primarily in a public way, public utility towers and poles, mechanical equipment not specifically regulated in this Code, and hydraulic flood control structures.
>
> Additions, alterations, repairs, and changes of use or occupancy in all buildings and structures shall comply with the provisions for new buildings and structures except as otherwise provided in Sections 104, 306 and 502 of this Code.
>
> Where, in any specific case, different sections of this Code specify different materials, methods of construction or other requirements, the most restrictive shall govern.

***Standard Building Code:*** In 1945, the Southern Building Code Congress adopted its *Standard Building Code*. The purpose of the *Standard Building Code,* which was specifically designed to meet the needs of southern municipalities, is as follows:[18]

> The purpose of this Code is to provide minimum requirements to safeguard life, health and public welfare and the protection of property as it relates to these safeguards by regulating and controlling the design, construction, alteration, repair, equipment, use and occupancy, location, maintenance, removal and demolition of all buildings or structures and appurtenances thereto.
>
> The *Standard Building Code* is dedicated to the development of better building construction and greater safety to the public and uniformity in building laws; to the granting of full justice to all building materials on a fair basis of the true merits of each material; and to development on a sound economic basis for the future growth of our Nation through unbiased and equitable dealing with building construction.

***BOCA Basic Building Code:*** In 1950, the Building Officials Conference of America (now the Building Officials and Code Administrators International) published its *Basic Building Code*. The scope and purpose of this Code, which has found favor in many cities (particularly in the midwest and east), is stated in the following excerpt:[19]

> *100.2 Scope:* These regulations shall control all matters concerning the construction, alteration, addition, repair, removal, demolition, use, location, occupancy, and maintenance of all buildings and structures and their service equipment as herein defined, and shall apply to existing or proposed buildings and structures in the (name of jurisdiction); except as such matters are otherwise provided for in the local jurisdictional charter, or other ordinance or statutes, or in the rules and regulations authorized for promulgation under the provisions of this Code.
>
> *100.4 Code Remedial:* This Code shall be construed to secure its expressed intent, which is to insure public safety, health, and welfare insofar as they are affected by building construction, through structural strength, adequate egress facilities, sanitary equipment, light and ventilation, and firesafety; and, in general, to secure safety to life and property from all hazards incident to the design, erection, repair, removal, demolition or use and occupancy of buildings, structures or premises.

***The Role of Standards in Building Codes:*** In the four model building codes and in other building codes, various national standards are adopted by reference, thus keeping the building codes to a workable size and eliminating much duplication of effort. Many building codes reference NFPA standards such as NFPA 101, *Life Safety Code* (see Bibliography), NFPA 13, *Standard for the Installation of Sprinkler Systems,*[20] and NFPA 30, *Flammable and Combustible Liquids Code.*[21] Advisory construction and performance standards by testing facilities such as Underwriters Laboratories Inc. and the Factory Mutual Engineering Corporation, as well as standards issued by technical groups such as the American Society of Mechanical Engineers and the American Society for Testing and Materials, may be referenced in building codes. Standards suggested by trade associations such as the American Gas Association, the American Iron and Steel Institute, the National Forest Products Association, and the American Society of Heating, Refrigerating, and Air Conditioning Engineers are sometimes incorporated in, or referenced in, building codes.

## FIRESAFETY STANDARDS-MAKING ORGANIZATIONS

In addition to the NFPA, other nongovernmental agencies have developed consensus standards and codes for use as guidelines in enforcing fire protection standards and laws. Two other recognized organizations involved in certain fire protection areas are the American National Standards Institute and the American Society for Testing and Materials. Additionally, there are several other types of organizations (testing laboratories, trade associations, and institutions) that develop safety standards. However, these organizations either do not operate under consensus procedures, or their range of representative interests is so limited that the resultant standards do not lend themselves for adoption by public regulatory authorities. Several of the standards developed by some of these organizations are subsequently subjected to the consensus processes of the American National Standards Institute; this, in effect, means that firesafety standards so submitted will be processed through the NFPA standards-making system.

This section discusses, as representative of standards-making organizations having firesafety concerns, the consensus standards-making activities of the American National Standards Institute, the American Society for Testing and Materials, and the National Fire Protection Association. Also included are descriptions of some of the codes and standards developed by these organizations that are incorporated in, or referenced in, many of our federal and state laws, local building codes, and fire prevention codes.

### American National Standards Institute (ANSI)

ANSI sets public requirements for national standards and develops and publishes them on a wide range of subjects. In order to achieve uniformity in voluntary and mandatory state and federal standards, it coordinates voluntary standardization activities of concerned organizations.

Proposals submitted to ANSI may be approved by any one of three methods that the proponent may elect to utilize. These methods are: (1) the Accredited Organization Method, (2) the American National Standards Committee Method, and (3) the Canvass Method. Details of these methods are somewhat involved and are beyond the scope of this text. A proponent choosing either the Accredited Organization Method or the American National Standards Committee Method must provide a copy of the standard and a brief history of its development, purpose, and any prior use it may have served. This information is referred first to the cognizant standards management board and then, within thirty days, to the Accredited Organization or the secretariat of the American National Standards Committee, depending upon the method the proponent chooses. These organizations are then requested to develop evidence of consensus and to report the results to the proponent within six months. The organization must include identification of those in favor of the standard's adoption, those opposed (with reasons), those not voting, and those not responding.

The proponent must then consider the results and attempt to resolve any objections and comments before again submitting the proposed standard to the ANSI. Evidence of the proponent's consideration of, and attempts to resolve, objections and comments must be included in the submission.

When a proponent elects to use the Canvass Method, a list of organizations and others known to have concern with, and competence in, the subject of the proposed standard must be submitted to the ANSI and reviewed by the appropriate standards management board. The review period lasts thirty days, within which time additions may be proposed by standards management board members. The proponent must transmit to the organizations and others listed for canvass a ballot, other pertinent material, and at least one copy of the standard. The ballot form provides opportunity for the canvassee to concur, object (in writing), or not vote. The balloting procedure may last no longer than six months.

The sponsor making the canvass must consider all objections and comments, and attempt to resolve them. Any changes made in the standard as a result of these objections or comments must be reported to the organizations canvassed so that they may have opportunity to accept or reject the changes. This second canvassing period cannot exceed three months. Final results are then reported to the ANSI.

ANSI standards cover a variety of products, materials, and equipment that are used both in highly specialized fields and in nearly all areas of modern life. The ANSI publishes standards on ceramic tiles, chemical process equipment, home appliances, electronics equipment, motion picture film and equipment, acids, refractory materials, oil burners, office machines and supplies, hospital supplies, and combustion engines, to name but a few.

## *American Society for Testing and Materials (ASTM)*

ASTM develops and publishes standards on finished products and on materials used in manufacturing and construction. Because some products and materials are used only within certain companies, industries, and government agencies, not all ASTM standards are developed by the full-consensus system. However, standards that deal with commodities used by the general public are developed by a full-consensus procedure, wherein all interested parties are fairly represented in the committee writing the standard. The standard committee is comprised of anyone technically qualified or knowledgeable in the area of the committee's scope. There are presently approximately 22,000 persons serving the ASTM from all segments of society.

The ASTM standardization procedure begins when a suggestion to prepare a new standard originates in, or is recommended to, an ASTM Technical Committee, which then refers it to a subcommittee for preparation. Unless usage and acceptance of the proposed standard are considered sufficient enough to warrant immediate adoption, the standard must go through a tentative procedure, wherein it is considered by the subcommittee and, if approved, is then proposed to the Technical Committee and voted upon. If approved by two-thirds of those voting, a report is prepared by the committee recommending the acceptance of the proposed standard. This report must contain a statement of facts regarding the origin and development of the standard, as well as the results of previous voting, listing affirmative and negative votes and those ballots marked "Not Voting." The report is submitted to the ASTM headquarters for review. If accepted, it is then published as a "Tentative" before being promulgated as a standard. (*N.B.:* Not all proposals need go through this tentative procedure. If they are deemed worthy of immediate adoption, they may be promulgated directly.)

After publication as a "Tentative," the proposal is then ready for promulgation as a standard. The proposal is first considered at a meeting of the Technical Committee on the recommendation of the responsible subcommittee. If approved by a two-thirds vote, it is referred to letter ballot, wherein it must receive another two-thirds vote before submission to headquarters. Here it once again goes through various voting procedures and, if not rejected, is either accepted or amended before publication as an ASTM standard.

ASTM standards are published annually. They cover a wide range of products and materials such as thermal insulations, electrical insulating materials, petroleum products and lubricants, environmental acoustics, textiles, building construction, paint, and resilient floor coverings.

### *National Fire Protection Association (NFPA)*

An earlier section of this chapter discussed the formation of codes and standards, using as an example the standards-making process followed by the NFPA. Once a code or standard has been adopted by the NFPA, it becomes available for adoption by any organization or jurisdiction having enforcement authority. A number of NFPA standards are widely used and commonly referenced in fire legislation. Descriptions of some representative codes and standards of the NFPA are contained in the following paragraphs.

NFPA codes and standards encompass the entire scope of fire prevention, fire protection, fire fighting, and fire hazards, ranging from the *National Electrical Code* (see Bibliography), believed to be the most widely adopted set of safety requirements in the world, to codes or standards of specific limited areas that are nevertheless important in controlling a life or fire hazard. Among such items is NFPA 1122L, *Code for Unmanned Rockets,*[22] the purpose of which is to prohibit the making and launching of dangerous homemade "rocket bombs" and to eliminate deaths and injuries from experimenting with such devices. Another typical standard of limited but important application is NFPA 656, *Standard for the Prevention of Dust Ignition in Spice Grinding Plants.*[23]

NFPA 101, *Life Safety Code* (see Bibliography), specifies measures for providing that degree of public safety from fire that can be reasonably required. It covers construction, protection, and occupancy features in order to minimize danger from fire, smoke, fumes, or panic before buildings are evacuated. The Code specifies the number, size, and arrangement of exit facilities sufficient to permit prompt escape of occupants from various types of buildings in case of fire or other conditions dangerous to life.

NFPA 30, *Flammable and Combustible Liquids Code* (see Bibliography), provides the rules for tank storage, piping, valves and fittings, and container storage for such liquids in industrial plants, bulk plants, processing plants, and service stations.

NFPA 13, *Standard for the Installation of Sprinkler Systems* (see Bibliography), is the national standard for the installation and testing procedures required for all types of fire protection sprinkler systems.

NFPA 54, *National Fuel Gas Code*[24] covers installation, operation, and maintenance of gas piping, gas appliances, and other gas equipment in residential, commercial, and industrial premises for use with either natural or manufactured gas.

NFPA 10, *Standard for Portable Fire Extinguishers,*[25] contains criteria for the selection, installation, inspection, maintenance, and testing of all types of portable fire extinguishers.

NFPA 1901, *Standard for Automotive Fire Apparatus,*[26] applies to all types of fire apparatus including those equipped with a fire pump, hose body, water pump, water tank, aerial ladder, elevating platform, or water tower.

NFPA 501C, *Standard for Recreational Vehicles,*[27] covers plumbing, heating and electrical systems, and firesafety and life safety provisions for travel trailers, motor homes, camping trailers, and truck campers.

*288   Principles of Fire Protection*

NFPA 74, *Standard for Household Fire Warning Equipment,*[28] covers the proper selection, installation, maintenance, and operation of fire warning equipment, such as smoke and heat detection devices, for use within family living units.

## Summary

Early building codes were primarily concerned with the prevention of building collapse. As civilization progressed and cities became crowded, regulations were formed to limit the types, numbers, and heights of buildings that could be constructed, and also to prevent the start and spread of fire in those buildings. As building and fire regulations developed throughout the United States, many were incorporated into the law at federal, state, and local levels of government.

Today's model building codes establish minimum requirements for construction and design, and fire protection codes and standards play an important part in their development. Fire codes that help ensure life safety in any structure where people congregate for a variety of purposes are formed by such nongovernmental consensus standards-making associations as the American National Standards Institute, the American Society for Testing and Materials, and the National Fire Protection Association.

## Activities

1. Discuss the major differences between fire prevention codes and building codes.
2. (a) Explain why there is difficulty in establishing responsibility for the enforcement of the safety provisions of fire prevention and building codes.
   (b) How could this difficulty be rectified?
3. Discuss the different types of authority that federal, state, and local governments have in enforcing codes and standards.
4. Standards may be adopted into law by reference or by transcription.
   (a) Discuss the differences between these two methods.
   (b) Which method do you prefer?
5. (a) What are the objectives of specification codes?
   (b) What do performance codes establish?
6. What features are fundamental to each of the four major model building codes discussed in this chapter?
7. (a) Describe the primary concerns of building and fire laws in 14th-century London and colonial America.
   (b) What was the first fire code to be established in America, and how was it enforced?
8. In outline form, present what you feel is an efficient procedure that could

be used by voluntary standards-making organizations involved in the development of firesafety standards.
9. What areas of firesafety do standards from the American National Standards Institute and the American Society for Testing and Materials cover?
10. (a) What criteria should a community use to adopt a fire prevention code?
    (b) How can a community best utilize a code that does not cover all fire hazards within its area?

## Bibliography

[1]*Recommended Building Code,* 1st Ed., National Board of Fire Underwriters, New York, 1905.

[2]*National Building Code,* 1976 Ed., American Insurance Association, Engineering and Safety Services, New York, 1976.

[3]*Fire Protection Handbook,* 13th Ed., NFPA, Boston, 1969, p. 3-7.

[4]NFPA 70, *National Electrical Code,* NFPA, Boston, 1978.

[5]NFPA 101, *Code for Safety to Life from Fire in Buildings and Structures,* NFPA, Boston, 1976.

[6]NFPA 1, *Fire Prevention Code,* NFPA, Boston, 1975.

[7]NFPA 58, *Standard for the Storage and Handling of Liquefied Petroleum Gases,* NFPA, Boston, 1976.

[8]*Fire Prevention Code,* American Insurance Association, Engineering and Safety Services, New York, 1976.

[9]*Uniform Fire Code,* 1976 Ed., International Conference of Building Officials and Western Fire Chiefs Association, Whittier, CA, 1976.

[10]BOCA *Basic Fire Prevention Code,* 2nd Ed., Building Officials & Code Administrators International, Inc., Chicago, 1970.

[11]*Southern Standard Fire Prevention Code,* 1974 Ed., Southern Building Code Congress, Birmingham, 1974.

[12]NFPA 206M, *Guide on Building Areas and Heights,* NFPA, Boston, 1976.

[13]*Uniform Building Code,* 1976 Ed., International Conference of Building Officials, Whittier, CA, 1976.

[14]*Standard Building Code,* 1976 Ed., Southern Building Code Congress International, Inc., Birmingham, 1976.

[15]BOCA *Basic Building Code/1975,* 6th Ed., Building Officials and Code Administrators, Chicago, 1975.

[16]*National Building Code,* 1976 Ed., American Insurance Association, Engineering and Safety Services, New York, 1976.

[17]*Uniform Building Code,* 1976 Ed., International Conference of Building Officials, Whittier, CA, 1976.

[18]*Standard Building Code,* 1976 Ed., Southern Building Code Congress International, Inc., Birmingham, 1976.

[19]BOCA *Basic Building Code/1975,* 6th Ed., Building Officials and Code Administrators, Chicago, 1975.

[20]NFPA 13, *Standard for the Installation of Sprinkler Systems,* NFPA, Boston, 1976.

[21]NFPA 30, *Flammable and Combustible Liquids Code,* NFPA, Boston, 1977.

[22]NFPA 1122L, *Code for Unmanned Rockets,* NFPA, Boston, 1976.

[23]NFPA 656, *Standard for the Prevention of Dust Ignitions in Spice Grinding Plants,* NFPA, Boston, 1971.

[24]NFPA 54, *National Fuel Gas Code,* NFPA, Boston, 1974.

[25]NFPA 10, *Standard for Portable Fire Extinguishers,* NFPA, Boston, 1978.

[26]NFPA 1901, *Standard for Automotive Fire Apparatus,* NFPA, Boston, 1975.

[27]NFPA 501C, *Standard for Recreational Vehicles (Travel Trailers, Motor Homes, Truck Campers, Camping Trailers),* NFPA, Boston, 1977.

[28]NFPA 74, *Standard for Household Fire Warning Equipment,* NFPA, Boston, 1978.

Chapter Twelve

# Fire Protection Organizations, Information Sources, and Career Opportunities

*A direct result of the modern world's complexity is an increase in fire hazards. The need for reliable information about fire protection has engendered a number of organizations and information sources whose basic function is ensuring firesafety for people and property. The increase in fire hazards has also brought about the need for specialized fire science careers.*

## THE SCOPE OF FIRE PROTECTION ORGANIZATIONS

The scope of fire loss in the United States has brought about the development of many organizations and agencies that are directly or indirectly involved with the national fire problem. These groups range from fire service organizations such as the International Association of Arson Investigators, which has a direct relationship with the fire problem, to federal agencies such as the Department of Agriculture and the Department of Health, Education, and Welfare, which have fire protection interests as an indirect part of their overall objectives.

Some of these organizations prepare materials for fire service personnel, fire science students, the general public, and for other organizations concerned with fire protection. Still others test materials for their firesafety and burning characteristics. The results of such testing and research efforts are used to help solve the national fire problem.

This chapter discusses the organizations that are involved in fire protection and fire prevention activities, presents career opportunities in the fire science field, and identifies some of the major sources of firesafety information. For purposes of this text, these organizations are divided into three major categories: (1) organizations with fire protection interests, (2) federal agencies involved in firesafety, and (3) fire testing and research laboratories.

## ORGANIZATIONS WITH FIRE PROTECTION INTERESTS

Private, national, and international organizations with fire protection interests provide a liaison between the public and private sectors, and the academic, professional, and technical communities. Because fire is a universal threat, it is natural that a large number of organizations take an active interest in fire protection and control. The degree of interest and activity varies widely among different organizations, and a large number of them can be included in one of the following classifications: educational and technical membership organizations; insurance organizations; fire service organizations; noncommercial, trade, professional, or labor organizations; and international organizations. Each type fulfills its objectives by providing some form of firesafety awareness or activity.

### The National Fire Protection Association

The National Fire Protection Association (NFPA), organized in Boston in 1896, is a scientific and educational membership organization concerned with the causes, prevention, and control of destructive fires. According to its Articles of Organization, the NFPA was organized "to promote the science and improve the methods of fire protection, to obtain and circulate information on the subject, and to secure the cooperation of its members and others in matters of common interest." Incorporated in 1930 under the laws of the state of Massachusetts, the NFPA is a private, voluntary, charitable, and tax-exempt organization that serves on a worldwide basis as a primary source of fire protection information. The Association's activities may be summarized as follows: (1) technical standards development, (2) information exchange, (3) technical advisory services, (4) public education, (5) research, and (6) services to public protection agencies.

*Technical Standards:* Technical standards are developed by the more than 150 committees of the NFPA. These committees provide a balanced representation of the organizations, companies, etc., that are affected by the standards. Committee members serve voluntarily and consist of more than 2,400 experts from many areas concerned with fire protection. Standards are approved by the NFPA membership at either the Association's Annual or Fall Meeting and, once approved, are published and made available for voluntary adoption by any organization or jurisdiction with power of enforcement. (The NFPA standards-making process is discussed in greater detail in the preceding Chapter Eleven, "Codes and Standards.")

*Information Exchange:* Information on significant fires, fire-related developments, and the latest techniques in fire protection, prevention, and suppression is communicated to NFPA members and interested persons by means of various publications, including: *Fire Journal, Fire Command, Fire Technology, Fire News, Technical Committee Reports, Technical Committee Documentation,* presentation of general and technical papers at the Fall and Annual Meetings of the Association, the annual *Yearbook and Committee List,* audiovisual materials, handbooks and texts, and the NFPA *Fire Protection Handbook.*

*Technical Advisory Services:* To promote the use of NFPA standards and to provide guidance for their application, special services are established by the NFPA in certain areas of firesafety. The advisory services program is supported by numerous publications, such as technical books, pamphlets, and periodicals, all of which cover a wide range of fire protection topics.

*Public Education:* Within its public education program, the NFPA develops, produces, and distributes educational materials that contain guidelines and instructions for the general public. In addition to such nationally known features as Sparky the Fire Dog, Fire Prevention Week, and Operation EDITH (Exit Drills in the Home), firesafety messages are carried on television, film, and in print. (See Fig. 12.1.)

Fig. 12.1. *Public firesafety educational materials of the National Fire Protection Association: (left) a still from the film, "When Your Clothing Burns. . ."; (center) Sparky the Fire Dog; and (right) a poster included in firesafety educational materials.*

## 294　Principles of Fire Protection

***Services to Public Protection Agencies:*** The NFPA provides information and advice to fire departments, state fire marshal offices, municipal, state, and federal authorities, and to other agencies and individuals involved in public fire protection. Informational records and data related to fire service operations, management, and fire prevention inspection and enforcement are also developed. The NFPA additionally encourages education, training, and professional advancement in the fields of public fire protection.

***Research:*** Data collection and analysis, as well as a number of applied research programs, make up the NFPA's research activities. Reports on fires are analyzed, then classified by property type, hazard and cause, and fire service-related information. NFPA fire data is utilized by members of standards committees and in nearly all Association activities. Agencies of the federal and state governments, as well as educational and research institutions, also rely on NFPA fire data.

The NFPA's applied research programs are usually funded by government agencies. These programs range from the development and implementation of training programs and seminars to the design of management models, including the design and implementation of data systems.

***Society of Fire Protection Engineers (SFPE):*** This society was organized in 1950. Membership is limited to qualified professional fire protection engineers. The Society promotes recognition of the art of fire protection engineering, sponsors local chapters, publishes a bulletin and technical reports for its members, and holds meetings and technical seminars. (See Fig. 12.2.)

## *Fire Service Organizations*

On the North American continent there are a large number of prominent associations that serve the needs of the fire service. Among them are:

***Fire Marshals Association of North America (FMANA):*** FMANA is an organization of state, provincial, and city fire marshals, heads of municipal fire prevention bureaus, fire investigators, and their staffs. Organized in 1906, the FMANA (see Fig. 12.2) became a section of the NFPA in 1927 and adopted its constitution and by-laws in 1947, wherein the following objectives are listed:[1]

    (a) To unite, for mutual benefit, those public officials engaged primarily in the control of arson or the prevention of fire, or both.

    (b) To act as a central agency for exchange of professional information among its members.

    (c) To assist fire marshals in the conduct of their professional activities.

    (d) To correlate the activities of fire marshals toward the reduction of fire waste.

*Fig. 12.2. Emblems of: (left) the Fire Marshals Association of North America, (center) the Society of Fire Protection Engineers, and (right) the National Fire Prevention and Control Administration.*

Members of the association are the statutory officials and their full-time salaried deputies and assistants, while its associate members are legally designated nonsalaried or part-time officials who may enjoy all the privileges of membership except that of elective office. The FMANA publishes a "Bulletin" and a *Yearbook* and holds two meetings annually. Quite often the FMANA and its members serve as advisors to fire-related projects, hearings, and code problems.

***Fire Service Section (A Section of the NFPA):*** Organized in 1973, this organization is open to any NFPA member who: (1) is a member of a fire department that provides public fire prevention and fire suppression services to a state, county, municipality, organized fire district, airfield, or military base, or (2) is engaged in training or educating fire department members. Its objectives are:[2]

- To unite for mutual benefit NFPA members who work in the fire service.
- To act as a means of information exchange among its members.
- To advance interest in such fields as fire protection, prevention, and suppression.
- To stimulate awareness of the need for management, training, and education programs.
- To encourage and assist its members in conducting meetings for the exchange of information.
- To encourage its members to participate on NFPA Technical Committees.
- To advance the development of fire suppression equipment.
- To encourage public authorities to purchase fire protection equipment on the basis of performance standards.
- To bring to its members' attention regulations and matters of legislation that would be of interest.
- To promote cooperation within the fire service.
- To help the fire service to protect life and property from fire.
- To provide a professional fire service society.

***International Association of Fire Chiefs (IAFC):*** This group was organized in 1873 ". . . to further the professional advancement of the fire service and to insure and maintain greater protection of life and property from fire."[2] The Association holds an annual conference and exhibit of fire apparatus and equipment, and publishes material on fire service matters for its members and other interested parties.

***International Association of Fire Fighters (IAFF):*** Since 1918, the IAFF has served permanent, paid employees of fire departments who are engaged in fire fighting and fire prevention. The Association is affiliated with the AFL and CIO, and membership totals between 160,000 and 170,000. The Association's official monthly publication is "The International Fire Fighter." Other publications include materials intended to promote the fire service, guide union fire fighters in their public relations, and present informative topics of special interest to fire fighters (such as the content of the publication titled "Annual Death and Injury Survey").

***International Association of Arson Investigators (IAAI):*** Membership in the IAAI is open to anyone, twenty-one years of age or over, who is actively engaged in some phase of arson suppression and whose qualifications are found suitable to the Membership Committee of the Association. The IAAI publishes a quarterly bulletin titled "The Fire and Arson Investigator," and holds an annual meeting in conjunction with an arson conference.

***International Association of Black Professional Fire Fighters (IABPFF):*** This association was organized in 1970 to encourage the recruitment and employment of qualified black people as fire fighters in paid fire departments, and to serve as a liaison between black fire fighters across the nation. As a life member of the National Association for the Advancement of Colored People (NAACP), the organization seeks to promote interracial progress throughout the fire service.

***International Fire Service Training Association (IFSTA):*** This educational alliance, founded in 1943, was organized to develop training materials for the fire service. Its manuals, published by the Fire Protection Publication Division of Oklahoma State University at Stillwater, are widely used in the United States and Canada.

***International Municipal Signal Association (IMSA):*** This educational, non-profit organization is dedicated to imparting knowledge, technical information, and guidance to municipal signal and communication department heads and their first assistants. The types of communications covered include traffic control, fire alarm, and police alarm. ISMA publishes a bimonthly titled *ISMA Signal Magazine.*

***International Society of Fire Service Instructors (ISFSI):*** This society provides a forum for the continuing exchange of ideas and discussion of training techniques among those persons involved with fire service training and education. The society's goals are: (1) to provide uniform standards for fire service instructors; (2) to help develop the instructors themselves by means of sound training and educational opportunities; and (3) to actively promote the role of the fire service instructor.

*Joint Council of National Fire Service Organizations (JCNFSO):* This is an alliance of the principal national fire service organizations. Represented in it are delegates from the fire service associations previously described (the FMANA, IAAI, IABPFF, IAFC, IAFF, IFSTA, IMSA, ISFSI, NFPA, and the Metropolitan Committee of the IAFC).

The Joint Council was formed to identify problems common to each organization, to review current developments of concern to all, and to establish areas of common interest where cooperative efforts of member organizations can be used to produce desirable results. One area of common interest in which national collective action was deemed desirable was the establishment of standards upon which the levels of competency of personnel within the fire service could be determined. Committees have been created to develop recommended minimum standards of professional competence for: (1) fire fighters, (2) fire inspectors and investigators, (3) fire service instructors, and (4) fire service officers.

The Joint Council has also established a National Professional Qualifications Board for the Fire Service to supervise the program. Its other functions are to identify and define levels of professional progression, to review the work of the Committees on professional standards, and to be responsible for the accreditation and supervision of national programs for certification. The Secretariat for the Committees and the Board is provided by the staff of the National Fire Protection Association.

## Fire Testing and Research Laboratories

There are many laboratories in the United States capable of performing, in varying degrees, fire tests of materials and/or equipment; many of these same laboratories, as well as other laboratories, have facilities for conducting fire-related research work. Generally, these laboratories can be classified into three categories: (1) private and industrial laboratories, (2) university laboratories, and (3) government laboratories.

Specific work currently being done by any of the laboratories mentioned herein will not be listed, since such a cataloging would be quickly outdated. Also, fire-related testing and research may be just one of a number of capabilities possessed by some of the laboratories named; for many, fire testing may just be a small part of their overall work. A *Directory of Fire Research in the United States*[3] has been published periodically by the National Academy of Sciences. This publication can be a valuable source for specific

## 298    Principles of Fire Protection

details about any of the organizations discussed here, as well as any other testing and research organizations in existence.

***Private and Industrial Laboratories:*** In the United States there are approximately sixty-five private and industrial laboratories that perform a wide range of fire tests. Space does not permit that each be described in detail. However, there are two — Underwriters Laboratories Inc. and Factory Mutual Laboratory Facilities — whose work warrants particular emphasis.

1. *Underwriters Laboratories Inc. (UL).* Underwriters Laboratories was founded in Chicago in 1894, with its sole objective being the promotion of public safety through conduct of ". . . scientific investigation, study, experiments, and tests to determine the relation of various materials, devices, products, equipment, construction, methods, and systems to hazards appurtenant thereto or to the use thereof affecting life and property and to ascertain, define, and publish standards, classifications, and specifications for materials, devices, products, equipment, construction, methods, and systems affecting such hazards, and other information tending to reduce or prevent bodily injury, loss of life, and property damage from such hazards."

Annually, UL publishes lists of manufacturers whose products, when tested, have proved acceptable under appropriate standards, and which are subjected

*Fig. 12.3.    Representative labels of Underwriters Laboratories Inc.*

Fig. 12.4. Representative labels of the Factory Mutuals.

to one of the follow-up services provided by the Laboratories as a countercheck. The word "listed" appears on UL labels attached to these products as authorized evidence that these products have been found to be in compliance with the Laboratories' requirements. (See Fig. 12.3.)

2. *Factory Mutual Systems (FM)*. Factory Mutual, founded in 1835 to provide mutual fire insurance coverage, functions to minimize fire and extended coverage losses and interruption of production, and to provide insurance at cost through loss prevention inspections of plants, research, and consultation services to its insureds. The Factory Mutual Research staff in Norwood, Massachusetts, consists of the Standards, Research, and Approvals Groups. Factory Mutual maintains testing facilities in Norwood, Massachusetts, and also conducts large-scale applied research in its one-acre, 60-foot-high FM Test Center in West Glocester, Rhode Island. Factory Mutual Laboratory Facilities are available on a contract basis through Factory Mutual Research. Some of Factory Mutual's labels are shown in Fig. 12.4.

***University and Federal Government Laboratories:*** More than forty American colleges and universities are equipped with laboratories for fire testing and research. In addition to the colleges and universities that serve primarily as institutions for fire science training and education, there are others, both private and state-supported, whose engineering, physics, or science departments engage in such activities.

Several departments of the federal government — those of Agriculture, Air Force, Army, Commerce, Navy, Transportation, and independent agencies — also have research laboratories located throughout the country. These facilities are a direct result of an increasing national interest in firesafety as well as other safety- and health-related issues.

## *Insurance Organizations*

Many important groups perform varied fire protection and inspection services on behalf of the insurance industry and its insureds. For example, the Association of Mill and Elevator Mutual Insurance Companies serves the mill and elevator industry's needs; the American Institute of Marine Underwriters is organized to serve the marine underwriters and to promote, advance, and protect their interests.

There are five large insurance organizations, however, that serve a wide range of casualty and property insurers and contribute to fire protection in many ways. They are: (1) American Insurance Association, (2) American Mutual Insurance Alliance, (3) Factory Mutual System, (4) Industrial Risk Insurers, and (5) Insurance Services Office.

*American Insurance Association (AIA):* This organization was created when the National Board of Fire Underwriters, the Association of Casualty and Surety Companies, and the former American Insurance Association merged in September, 1964. AIA is a trade association that services a large number of companies operating in the property and casualty insurance fields for which it provides legislative services, engineering and safety services, and research and review of claim and loss adjustment functions.

The Engineering and Safety Service of the American Insurance Association offers guidance in fire protection, environmental protection, industrial hygiene, product safety, building code requirements, and accident prevention. The Service also issues eighteen series of bulletins for subscribers; one series, known simply as "Special Interest Bulletins," is available to all fire protection interests. AIA publishes the *National Building Code,* a *Fire Prevention Code,* other suggested codes and standards, and general educational materials dealing with fire prevention.

*American Mutual Insurance Alliance (AMIA):* Organized in 1922, the AMIA is a national organization whose members are mutual fire and casualty insurance companies. It provides engineering, legal, legislative, educational, and public relations services for its member companies. Fire prevention studies are carried out through the Alliance's Accident and Fire Prevention Department, and technical training sessions on different aspects of fire protection engineering are organized by the Alliance for instructing new company inspection and engineering personnel. A great deal of firesafety educational material, as well as a quarterly *Journal,* are issued by the Alliance.

*Factory Mutual System (FM):* Founded in 1835 to provide fire insurance coverage for large industrial and commercial properties in the United States and Canada, the Factory Mutual System is composed of four mutual property insurance companies and two System associates. It is now multinational in scope and works to minimize fire and extended coverage losses, interruption of production, and to provide insurance at cost through loss prevention inspections. The System is also well-known for its testing and approval services. ("Approval" refers to a product's having been tested by Factory Mutual Standards and found suitable for general application.)

Factory Mutual System issues *Loss Prevention Data* books that contain recommendations and information for preventing fire, explosions, wind damage, electrical breakdowns, and boiler and machinery accidents. Factory Mutual also publishes numerous other loss prevention materials and the bimonthly *Record,* a publication that covers property conservation engineering and management.

***Industrial Risk Insurers (IRI):*** This group represents a merger of Factory Insurance Association and the Oil Insurance Association, which became effective on December 1, 1975. The IRI writes insurance against fire and allied perils for selected industrial properties. Its inspectors visit plants on a scheduled basis to recommend improvements in loss prevention and control. A training laboratory is maintained to benefit its engineers and to familiarize representatives of its policyholders with fire protection equipment. Material on industrial fire protection and a magazine called *Sentinel* are published.

***Insurance Services Office (ISO):*** This association, founded in 1971, is a national association established by the property and liability insurance industry to provide the insuring public, insurance regulatory bodies, and the insurance industry with a full range of insurance services. ISO is licensed in all 52 states and territories. It is licensed as a fire-rating organization in most of the jurisdictions, and operates as an advisory organization to state fire-rating organizations in the other jurisdictions. It also acts as an advisory organization for other property-liability lines of insurance in those states in which statutory bureaus are operative. ISO renders a wide range of advisory, actuarial, rating, statistical, research, and other types of services covering thirteen lines of insurance. Most of the former state and regional fire-rating organizations are now a part of the ISO.

## Other Organizations Having Fire Protection Interests

According to one estimate, in the United States alone there are over 5,000 noncommercial, trade, professional, and labor organizations. A large number of them have some degree of fire protection concerns related to their activities. These organizations cover an extraordinarily wide range of interests — hospitals, transportation, construction, chemistry, materials, electronics, and education, to mention but a few.

The scope of interest in firesafety is worldwide. Outside the United States there are many national and international fire organizations, far too numerous to list within the confines of this textbook. Although the political ideologies of some of these organizations may vary and even seem to be at odds with one another, all of them are united in their efforts to protect human life and property from fire.

# FEDERAL AGENCIES INVOLVED IN FIRE PROTECTION

For many years, federal agencies have been involved in fire protection, usually in such nonregulatory roles as providing for protection of federal property, employees, and the public while on federal property. In recent years, though, federal regulatory authority has been increased. Assistance programs that mandate compliance with regulations as a condition of participation have been expanded significantly, and new federal authority designed to protect the mass public or certain segments of society is having a major impact. The various federal agencies directly involved with firesafety have headquarters in the Washington, DC area, and most oversee eight to ten regional and field offices in major cities throughout the country.

## *Department of Commerce*

In 1974 Congress established the first major federal program aimed at firesafety for the general public, with direct concern for fire departments. The Federal Fire Prevention and Control Act of 1974 also provides for the reimbursement to communities and local fire departments of expenses incurred in fighting fires on federal property. Two of the agencies within the DOC concerned with firesafety are the National Fire Prevention and Control Administration and the National Bureau of Standards.

***National Fire Prevention and Control Administration (NFPCA):*** Fire service education and training, establishment of a national fire academy, and support to state and local training programs are considered top priorities by the NFPCA. (See Fig. 12.2.) The report, "America Burning," emphasizes that fire prevention and control are primarily local responsibilities, and that through codes and firesafety laws, local governments ". . . have shouldered the major burden of protecting citizens from fire."[4] The report recommends that the federal government should lend technical and educational assistance to state and local governments.

The NFPCA was created to perform these functions. Also, it operates the Federal Fire Council, which for many years has provided a focal point for data collection, information dissemination, and cooperation among federal agencies in reducing losses on federal property.

***National Bureau of Standards (NBS):*** Since the early 20th century, the Commerce Department's National Bureau of Standards has been engaged in fire research primarily directed toward building technology and fire tests of materials. The NBS serves as the Secretariat for the National Conference on State Building Codes and Standards, which promotes adoption and enforcement of uniform standards.

## *Department of Labor*

In 1970 the DOL established the Occupational Safety and Health Administration (OSHA) to adopt, develop, promulgate, and enforce mandatory standards that provide for the safety and health of employees in places of employment. OSHA adopted by reference fifty NFPA standards as mandatory firesafety provisions under early authority of the Occupational Safety and Health Act of 1970. OSHA has engaged in a wide variety of educational and training programs for federal and state compliance officers and industry personnel through the OSHA Training Institute near Chicago, and through field training programs and short courses at colleges and universities.

## Department of Agriculture

The fire protection programs promulgated by the Department of Agriculture are carried out mainly by the U.S. Forest Service and the Farmers' Home Administration. Both of these agencies play an important role in fire prevention education programs in rural areas.

*U.S. Forest Service (USFS):* Known to millions through its familiar symbol of Smokey the Bear, this agency provides protection to over 200 million acres of forests, grasslands, and nearby private lands, as well as technical and financial assistance to all states for protecting federal and nonfederal land. The Forest Service conducts and sponsors research in fire prevention and suppression, fuel management and modification, and weather modification. Research is also conducted on wood and wood-based products.

*Farmers' Home Administration:* The Farmers' Home Administration makes loans to rural areas for improving necessary community facilities such as water supplies, fire stations, and apparatus and equipment.

## Department of Housing and Urban Development

One- and two-family dwellings, multifamily housing, and care-type housing are covered by the *Minimum Property Standards* promulgated and enforced by the Department of Housing and Urban Development (HUD). HUD insurance on mortgages and loans and direct assistance programs are based on compliance with these standards. HUD has sponsored research in support of its standards-making activities and improving community services, such as development of Fire Station Location Models and the *Uniform Fire Incident Reporting System* (UFIRS), the latter with the NFPA.

HUD's Federal Insurance Administration (FIA) has cooperated with the insurance industry to provide fire insurance, particularly in riot-prone areas where commercial insurance was not available. Also under HUD comes the Federal Disaster Assistance Administration, which provides assistance to public agencies and individuals who have suffered property damage in emergencies and major disasters, including fires and explosions, when such incidents are declared major disasters by the President.

## Department of the Interior

Within the Department of the Interior are three agencies — the Bureau of Land Management (BLM), the Bureau of Indian Affairs (BIA), and the National Park Service (NPS) — under whose jurisdiction is fire protection for some 545 million acres of land. On these lands, the majority of which lie in remote areas of western states, the department provides for all management, surveillance, and suppression services.

The control of fires in mines is a major responsibility of the Bureau of Mines (BUMINES), another agency within the Department of the Interior. Research carried out by BUMINES has contributed a great deal to mine safety and firesafety in mines. BUMINES is also involved in testing and approving safety equipment for mine use, such as gas detectors, breathing apparatus, and fire extinguishing equipment.

## Department of Defense

All military departments provide for fire protection of DOD facilities, such as vehicles, ships, aircraft, and personnel. The military operates a number of fire training schools, and has made major contributions to fire protection by means of in-house and contract training programs.

## General Services Administration

This agency provides supplies and buildings to federal agencies. GSA has made major advances in the development and adoption of a systems approach to firesafety in buildings. Its standards are likely to be a major factor in fire protection features of buildings in the future.

## Department of Transportation

Major regulatory authority for all modes of transportation comes under the Department of Transportation (DOT), specifically under the *Department of Transportation Act of 1966* and the *Transportation Safety Act of 1974*. The authority for the construction of both private and commercial vehicles, the operation of commercial vehicles, the collection of accident data, the establishment and enforcement of standards and regulations, and the regulation of the transportation of hazardous materials is provided for by laws that establish programs within the Department of Transportation. For example, the Materials Transportation Bureau has regulatory authority for the transportation, storage, and labeling of hazardous materials, and the Department's Urban Mass Transportation Administration conducts research and provides financial assistance in establishing mass transit systems.

## Other Federal Agencies

Among the other federal agencies concerned with fire prevention and protection are: the Maritime Administration, which sponsors research in shipboard fire hazards and helps to improve the fire protection of port cities; the Federal Communications Commission, which exercises federal regulatory authority over assignment of frequencies and operation of fire, rescue, and medical services communications; the Federal Trade Commission, which investigates safety claims for products; the National Aeronautics and Space Administration, which does useful research on detection devices, fire-retardant and fire-resistive materials, insulating materials, and personnel protective equipment; the Treasury Department, which provides revenue-sharing funds for public safety projects including fire equipment for volunteer fire departments; and the Veterans Administration, which looks after firesafety in its hospitals and has sponsored research in the flammability of materials, fire test methods, and in the treatment of burn victims. Table 12.1 shows a comprehensive listing of federal agencies that are involved with fire protection.

## CAREER OPPORTUNITIES IN THE FIRE SCIENCES

Fire can be one of the most dangerous and costly of all destructive forces. Because of the hazard that fire presents to both life and property, the scientific study of fire protection and the methods of fire prevention have grown since the early 19th century. From this growth, and from recent technological advances in building design and the increasing use of hazardous materials, the fire protection profession has become highly specialized and challenging.

It is beyond the scope of this chapter to detail each and every career that a student of the fire sciences may enter. Many types of persons with fire science backgrounds are hired by insurance companies, building departments, and divisions of the state government, whose responsibility is to provide bases for evaluating danger to life from fire and for evaluating and determining ways for minimizing fire danger to buildings and their contents. Architectural firms often either hire or consult fire science professionals to examine the firesafety of proposed building designs, and a great many industries have their own fire protection staffs and private fire brigades.

The increase in fire science specialty fields has brought about a greater need for education and training. Since the early 1960s, there has been a rapid increase in the number of two-year community college fire science programs in the United States. Community college fire science programs are usually designed to supplement, but not replace, fire training provided by fire departments and state training agencies.[5] Four-year fire science programs offered by universities, however, expand upon the traditional career of fire fighter by offering degrees in fire protection engineering and technology that can help qualify individuals for work in fire protection fields other than fire depart-

## Table 12.1 Federal Agencies Involved in Fire Protection*

| Department or Agency | Key Organizational Elements |
|---|---|
| Department of Agriculture | Farmers Home Administration<br>U.S. Forest Service (USFS) |
| Civil Service Commission (CSC) | |
| Department of Commerce (DOC) | Maritime Administration (MARAD)<br>National Bureau of Standards (NBS)<br>National Fire Prevention and Control Administration (NFPCA)<br>Economic Development Administration (EDA) |
| Consumer Product Safety Commission (CPSC) | |
| Department of Defense (DOD) | Civil Defense Preparedness Agency (CDPA)<br>Military Departments (Air Force, Army, Marine, Navy) |
| Environmental Protection Agency (EPA) | |
| Federal Communications Commission (FCC) | |
| Federal Trade Commission (FTC) | |
| General Services Administration (GSA) | Federal Supply Service<br>Public Buildings Service |
| Department of Health, Education, and Welfare (HEW) | National Institutes of Health (NIH)<br>National Institute of Occupational Safety and Health (NIOSH)<br>Office of Education<br>Public Health Service (PHS)<br>Social Security Administration (SSA) |
| Department of Housing and Urban Development (HUD) | Federal Housing Administration (FHA)<br>Federal Disaster Assistance Administration |
| Department of the Interior | Bureau of Indian Affairs<br>Bureau of Land Management (BLM)<br>Bureau of Mines (BUMINES)<br>Mining Enforcement and Safety Administration (MESA)<br>National Park Service (NPS) |
| Department of Justice | Law Enforcement Assistance Administration (LEAA) |
| Department of Labor (DOL) | Bureau of Labor Statistics (BLS)<br>Wage and Manpower Division<br>Occupational Safety and Health Administration (OSHA) |

*(Continued)*

---

*From *Fire Protection Handbook,* 14th Ed., NFPA, Boston, 1976, p. A-25.

Table 12.1    Federal Agencies Involved in Fire Protection* *(Continued)*

| Department or Agency | Key Organizational Elements |
|---|---|
| National Academy of Sciences (NAS) (Quasi-government) | |
| National Academy of Engineering (NAE) (Quasi-government) | |
| National Research Council (Quasi-government) | |
| National Aeronautics and Space Administration (NASA) | |
| National Science Foundation (NSF) | |
| National Transportation Safety Board (NTSB) | |
| Nuclear Regulatory Commission (NRC) | |
| Energy Research and Development Administration (ERDA) | |
| Small Business Administration (SBA) | |
| Department of Transportation (DOT) | Federal Aviation Administration (FAA) Federal Highway Administration (FHWA) Federal Railroad Administration (FRA) Materials Transportation Bureau (MTB) National Highway Traffic Safety Administration (NHTSA) U.S. Coast Guard (USCG) Urban Mass Transportation Administration (UMTA) |
| Department of the Treasury | Bureau of Alcohol, Tobacco, and Firearms |
| Veterans Administration (VA) | |

ments. The requirements for a fire science degree and the career opportunities in the fire sciences are presented in the following paragraphs.

## *Fire Science Education*

Many community colleges throughout the country offer associate degrees in the fire sciences. Three universities — Illinois Institute of Technology, the University of Maryland, and Oklahoma State University at Stillwater — offer four-year bachelor of science programs aimed at preparing individuals for careers in fire protection engineering and technology.

The Illinois Institute of Technology offers a bachelors degree in Fire Protection and Safety Engineering, which many students supplement with a Master of Business Administration degree. The fire protection engineer (FPE) works with architects, city planners, ecologists, and sociologists in trying to solve fire hazards in areas such as large cities, rural towns, suburban communities, space platforms, and ocean depths. FPEs can find careers in three major areas: (1) industry, (2) research and product design, and (3) sales and insurance.

A degree in Fire Protection Engineering is offered by the University of Maryland upon completion of a curriculum that provides the student with an understanding of engineering, physics, mathematics, and the major aspects of fire protection engineering. Among the various fields in which fire protection engineers can work after graduation are industry, fire protection organizations, insurance companies, automatic sprinkler companies, the military, and the federal government.

Oklahoma State University at Stillwater provides a Bachelor of Science degree in Fire Protection and Safety Technology. During the first two years of study, emphasis is placed on the recognition, evaluation, and correction of fire hazards. After this two-year study, an associate degree is awarded. Junior and senior years of study concentrate on industrial safety, hygiene, and security, as well as other advanced subjects that are intended to ensure the competence of the graduate in the professional world.

## Fire Protection Careers

Safeguarding people and property from the ravages of fire offers rewarding career opportunities to large numbers of people with many and varied skills. Thus, for today's fire science student, there are many types of careers in fire protection available.

People are needed to fight fires in homes, industrial and mercantile properties, farms and forests, planes and ships and tank trucks, hospitals and nursing homes, and all other places where people live and work and play.

People are needed to prevent fires from starting by making fire inspections, and by enforcing the laws and codes enacted at all levels of government.

People are needed for fire prevention education of adults and school children. They are needed to investigate fires and to control arson, and to train others in fire fighting procedures and fire prevention measures.

People are needed to conduct the fire insurance business, to make rates, to write policies, and to remove hazards.

And still others are needed to design, manufacture, and sell fire apparatus, automatic sprinklers and other extinguishing systems, alarm systems, fire doors, first aid fire appliances, fire-retardant materials, etc.

Fire protection engineers are needed in ever-increasing numbers to cope with ever-greater fire protection problems such as those involving high-rise buildings, new chemical hazards, rapid increase in the use of plastics with accompanying smoke and toxic fume hazards, automated warehouses, etc.

Among the many professionals who play a major role in solving the growing number of fire protection problems is the fire protection engineer. Most fire protection engineers receive their undergraduate educations from accredited colleges or universities. The principal areas in which a fire protection engineer may work are industrial engineering, research and product design, and sales and insurance.

For the fire protection engineer, industrial engineering usually involves direct problem solving; the engineer evaluates information to establish the extent of fire and explosion potential, and may also design systems such as automatic sprinklers, water supplies, fire alarm and control systems, and municipal fire defenses for large cities. In the areas of research and product design, the fire protection engineer is concerned with such diverse problems as fabric flammability, firesafety in high-rise buildings, and protection of automated storage warehouses. Fire protection engineers are also utilized by the fire insurance industry where they are involved with evaluating fire risks, studying fire losses, underwriting analysis, and with management concerns. Some fire protection engineers become involved in selling fire protection and extinguishing equipment systems.

## Careers in a Fire Department

In addition to the various levels of fire fighters and the various ranks of officers, today's fire departments have the need for individuals trained to perform specialized functions that require mechanical, automotive, and communications skills. Such positions include: motor pump operator, chief dispatcher, fire alarm dispatcher, automotive mechanical supervisor, fire equipment mechanic, and emergency medical technician (EMT). Depending upon the geographical location of the fire department and the specific fire hazards of a community, there may be a need for fireboat pilots, marine engineers, airport fire investigators, and airport fire fighters.

Different communities have specific fire protection needs; thus, qualifications for the preceding positions are determined at the local level. National standardized qualifications do exist, however, for the fire fighter and the apparatus driver/operator, and are published in NFPA 1001, *Fire Fighter Professional Qualifications*,[6] and NFPA 1002, *Fire Apparatus Driver/Operator Professional Qualifications*,[7] details of which are given in the remaining paragraphs of this chapter. The qualifications are intended to ensure that any individual measured by them truly possesses the skills needed to perform the tasks required by each position. Because nationally recognized qualifications exist for fire fighter and apparatus driver/operator, these two careers will be examined in detail in the following paragraphs, as will be the position of fire department paramedic which, because of its recent conception, has become the subject of some controversy.

**Fire Fighter:** A fire fighter's main responsibility is to protect the public from the dangers of fire, to prevent loss of life from fire, and to minimize the property damage that fire can cause. This is accomplished through highly organized teamwork under the supervision of a commanding officer. Thus, the career of the fire fighter involves a group work activity: some fire fighters are assigned to forcible entry tasks, others to pumping apparatus, and others to the operation and placement of ladders. After the extinguishment of a fire, fire fighters remain on the scene to ensure that no further danger exists from rekindling. Therefore, while fire fighters may have specialized tasks on the fire scene, it is through the work of each individual that the entire unit accomplishes its task.

Between fires, fire fighters may be assigned to inspect public buildings for compliance with fire codes and safety regulations. During these inspections, fire fighters may apply their knowledge of firesafe buildings to ensure that exits, passageways, and hallways are not blocked, that combustibles are stored safely, and that any hazards or inoperable firesafety devices are corrected.

Fire fighter applicants must be at least eighteen years old and have a high school diploma or recognized equivalent. Because of the physical demands of the job, candidates must possess strength, endurance, dexterity, and the ability to transport heavy equipment under adverse conditions, and must be able to deal with injured and panic-stricken individuals and guide such persons to safety. The physical and mental demands of the job mandate a physical examination and an evaluation of the candidate's character before an applicant may be accepted into the fire department. NFPA 1001, *Fire Fighter Professional Qualifications,* details the medical requirements that should be met by fire department candidates.

In 1973 and 1974, the Fire Service Professional Standards Development Committee for Fire Fighter Qualifications concluded that fire fighter standards also relate to the airport fire fighter, the emergency medical technician, the alarm operator, the master mechanic, and the driver/operator.

*Fire Apparatus Driver/Operator:* NFPA 1002, *Fire Apparatus Driver/ Operator Professional Qualifications,* outlines the minimum requirements of professional competence that the individual seeking this position must possess. Before being certified as a driver/operator, the individual must already be a member of a fire department or fire brigade and must have fulfilled the requirements of Fire Fighter II, defined in NFPA 1001, *Fire Fighter Professional Qualifications,* as follows:[8]

> ... at the second level of progression in the fire department, [one] who has demonstrated the knowledge of, and the ability to perform the objectives specified in this Standard for that level, and who works under minimum supervision.

The driver/operator must be able to demonstrate the performance of routine tests, inspections, and servicing functions of fire department vehicles to assure that they are in good working order. This preventive maintenance includes such component parts as the battery; the braking, coolant, and electrical systems; engine and hydraulic oil levels; and tire care for all department vehicles. A knowledge is also needed of conditions that may result in pump damage or unsafe operation, and the ability to identify measures for the correction of such conditions.

The driver/operator must be legally licensed for the particular class of vehicle, and must have a thorough knowledge of the state and local laws governing the safe operation of the vehicles. A driving test must be passed in which the candidate displays the ability to maneuver the vehicle as prescribed in NFPA 1002.[9] The vehicles with which the driver/operator must be familiar include all fire department apparatus equipped with fire pumps, aerial ladders, tillers, and elevating platforms.

*Paramedic:* The use of fire fighters as mobile intensive care paramedics is a relatively new concept, and as such is not without a certain amount of controversy. Fire fighter-paramedic programs were initially developed because they were felt to be a less expensive alternative to medical doctors on ambulance and rescue units. However, there are administrators who feel that even fire fighters are too expensive and that the positions should be given to civilians who might not demand the salaries paid to fire fighters. Such feelings are often based on the consideration that a fire fighter is likely to pass a promotion examination and leave the paramedic position within a few years or even within several months. A contrasting viewpoint was put forth in an article in *Fire Command!:*[10]

> It is anticipated that the study habits acquired by the fire fighter-paramedics during their intensive twelve-week training program will pay big benefits as they prepare for future promotional exams. As they pass those exams and are elevated to company officer positions, they will take their paramedic skills and knowledge with them. They will thus be competent to train their subordinates in advanced lifesaving and first aid skills. In Los Angeles City and County, every engine company is equipped with a resuscitator, inhalator, aspirator, and other lifesaving and first aid equipment. If a medical emergency is reported at a location that is closer to the engine company quarters than to a rescue squad, the engine will be dispatched along with the rescue unit. Thus, the company officer must take initial action pending arrival of the rescue squad paramedics. If the officer has been trained and certified as a paramedic, there is a much greater chance of accurately diagnosing the appropriate first aid. . . . It can easily be seen that diffusing paramedic skills throughout all ranks and geographic areas of the department will greatly enhance the level of service in this significant portion of the department's emergency activity.

The paramedic must be able to determine the type and severity of the injuries present in order to establish priority for the emergency care required. Certain medical knowledge is needed, as the paramedic may be required to open and maintain an airway, control hemorrhages, assist at childbirth, perform initial care for burn victims, and treat shock. Under the auspices of a physician, the paramedic may also be required to administer drugs, including intravenous fluid.

## Summary

The basic goal of fire protection organizations is to help solve the country's fire problems. This can be accomplished in many ways, as reflected in the vast number of such organizations and in the various roles they play. While some groups primarily sponsor educational activities, both for members of the fire service and for the public-at-large, other associations provide insurance for property damages that may result from fires or explosions. Within the fire service itself are several national and international organizations that seek to unite fire officers and fire fighters, to provide special training for them, and to establish a forum for the exchange of ideas relating to fire protection, fire prevention and fire suppression.

Under many departments of the federal government are agencies with fire protection interests. These agencies perform a variety of services, from developing and enforcing standards and codes to regulating the transportation of hazardous materials. The federal government also sponsors fire testing and research laboratories, as do private industries, colleges, and universities throughout the country.

The diversity and specialization of careers available to those interested in the fire sciences is a reflection of the need for a thorough knowledge of fire and its properties. Additionally, the types of careers available in the fire service have expanded considerably in scope in order to keep pace with the staggering losses in life and property from fire.

## Activities

1. What are some of the ways that organizations with fire protection interests contribute to helping solve the national fire problem?
2. From the following list of fire protection careers, identify: (1) the training and/or qualifications needed for that particular career, and (2) the responsibilities of each job.
   (a) Fire protection engineer.
   (b) Fire fighter.
   (c) State fire marshal.
   (d) Paramedic.
   (e) Fire apparatus driver/operator.
3. The following insurance organizations serve a wide range of casualty and property insurers, and contribute to fire protection in many ways. Identify the types of fire protection services supplied by each.
   (a) American Mutual Insurance Alliance.
   (b) Industrial Risk Insurers.
   (c) Insurance Services Office.
   (d) Factory Mutual System.
   (e) American Insurance Association.

4. The Joint Council of National Fire Service Organizations has acted to establish standards upon which the levels of personnel competency within the fire service can be determined. What are the benefits of such action?
5. What federal agencies are responsible for firesafety in the following areas?
   (a) Mines.
   (b) Forests and grasslands.
   (c) Testing of materials.
   (d) Employee safety.
   (e) Rural fire prevention and protection.
   (f) Transportation of materials.
6. (a) In what ways has the Federal Insurance Administration cooperated with private insurance associations?
   (b) In what ways have colleges and universities kept pace with the specialized needs of the fire services?
7. What fire protection services do the following agencies provide? How do their services affect the general public?
   (a) The Bureau of Land Management.
   (b) The Maritime Administration.
   (c) The Veterans Administration.
   (d) The National Aeronautics and Space Administration.
8. (a) What are the types of research performed by the National Fire Protection Association?
   (b) How is this research utilized?
9. What services does Underwriters Laboratories Inc. provide to ensure the firesafety of consumer goods?
10. Explain why you feel that private fire protection organizations should or should not be replaced by similar organizations on the federal level, including in your explanation consideration of each of the following items:
    (a) Fire as a national problem.
    (b) The basic concerns of private fire protection organizations and of federal fire protection organizations.
    (c) The contributions of private fire protection organizations and of federal fire protection organizations.
    (d) The membership makeup of private fire protection organizations and of federal fire protection organizations.
    (e) The accomplishments of private fire protection organizations and of federal fire protection organizations.

## Bibliography

[1]"State Fire Marshals Conference Report," June 1977, prepared by the Fire Marshals Association of North America for National Fire Prevention and Control Administration, NFPA, Boston, p. 7.

[2]*Fire Protection Handbook,* 14th Ed., NFPA, Boston, 1976, p. A-6.

[3]*Directory of Fire Research in the United States,* Committee on Fire Research of the Division of Engineering, National Research Council, National Academy of Sciences, Washington, DC, 1975.

[4]"America Burning," May 1973, The National Commission on Fire Prevention and Control, Washington, DC, p. x.

[5]*Fire Protection Handbook,* 14th Ed., NFPA, Boston, 1976, p. 9-38.

[6]NFPA 1001, *Standard for Fire Fighter Professional Qualifications,* NFPA, Boston, 1974.

[7]NFPA 1002, *Standard for Fire Apparatus Driver/Operator Professional Qualifications,* NFPA, Boston, 1976.

[8]NFPA 1001, *Standard for Fire Fighter Professional Qualifications,* NFPA, Boston, 1974, p. 6.

[9]NFPA 1002, *Standard for Fire Apparatus Driver/Operator Professional Qualifications,* NFPA, Boston, 1976, pp. 24-27.

[10]James O. Page, "Why Fire Fighters?", *Fire Command!,* Vol. 39, No. 8, Aug. 1972, p. 31.

# SUBJECT INDEX

## A

Acrolein, as a fire gas, 64
Administration of fire departments, 248–253
Aerial ladders
  early use of, 12
  first, 12
*A History of the British Fire Service*, by G. V. Blackstone, 3, 7, 8
AIA (*see American Insurance Association*)
Alarms
  additional, Table 9.2
  fire department response to, 232
Alarm systems, provisions for, in *Life Safety Code*, 34
"America Burning"
  estimated building fire causes in, 24, Table 2.2
  goals of master planning in, 734
  property loss from fire in, 23, Table 2.1
  weaknesses in building design in, 142
American Insurance Association (AIA), 300
American Mutual Insurance Alliance (AMIA), 300
American National Standards Institute (ANSI), 284
American Society for Testing and Materials (ASTM), 285
AMIA (*see American Mutual Insurance Alliance*)
Ammonia
  as a fire gas, 63
  properties of, 63
Apartment buildings
  as light hazard occupancies, 171
  as residential occupancies, 38, 39
  definition of, 39
Apparatus
  basic unit for, 254
  breathing, 198
  electric lights and generators as, 197
  equipment carried on, 196–197
  for fire department, 232
  multiple, 227
  nineteenth century, 11
  power demands for, 197

rural, 228
*Aquarii*, duties of, 3
Arcing
  as source of ignition, 58
  description of, 58
Arson
  a growing problem, 132–133, Table 5.5
  criminal procedure for, 136
  definition of, 132
  degrees of, 132
  details for investigation of, 135
  determining fire by, 134
  Federal Fugitive Felon Act in relation to, 137
  fire cause investigation of, 204
  first degree, 136
  fourth degree, 137
  intent as component of, 134
  International Association of Arson Investigators (IAAI) for, 296
  investigation of, 133
  Model Arson Law for, 136
  needs areas for combatting, 138
  other laws related to, 137
  prevention of, 137–138
  proof of, 136
  requirements for conviction of, 136
  responsibility for investigations of, 135
  second degree, 136
  Statute of Limitations for, 137
  suppression of, 277
  third degree, 136–137
Arsonists, classification of, 134
Assembly occupancies
  fire hazards in, 41
  examples of, 41
Atkins, Thomas, as fire chief, 6
Augustus Caesar, fire protection during reign of, 3
*Automotive Fire Apparatus, Standard for*, 196–198
Automatic fire extinguishing systems
  carbon dioxide as, 183
  combustible metal as, 191
  dry chemical extinguishing systems as, 178
  foam extinguishing systems as, 181

*315*

halogenated agents and systems as, 184
sprinkler systems as, 168
standpipes as, 177
water as, 164
water spray fixed systems as, 179
*Automatic Sprinkler and Standpipe Systems*, by John L. Bryan, 9
Automatic sprinkler heads, by Henry S. Parmelee, 10, Fig. 1.2
*Automatic Sprinkler Protection*, by Gorham Dana, 9
Automatic sprinkler system
 by Philip W. Pratt, 9
 conception of, 9
 definition of, 9
 development of, 10
 first American, 9
 first patented, 9
 multicycle, 10
 operational definition of, 9
 Parmelee
  design of, 10
  installation of, 10
Automatic sprinkler systems (*see also sprinkler systems*)
 human action as reason for failure of, 125
 in large-loss fires, 124–125
 reasons for failure of, 125
 use of, in business occupancies, 43
Auxiliary equipment for fire department, 232
Auxiliary signaling systems, 211, 213

## B

"Babcock" fire engines, 12
Basic Casualty Report, information included in, 118
Basic Incident Report, information included in, 118
Beam-type smoke detectors, 220
Blackstone, G. V., *A History of the British Fire Service* by, 3, 7, 8
BLM (*see Bureau of Land Management*)
BLEVE (Boiling Liquid-Expanding Vapor Explosion)
 description of, 90
 fireball formed by a, 90
*BOCA Basic Building Code*, 281–283
Boiling point, of a liquid, 84

Boston
 conflagrations in, 5
 fire service in, 5
 fires in, 5
 first paid fire department in, 5
 mutual fire societies in, 7
Boyle's Law, 89
Braidwood, James
 as chief of Edinburgh Fire Brigade, 4
 first handbook on fire department operations by, 4
Breathing apparatus, 198
British thermal unit (Btu), definition of, 55, 164
Bromine, as a halogen, 94
Bromochlorodifluoromethane (Halon 1211), 186
Bromotrifluoromethane (Halon 1301), 187
Bryan, Dr. John L.
 *Automatic Sprinkler and Standpipe Systems* by, 9
 definition of automatic sprinkler system by, 9
Building codes
 development of, 268–270
 factors in, 271
 in relation to fire codes, 270–271
 in relation to firesafety, 24
 *Life Safety Code*, in relation to, 35
 model, 280–284
  *BOCA Basic Building Code* as, 281, 283
  *National Building Code* as, 281–282
  *Standard Building Code* as, 281, 283
  *Uniform Building Code* as, 281, 282
 role of standards in, 283–284
Building construction, firesafe, 141
Building contents, improper storage of, 124
Building department
 relation of, to fire departments, 251–252
Building design (*see also building firesafety, elements of*)
 as determinant of fire spread, 145
 changes in, 41
 continuity of operations decisions in, 142
 decisions for, 145
 effect of, on fire department operations, 145–146
 exposure protection, as consideration for, 149

features that influence safety of, Table 6.1
fire department response time as a factor for, 149
firesafe, 37, 141
for fire protection, 148
for multiple-purpose buildings, 29
fundamentals of, 28–29
human characteristics for, 26
interior finish in, 150
life safety decisions in, 142
major goal in, 28
of one- and two-family dwellings, 29
problems presented in, 29
property protection decisions in, 142
Building firesafety, elements of
fire prevention, Table 6.1
good housekeeping, 145
keeping fuel load down, 145
Building laws, early, 267
Building ordinances, 268
Buildings
exit design in, 31
fire hazards in, 143
fire loading in, 157
firesafety planning for, 145
high-rise, smoke in, 160
interior of, fire fighting accessibility to, 146
smoke control in, 159
test procedures for fire resistance of, 154
Building services, as condition for firesafety, 145
BUMINES (*see Bureau of Mines*)
Bureau of Indian Affairs, 304
Bureau of Land Management (BLM), 304
Bureau of Mines (BUMINES), 304
Burns
as products of combustion, 65
classification of, 65
degrees of, 65
first degree, 65
second degree, 65
third degree, 65
from clothing ignition, 23
Business occupancies
definition of, 42
examples of, 42
fire hazards in, 42
ignition sources in, 42
risk hazards in, 42

## C

Calorie, definition of, 55
"Capacity Method," for determining exit widths, 32, 33
Carbon
as a component of cotton, 78
as a component of wood, 73
Carbon dioxide
as a fire gas, 63
characteristics of, 183
cooling capacity of, 183
cooling properties of, 184
effectiveness of, on dry chemicals, 196
extinguishing properties of, 184
portable fire extinguishers, 193
properties of, 63
smothering properties of, 184
use of, in fire control, 183
Carbon dioxide systems, 183
Carbon monoxide
as a cause of death, 143
as a fire gas, 62–63
as a product of flashover, 75
properties of, 63
Carbon tetrachloride, in portable fire extinguishers, 193
Career opportunities
in a fire department
fire apparatus driver/operator, 310
fire fighter, 309
paramedic, 310
in fire protection
fire protection engineer, 308
state fire marshal, 308
in fire sciences, 307
Celsius degree, definition of, 55
Central station systems, 210, 211, Fig. 8.4
Chain-reaction theory of combustion, 190
Charles' Law, 89
Chemical emergencies
extinguishment of threat of, 102
procedures for, 101
Chemicals, corrosive
inorganic acids as, 92–93
halogens as, 93–94
storage of, 94
Chicago Fire, 9
Chlorine, as a halogen, 94
Churches, as light hazard occupancies, 171

## 318  Principles of Fire Protection

Circulation patterns, as condition for firesafety, 145
Climate, as factor for fire protection, 226
Clothing ignition, burns from, 23
Cocoanut Grove fire, results of, 126
*Code for Safety to Life from Fire in Buildings and Structures* (*see Life Safety Code*)
Code of Hammurabi, 267
Codes (*see also specific codes*)
  building, 227
  enforcement of, 113, 263, 275–279
  fire, 227
  fire prevention, 227
  formation of, 272–279
  "model" fire prevention, 280
    *BOCA Basic Fire Prevention Code* as, 280
    *Fire Prevention Code* (AIA) as, 280
    *Fire Prevention Code* (NFPA) as, 280
    *Southern Standard Fire Prevention Code* as, 280
    *Uniform Fire Code* as, 280
  types of, 279–284
Colonial America
  fire engines in, 5
  fire protection in, 5, Fig. 1.1
  "fire wardens" in, 5
  first fire ordinance in, 5
  night watch fire service in, 5
  ordinances in, 5
Combination systems, for water distribution, 240
Combined dry-pipe and preaction systems, 175
Combined rate-of-rise and fixed-temperature detectors, 219
Combustible liquids
  characteristics of, 83–84
  classification of
    Class II, 83
    Class IIIA, 83
    Class IIIB, 83
  fire fighting involving, 81
  flash point of, 83
  smothering effect of water on, 167
  storage of, 85
Combustible metal extinguishing systems, 191
Combustible metals
  explosions involving, 191
  extinguishing agents for, 191
  fires involving, 191

  labels and placards for, 191
  transportation of, 192
Combustible solids
  plastics as, 76
  textiles as, 78
  wood as, 72
Combustion
  burns, 65
  chain-reaction theory of, 190
  essentials for, 54
  fire gases, 62–64
  flame, 64
  flameless surface mode, 66
  flaming mode, 66
  fuel as a component of, 71
  heat, 64–65
  products of, 62–66
  smoke, 65–66
Combustion explosions, safeguards against, 91
Combustion, products of (*see products of combustion*)
Community relations
  fire department activities in, 48
  news media role in, 48
  press releases in, 48
Communication center
  illustrated, Fig. 8.3
  location of, 208
  procedure for, 208
Communications
  for insurance grading schedule, 234
  public fire service, 202
Communications equipment, two-way radios as, 197
Composite doors, 159
Compressed gases, 87
Concrete
  as an interior finish, 153
  fire resistance of, 153
  prestressed, 153
  reinforced, 153
Conduction, as a method of heat transfer, 56
Confinement
  as method of smoke control, 160
  as tactical operation, 256
  of smoke and fire, 156
Conflagrations
  analysis of, 128
  application of term, 128
  factors contributing to spread of, Table 1.3
  post World War II, 16

recent, Table 1.3
types of, 128–131
wood-shingle, Table 1.2
Consumer Product Safety Commission (CPSC), 25
Continuity of operations
as a factor of firesafety design, 143
considerations for, 143
design considerations for, 142
Convection, as a method of heat transfer, 57
*Corps of Vigiles*, 3
Corridors, exit requirements for, 30
Cotton
carbon as a component of, 78
hydrogen as a component of, 78
ignitability of, 78
oxygen as a component of, 78
Crosby, Uberto C., address by, 1
Crowd movement, 32
Cryogenic gases, 87
Cryogenic liquids
action at emergencies involving, 105–106
handling emergencies involving, 106–108
Curfew
as a fire protection regulation, 3
in Oxford, England, 4
meaning of, 3
Curtain-type doors, 159

## D

Dana, Gorham, *Automatic Sprinkler Protection* by, 9
Day-care centers (*see day-care facilities*)
Day-care facilities
as educational occupancies, 41
fire protection considerations for, 41
*Life Safety Code*, classification of, 41, 42
life safety hazards in, 41
Data processing, for fire departments, 253
Deflagration
as a form of thermal explosion, 61
definition of, 61
Deluge systems, 175
Department of Agriculture
Farmers Home Administration (FHA) of, 303
U.S. Forest Service (USFS) of, 303

Department of Commerce (DOC)
National Bureau of Standards (NBS) of, 302
National Fire Prevention and Control Administration (NFPCA) of, 302
Department of Defense (DOD), 304
Department of Housing and Urban Development (HUD), 303
Department of the Interior, 304
Department of Labor (DOL), 302–303
Department of Transportation (DOT), 304–305
Design, building (*see building design*)
Design features of dormitories, 39
Detection devices, uses of, 222
Detection equipment
location of, 222
spacing of, 222
standard for, 222
Detectors
beam-type, 220
combined rate-of-rise and fixed-temperature, 219
fixed-temperature, 217
flame, 221
gas, 222
heat, 216–219
ionization, 220
photoelectric, 220
rate compensation, 219
rate-of-rise, 218
resistance bridge, 221
sampling, 221
smoke, 219, 221
Detonation
as a form of thermal explosion, 62
definition of, 22
Dilution, as method of smoke control, 160
DOC (*see Department of Commerce*)
DOD (*see Department of Defense*)
DOL (*see Department of Labor*)
Doors, fire (*see fire doors*)
Doorways, exit requirements for, 30
Dormitories
as residential occupancies, 38, 39
design features of, 39
DOT (*see Department of Transportation*)
Dry chemical extinguishing systems
chemical properties of, 188
monoammonium phosphate in, 188
particle size in, 189
physical properties of, 188
portable fire extinguishers, 194

320  Principles of Fire Protection

potassium bicarbonate in, 188
potassium chloride in, 188
sodium bicarbonate in, 188
stability of, 188–189
toxicity of, 189
types of, 188
urea-potassium bicarbonate in, 188
Dry chemicals
chain-breaking reaction of, 190–191
cooling action of, 190
extinguishing properties of, 189
radiation shielding of, 190
stability of, 189
toxicity of, 189
uses in extinguishment, 189
Dry powder portable fire extinguishers, 194
Dust explosions, Table 3.1

## E

Edinburgh's Fire Brigade, 4
Educational occupancies
day-care facilities as, 41
fire drills for, 41
flexible-beam buildings as, 41
*Life Safety Code*, provisions for, 41
life safety considerations for, 41
life safety hazards in, 41
open-plan buildings as, 41
types of, 41
Egress design
concepts of, 30–34
factors for, 31
flow rates as factor of, 31, 32–33
in high-rise buildings, 31
Egress, numbers of, 33–34
Electrical energy
as a source of ignition, 57–58
heat from, 57
relationship of, to fire, 58
types of
arcing, 58
lightning, 59
resistance, 58
sparking, 58–59
static sparking, 59
Electrical equipment, as cause of fires, 127
Electric lights and generators, for apparatus, 197
Elevators, as means of egress, 31
EMT (*see paramedic*)

Energy, definition of, 71
Engine company, as organization for fire suppression, 254
Equipment
carried on apparatus, 196–197
for high-hazard buildings, 232
nineteenth century, 11
protective, 198–199
used by mutual fire societies, 7
Escape plan, as means for life safety, 37
Evacuation, for elimination of fire deaths, 132
Evaporation rate, of a liquid, 84–85
Exhaust, as method of smoke control, 100
Exit drills
*Life Safety Code*, requirements for, 35
procedure for, 45
Exit Drills In The Home (*see Operation Edith*)
Exit safety, principles of, Fig. 2.2
Exit widths
"Capacity Method" for determining, 32, 33
"Flow Method" for determining, 32, 33
Exits
*Life Safety Code*, requirements for, 34
requirements for
corridors used as, 30
doorways used as, 30
stairways used as, 30
windows used as, 30
Explosion control
flame inhibition as a means of, 66
removal of fuel as a means of, 66
removal of oxygen as a means of, 66
water as a means of, 66
Explosions
atomic, 61
chemical, 60
mechanical, 61
thermal, 61–62
deflagration, 61
detonation, 62
types of, 60–62
violence of, 81
Explosive plants, as extra hazard occupancies, 171
Exposure
as tactical operation, 255
damage from, 150
factors for, 150
severity of, 150

Exposure protection
  effectiveness of, Fig. 6.2
  standard for, 150
External traffic control, 45
Extinguishing agents (*see also* extinguishing systems)
Extinguishing properties
  of dry chemicals, 189
  of water, 165
Extinguishing systems
  carbon dioxide, 183
  combustible metal, 191
  dry chemical, 188
  foam, 181
  halogenated, 184
  sprinkler, 168
  standpipes, 177
  water spray, 171
Extinguishment
  as a tactical operation, 256
  at chemical emergencies, 102–103
  at radiation emergencies, 100
Eyewitnesses, role of, during investigations, 116

# F

Fact finding, in fire reporting systems, 121–122
Factory Mutual Laboratory Facilities, 299
Factory Mutual Systems (FM), 10, 300–301
Fact processing
  "Fire Fact File" as element of, 122
  in fire reporting systems, 122
  steps in, 122
Fact use, in fire reporting systems, 122–123
Fahrenheit degree, definition of, 55
*Familia Publica*, 3
Farmers Home Administration (FHA), 303
Federal Fire Prevention and Control Act, 302
Federal Fugitive Felon Act, in relation to arson, 137
Fiberglass, 153
Finance department, of fire departments, 252
Fire
  as a chemical reaction, 57
  as a failure of code enforcement, 113
  as a failure of public education, 113
  as a "two-sided" god, 1
  basic definition of, 2
  behavior of, 51
  caused by ignition of bedding, 80
  causes of, in U.S., Table 2.2
  characteristics of, 51
  Class A, 194–195
  Class B, 195
  Class C, 195
  Class D, 195
  classification of, 194–195
  Cocoanut Grove, 126
  concepts of egress design related to, 30
  confinement of, 156
  control of
    by flame inhibition, 66
    by removal of fuel, 66
    by removal of oxygen, 66
    by water, 66
  deaths resulting from, 22
  definition of, 2, 52
  described by Haessler, 2
  determining arson as cause of, 134
  effect of, on building components, 154
  electrical, use of water on, 167
  electrical equipment as a cause of, 127
  elements of, 52
  estimation of damage by, 154
  extinguishment of, 53, 163
  growth and development of, 114
  high-rise, Table 6.3
  human behavior during, 22, 23
  human characteristics regarding, 22, 28
  in a shopping mall, 29
  in automotive transmission plant, 24
  in Chicago, 9
  in Cocoanut Grove nightclub, 126
  in complex buildings, 147
  influence of behavior of, 25
  in high-rise buildings, 31
  injury from, 23
  in nursing home, 17
  in one- and two-family dwellings, 39
  in Our Lady of the Angels Grade School, 16, Fig. 1.5
  in prehistoric times, 2
  in San Francisco, 269
  investigation of, 263–264
  involving combustible metals, 192
  involving fireworks and explosives, 128
  involving flammable liquids, 128

involving wood, effect of water on, 167
life loss from, 23
major causes of, 24–25, Table 2.2
major, in early times
  in China, 4
  in Constantinople, 4
  in India, 4
  in Japan, 4
  in London, 5
  in Moscow, 4
  in Rome, 5
open flames and sparks, 127–128
Our Lady of the Angels Grade School, 126
oxidation in, 53
people actions in, 26
people activities in, 27
people factors in, 25
people reactions in, 26
possible gains from, 24
principle causes of, 125
property loss from, 23
San Francisco, 269
severity of, 150
smoking and related, 126
tetrahedron, 53, Fig. 3.2
three elements for, 52
triangle, 53, Fig. 3.1
unpredictability of, 51
wood-shingle, 13
worship of, 2
Fire alarm box
  combination telephone-telegraph-type, 207
  distribution of, 204–205
  parallel arrangement, 207
  radio-type, 207–208
  series arrangement, 207
    telegraph-type, 205
    telephone-type, 206
Fire alarm system
  first municipal, 202
  function of, 202
  installation of, 12
  municipal, 9, 202–204
  radio-type, 203, 207–208
  telegraph-type, 203, 205–206
  telephone-type, 203, 206–207
  Type A (manual), 203, 204, Fig. 8.1
  Type B (automatic), 203, 204
  use of, 12, 205
Fire analysis
  fire ignition sequence in, 123
  "group fire," 123

large-loss, 123–124
types of, 123
Fire and explosion control
  flame inhibition as means of, 68
  fuel removal as a means of, 68
  means of, 66–68
  removal of oxygen as a means of, 67
  use of water as a means of, 67–68
Fire apparatus
  advances in, 12
  aerial ladders as, 12
  "Babcock" fire engines as, 12
  introduction of automobile as, 12
  steam fire engines as, 5, 12
Fire apparatus driver/operator
  minimum requirements for, 310
  skills of, 310
  standard for, 310
*Fire Attack 2*, by Warren Y. Kimball, 229
Fire behavior, of plastics, 77
Fire brigade chief, 44
Fire brigades (*see industrial fire brigades*)
Fire casualties, in investigations, 114
Fire chief
  as administrator of department, 251
  Benjamin Franklin as first, 8
  James Braidwood as, 4
  management responsibilities of, 248
  responsibilities of, 19
Fire codes
  development of, 268–270
  factors in, 271
  in relation to building codes, 270–271
  in relation to firesafety, 24
Fire control
  by flame inhibition, 66
  by removal of fuel, 66
  by removal of oxygen, 66
  by water, 66
Fire deaths
  carbon monoxide as cause of, 143
  from entrapment, 23
  heat and flames as cause of, 143
  smoke as cause of, 143
  studies of, 143
Fire defenses, municipal (*see municipal fire defenses*)
Fire defenses, public (*see public fire defenses*)
Fire defenses, rural (*see rural fire defenses*)
Fire defenses, urban (*see urban fire defenses*)

Fire department
  access of, to site, 149
  administration and management of, 248–253
  apparatus for, 232
  areas of responsibility in, 19
  assessment of, 232
  auxiliary equipment for, 232
  careers in a
    fire apparatus driver/operator, 310–311
    fire fighter, 309–310
    paramedic, 311
  communications center in, 208
  communications in, 202, 234
  community relations by, 48
  cost of staffing in, 251
  county, 245
  data processing for, 253
  establishment of first paid, 5–6
  evaluation of, Table 9.2
  evolution and scope of, 18–19
  finance department for, 252
  fireground operations by, 232
  fire prevention activities of, 17
  first municipal-type, 3
  growth of paid, 7
  history of, 243–244
  in large cities, 245
  intergovernmental relations, 251
  investigation as a function of, 113
  line functions in, 247
  lines of authority in, 247
  local, 245
  medical professionals in, 244
  minimum requirements of, 231
  NFPA information for, 294
  objectives of, 19, 244–245
  officers for, 232
  organization of, 19, 245
  organizational plans of, 248
  paid, 7
  personnel department in, 252
  promotional practices in, 250
  purchasing department of, 252
  relation of building department to, 251–252
  relation of police to, 252
  relation of water department to, 252
  response time, 149
  response to alarms by, 232
  role of, in combustible metal fires, 192
  role of, in fire protection, 226
  scope of responsibility of, 19–20
  staff functions in, 248
  staffing of, 246
  staffing practices of, 250–251
  strength of, in *ISO Grading Schedules*, 231
  structure of, 245–248
  traditional objectives of, 19
  training provided for, 232
  use of two-way radio equipment by, 209
  voluntary, 245
  work assignments for, 246–247
Fire department objectives, 244–245
Fire department operations
  effect of building design on, 145–146
  fire prevention as, 259–263
    enforcement of codes, 263
    inspections, 261–262
    organizations for, 260–261
    public education, 262–263
    seasonal activities, 263
  investigation of fires, 263–264
  mutual aid, 262–263
  nontactical operations, 257–258
    prefire planning, 257
    training, 257–258
  organizations for fire suppression, 254–255
    engine company, 254
    ladder company, 254
    rescue company, 254
    tactical control units, 254
    task forces, 254–255
  tactical operations, 255–257
    confinement, 256
    exposures, 255
    extinguishment, 256
    overhaul, 257
    rescue, 255
    salvage, 256
    size-up, 255
    ventilation, 256
  transportation incidents, 258
Fire department organization, 243–248
Fire department strength
  for *ISO Grading Schedule*, 231
Fire department structure, 245–248
Fire detection equipment
  combined rate-of-rise and fixed-temperature, 219
  fixed-temperature, 219
  flame detector, 221

gas detector, 222
heat detector, 216
rate compensation device, 219
rate-of-rise, 218
smoke detectors, 219, 221
Fire detectors, requirements for, 211
Fire development
as an area of investigations, 114
role of time factors in, 114
structural compartmentation as a factor of, 114
Fire districts, organization of, 245
Fire doors
classifications of
composite, 159
curtain-type, 159
hollow-metal, 159
metal clad, 159
rolling steel, 159
sheet-metal, 159
tin-clad, 159
evaluation of, in large-loss fire prevention, 131
standard for, 158
suitability of, 158–159
types of, 158–159
uses of, 158
Fire drills, for educational occupancies, 41
Fire engine
"Babcock," 12
in Boston, 5, 6
in New York City, 11
steam, 12
Fire extinguishers, portable, 192–196
Fire extinguishing systems, automatic, 215
"Fire Fact File," as element of fact processing, 122
Fire fighters
applicants for, 310
first level, 249
group work activity of, 310
International Association of Fire Fighters (IAFF) for, 296
national qualifications for, 249
paid, 249
physical requirements for, 310
protective clothing for, 199
protective equipment for, 198–199
responsibility of, 309
role of
during investigations, 115
in size-up, 255

second level, 250
selection of, 249
standards for, 249, 310
third level, 250
work week for, 251
Fire fighting
accessibility to building's interior for, 146–148
as a fire department responsibility, 19
at radioactive emergencies, 98–100
basic unit of, 254
involving flammable or combustible liquids, 81
water for, 228
water pressure for, 238–239
Fire flow, definition of, 230
Fire frequency, as a factor for fire protection, 226
Fire gas
acrolein as a, 64
ammonia as a, 63
as a cause of death, 62
as a product of combustion, 62–63
carbon dioxide as a, 63
carbon monoxide as a, 62–63
hydrogen chloride as a, 64
hydrogen cyanide as a, 63
hydrogen sulfide as a, 63
nitrogen dioxide as a, 64
phosgene as a, 64
sulfur dioxide as a, 63
Fireground, fire department operations at, 232
Fire hazard
as a factor in fire protection, 226
building elements and contents as a, 143–144
gas as a, 143
heat and flames as a, 143
in buildings
building elements and contents, 143
heat and flames, 143
smoke and gas, 143
of materials, 71
smoke as a, 143
Fire hose
early use of, 11
effective use of, 19
in American colonies, 11
in Boston, 11
Fire hydrants
early use of, 11
first installed, 11
inadequacies of, 11

Fire ignition sequence
  as an area of investigations, 113–114
  estimated number of building fires by, Table 5.3
  factors involved in, 113
  fuel as a factor of, 113
  heat as a factor of, 113
  human action as a factor of, 113
  natural act as a factor of, 113
Fire injuries, as symbols of fire, 23
Fire inspector, role of in fire prevention, 261
Fire insurance companies
  formation of, 7–8
  insurance brigades formed by, 8
Fire laws, early, 267
Fire loading
  by occupancy, 157–158
  determination of, 157
  of wood, 76
  role of in exposures, 150
  severity of, Table 6.4
Fire loss
  analysis of, 123
  by wood-shingle conflagrations, 13, Fig. 1.3, Table 1.1
  during early 20th century, 12–14
  during World War II, 14–15
  estimates of, 22
  inspections as a means to prevent, 131
  post World War II, 16, Fig. 1.5
  prevention of, 131
  public education as a means to prevent, 131
  scope of, 291
Fire Marshals Association of North America (FMANA), 294
Fire prevention
  as fire department operation, 253–264
  as fire department responsibility, 19
  as state fire marshal responsibility, 277
  evaluation of fire doors in, 131
  evaluation of fire walls in, 131
  evaluation of stairways in, 131
  housekeeping for, 145
  large-loss, 131
  modern methods of, 17
  persons responsible for, 145
  programs for, 46–47
Fire prevention codes, 227
Fire Prevention Day, establishment of, 9
Fire prevention programs
  annual savings from, Table 2.4
  benefits from, 46
  costs of, Table 2.4
  importance of, 46
Fire Prevention Week, 126, 293
Fire protection
  adequacy of, 223
  building design for, 148
  building makeup as factor for, 227
  careers in
    fire protection engineer, 308–309
    state fire marshal, 308
  climate as a factor for, 226
  community needs for, 226–227
  costs of, 227
  defenses and agencies, 17–20
  developing master plan for, 235–236
  during Roman Empire, 3
  early regulations for, 3–4
  engineering, 308–309
  factors for evaluating, 226
  federal agencies involved in, 301–305, Table 12.1
    Department of Agriculture, 303
    Department of Commerce (DOC), 302
    Department of Defense (DOD), 304
    Department of Housing and Urban Development (HUD), 303
    Department of the Interior, 304
    Department of Labor (DOL), 302–303
    Department of Transportation (DOT), 304–305
    General Services Administration (GSA), 304
  fire testing laboratories for, 297–299
  first organized, 3
  for radioactive materials, 96
  for rural areas, 228–229
  geography as factor for, 226
  in colonial America, 5
  in 19th century, 8
  insurance organizations for, 299–301
  levels of, 227
  life safety as factor for, 226
  master planning for, 235
  methods of, 17
  need for, 112
  nineteenth century progress in, 8
  organizations for, 291
  personnel for, 228
  population distribution as factor for, 226
  principles of, 54
  public, 226–236

requirements for in water systems, 238–239
role of fire department in, 226
role of portable fire extinguishers in, 192
science of, 54
water for, 236–239
Fire protection agencies, 20
Fire protection careers
  fire protection engineer, 308
  state fire marshal, 308
Fire protection districts, 245
Fire protection equipment, 196–199
Fire protection, industrial, establishment of, 9
Fire protection organizations, scope of, 291–292
Fire reporting systems
  elements of, 121–123
  fact finding as element in, 121–122
  fact processing as element in, 122
  fact use as element in, 122–123
  objectives of, 120–121
  three elements of, Table 5.1
Fire reports
  as legal record of fire incident, 117
  Basic Casualty Report as, 118
  Basic Incident Report as, 118
  information included in, 117
  information provided by, 116–117
  investigations for, 113
  purposes of, 117
  types of, 118–119
  uniformity in, 121
Fire resistance, of concrete, 153
Firesafe building construction (see building design)
Firesafe building design (see building design)
Firesafety
  building and site planning for, 145–149
  building codes in relation to, 29
  considerations for, 145
  elements of, 145, Table 6.2
  federal regulations regarding, 276
  fire codes in relation to, 24
  Fire Prevention Week, 126
  for people, 22
  for property, 22
  high-rise building, 37
  human characteristics in relation to, 26
  in relation to investigations, 115

in residential occupancies, 28
local authority for, 278
mobile homes as hindrance to, 37
planning for, 145–146
planning of sites for, 148–149
public education in, 46–48
standards-making organizations for 284–288
state authority for, 276–278
Firesafety design (see also building design)
  approaches toward, 143
  continuity of operations as a factor of, 143
  decisions concerning, 142–143
  fundamentals of, 141–145
  life safety as a factor of, 142
  objectives of, 142
  property protection as a factor of, 142–143
*Firesafety in the Home*, 31
Firesafety planning
  considerations for, 28
  in one-family dwellings, 29
  in multipurpose buildings, 29
  in shopping malls, 29
  traffic and transportation in, 149
Fire scene
  analysis of, 116
  condition of doors at, 116
  condition of windows at, 116
Fire scene evidence
  removal of, 115
  role of, in investigations, 115
Fire science education, 307–308
Fire sciences, career opportunities in, 305–311
Fire service, role of, in public protection, 46
Fire Service Section (of NFPA), 294–295
Fire severity
  expected by occupancy, Table 6.5
  for light commercial occupancies, 156
  for offices, Table 6.4
Fire spread, factors responsible for, Table 5.2
Fire suppression
  basic unit for, 254
  organization for, 254–255
Fire tests
  full-scale, 154
  laboratories for, 297–299
  standard, 154
Fire tetrahedron, 53, Fig. 3.2

Fire triangle, 53, Fig. 3.1
Fire walls, evaluation of, in large-loss fire prevention, 131
Fire waste, Table 2.1
Fireworks and explosives
  fire involving, 128
  incidents involving, Table 5.4
Fixed-temperature detectors, 217, Fig. 8.5
Flame, as a product of combustion, 64
Flame detectors
  flame flicker, 221
  infrared, 221
  photoelectric, 221
  ultraviolet, 221
Flame flicker flame detector, 221
Flame inhibitions, as means of fire and explosion control, 68
Flame retardant textiles, 80
Flame spread
  as a characteristic of wood, 75–76
  ratings for, 155
  tunnel tests for, 155
Flammability limit
  of gas, 53
  of vapor, 53
Flammability range, of gasoline, 81
*Flammable and Combustible Liquids Code*, 287
Flammable gases, 86
Flammable limits
  of common gases, Table 4.4
  of common liquids, Table 4.4
Flammable liquids
  characteristics of, 83–84
  classification of, 82
  fire fighting involving, 81
  fires involving, 128
  flash point of, 82
  storage of, 85
Flashover, 75
Flash point
  of combustible liquids, 83
  of common gases, Table 4.4
  of common liquids, Table 4.4
  of flammable liquids, 82
Flexible-plan buildings, 41
"Flow Method," for determining exit widths, 32, 33
Flow rates
  as factor of egress design, 31, 32–33
  design of passageways as a factor of, 32
  in high-rise buildings, 32

Fluorine, as a halogen, 94
FM (*see Factory Mutual System*)
FMANA (*see Fire Marshals Association of North America*)
Foam
  characteristics of, 182
  classes of, 181
  mechanical, 182
Foam extinguishing systems, 181–183
Foam systems, use of, 182–183
Forcible entry tools, 196
Franklin, Benjamin
  as America's first fire chief, 8
  as organizer of Philadelphia Contributionship, 8
  as organizer of Union Fire Company, 8
Freeman, John R., hydraulic studies by, 12
Friction, as a form of mechanical heating, 59
Fuel
  as a component of combustion, 71
  as a factor of fire ignition sequence, 113
  as an element of fire, 52
  removal of, 68
Fuel gases, 88

## G

Gas
  as a fire hazard, 143
  as a form of matter, 71
  classification of, by properties
    flammable, 86
    nonflammable, 86
    reactive, 86
    toxic, 86
  classification of, by usage
    full, 88
    industrial, 88
    medicinal, 88
  flammability limit of, 53
  physical properties of
    compressed, 87
    cryogenic, 87–88
    liquefied, 87
  ventilation for, 148
Gas detectors
  uses of, 222
  value of, 222

Gas laws, 89-90
  Boyle's Law, 89
  Charles' Law, 89
  Gay-Lussac's Law, 89-90
Gasoline
  as cause of flammable liquids fires, 128
  flammability range of, 81
  storage of, 81
Gas standards, 92
Gay-Lussac's Law, 89-90
General Services Administration (GSA), 304
Geography, as a factor for fire protection, 226
G-1 powder, 192
Glass
  as interior finish, 153
  uses of, in building construction, 153
Goals, for master planning, 234
Godfrey, Ambrose, first patented sprinkler system by, 9
Gravity system, for water distribution, 240
Great Britain, fire protection regulations in, 3-4
Great Chicago Fire, 9, 268
Great London Fire, 5, 8
Ground water supply, for water distribution systems, 23
"Group fire," prevention of, 132
GSA (*see General Services Administration*)
Gypsum, as interior finish, 153

# H

Haessler, Walter, *Extinguishment of Fire* by, 2
Halogenated extinguishing agents, 184-188
Halogens
  as corrosive chemicals, 93-94
  bromine as a, 94
  chlorine as a, 94
  description of, 93
  fluorine as a, 94
  iodine as a, 94
  properties of, 93
  storage of, 94
Halon
  definition of, 184
  extinguishing characteristics of, 187-188

principal types of, Table 7.3
types of, 186-193
Halon 1211
  as a liquefied gas, 193
  toxicity level of, 186
  weight of, 187
Halon 1301
  as liquefied gas extinguisher, 193
  commercial use of, 187
  toxicity of, 187
Halon 2402, 193
Hazardous materials
  corrosive chemicals, 92
  radioactive materials, 92, 95
  transportation emergencies involving, 103-104
Health care occupancies
  ability of occupants to escape in, 40
  categories of, 40
  definition of, 40
  fire hazards in, 40
  firesafety considerations for, 40
  life safety considerations for, 40
Heat
  as a factor in fire ignition sequence, 113
  as an element of fire, 52
  as a product of combustion, 64-65
  as a product of fire, 64
  caused by mechanical energy, 59
  definition of, 55
  exposure to, 65
  importance of determining intensity, 54-55
  results of exposure to, 65
Heat detectors
  categories of, 216
  line-pattern type, 216
  spot-pattern type, 216
Heat energy, sources of, 57
Heat measurement, 54-55
Heat transfer
  by conduction, 56
  by convection, 57
  by radiation, 56
  in cooling action of water, 166
  methods of, 56
Heat units
  Celsius degree, 55
  Fahrenheit degree, 55
  Kelvin degree, 55
  Rankine degree, 55
High hazard occupancies, Table 9.2

High-rise buildings
  as hindrance to firesafety, 37
  egress design in, 31
  fighting fires in, 31
  fires in, Table 6.3
  flow rates in, 32
  movement of smoke in, 31
  smoke in, 160
  "stack effect" in, 31
Hollow-metal doors, 159
Hotels, definition of, 39
Household fire warning systems, 214–215
HUD (*see Department of Housing and Urban Development*)
Human action
  as cause for unsatisfactory performance of sprinklers, 125
  as factor of fire ignition sequence, 113
Hydrants
  considerations for, in site planning, 149
  spacing of, 238
Hydraulic studies
  by John R. Freeman, 12
  by William Jackson, 12
Hydrochloric acid, 93
Hydrofluoric acid, 93
Hydrogen
  as a component of cotton, 78
  as a component of wood, 73
Hydrogen chloride, as a fire gas, 64
Hydrogen cyanide
  as a fire gas, 63
  properties of, 63
Hydrogen sulfide
  as a fire gas, 63
  properties of, 63

# I

IAAI (*see International Association of Arson Investigators*)
IABPFF (*see International Association of Black Professional Fire Fighters*)
IAFC (*see International Association of Fire Chiefs*)
IAFF (*see International Association of Fire Fighters*)
IFSTA (*see International Fire Service Training Association*)
Ignition
  of wood, 74–75
  prevention of, 131
  sources of, 52, 57
    chemical, 57
    electrical, 57–59
    mechanical, 59
    nuclear, 59
Ignition sources, in business occupancies, 42
Ignition temperature, of a liquid, 85
Illinois Institute of Technology, 307–308
IMSA (*see International Municipal Signal Association*)
Industrial fire brigades
  chief's responsibilities in, 44
  duties of, 44
  fire brigade chief of, 44
  industrial fire risk manager in, 43
  operating units in, 44
  operation of, 44
  organization of, 44
  personnel in, 44
  requirements for members in, 44
  role of, in combatting metal fires, 192
  training for, Fig. 2.3
Industrial fire emergencies
  external traffic control for, 45
  internal traffic control for, 45
Industrial fire risk manager
  communications by, 44
  evacuation drills by, 45
  function of, 44
  inspection of exits by, 45
  responsibilities for evacuation drills, 45
  responsibilities of, 43–44
Industrial gases, 88
Industrial occupancies
  fire in, 45
  fire records for, 43
  firesafety in, 43
  industrial fire brigades in, 43
  life loss in, 43
  life safety in, 43
  property loss in, 43
  use of automatic sprinkler systems in, 43
Industrial Risk Insurers (IRI), 301
Information analysis, in prefire planning, 257
Information dissemination, in prefire planning, 257
Information gathering, in prefire planning, 257
Infrared flame detectors, 221

Injury, from fire, 23
Inorganic acids
  hydrochloric acid, 93
  hydrofluoric acid, 93
  nitric acid, 93
  perchloric acid, 93
  storage of, 94
Inspections
  as a means to prevent fire losses, 131
  building, 262
  home, 261
  of fires, 263–264
  of residential occupancies, 131–132
  role of fire marshal in, 263
  steps in, 264
Insurance brigades
  authority of, 8
  maintenance of, 8
Insurance grading schedule (*see also ISO Grading Schedule*)
  communications for, 234
  fire department strength for, 231–233
  water supply, 230–231
Insurance organizations
  American Insurance Association (AIA), 300
  American Mutual Insurance Alliance (AMIA), 300
  Factory Mutual System (FM), 300–301
  Industrial Risk Insurers (IRI), 301
  Insurance Services Office (ISO), 301
Integrity, structural, 152
Interior finish, 150–156
  cellular plastics as, 151
  common materials of, 151
  concrete as, 153
  floor coverings as, 151
  functions of, 150
  glass as, 153
  gypsum as, 153
  masonry as, 154
  plastics as, 154
  potential fire hazards of, 150
  steel as, 152
  test procedures for, 155
  types of, 150–154
  wood as, 152
Interior layout, as consideration for firesafety, 145
Internal traffic control, 45
International Association of Arson Investigators (IAAI), 296

International Association of Black Professional Fire Fighters (IABPFF), 296
International Association of Fire Chiefs (IAFC), 295–296
International Association of Fire Fighters (IAFF), 296
International Fire Service Training Association (IFSTA), 296
International Municipal Signal Association (IMSA), 296
International Society of Fire Service Instructors (ISFSI), 296–297
Investigations
  as a fire department function, 113
  conducting of, 115
  details for arson, 135
  earliest recorded instances of, 112
  fire casualties as an area of, 114
  fire development as an area of, 114
  fire ignition sequence as an area of, 113–114
  firesafety in relation to, 115
  fire scene analysis as an area of, 116
  for arson, 133, 135
  general guide for, 115
  need for, 112
  preliminary, 115–116
  purpose of, 113
  role of eyewitnesses during, 116
  role of fire fighter during, 115
  role of fire scene evidence in, 115
  role of room contents in, 114
  role of state fire marshal in, 116
  scope of, 113–114
  time spent for, 115
  to determine criminal activity, 113
  to determine what happened, 113
  to provide information for fire report, 113
Iodine, as a halogen, 94
Ionization smoke detectors, 220
IRI (*see Industrial Risk Insurers*)
ISFSI (*see International Society of Fire Service Instructors*)
ISO (*see Insurance Services Office*)
*ISO Grading Schedule*
  communications in, 234
  deficiency point system in, 230, Table 9.1
  description of, 230
  fire department assessment in, 232
  fire department response capability in, Table 9.2

Subject Index 331

fire department strength in, 231
other important factors in, 233
primary concerns of, 235
water supply in, 237

## J

Jackson, William, hydraulic studies by, 12
JCNFSO (*see Joint Council of National Fire Service Organizations*)
Joint Council of National Fire Service Organizations (JCNFSO), 297
Junior Fire Marshal Program, 47

## K

Kelvin degree, definition of, 55
Kimball, Warren Y., *Fire Attack 2* by, 229

## L

Laboratories
  for fire testing, 297–299
  government, 297–299
  private and industrial, 297, 298–299
  university, 297–299
Ladder company, as organization for fire suppression, 254
Large-loss fires
  causes of, 123
  evaluation of fire doors in, 131
  evaluation of fire walls in, 131
  evaluation of stairways in, 131
  prevention of, 131
  role of automatic sprinkler systems in, 124–125
  structural defects as cause of, 123, Table 5.2
Learn Not to Burn, 47
Life guns, 199
Life loss
  determination of, 227
  from fire, 23
  prevention of, 132
Life nets, 199
Life safety
  as factor for fire protection, 226
  as factor in firesafety design, 142
  assessing, 22

design considerations of, 142
effect of sprinklers on, 109
escape plans for, 37
measure of, in one- and two-family dwellings, 37
planning for, 29
questions concerning, 142
smoke detectors for, 37
ventilation for, 148
windows for, 37
Life Safety Code
  classification of day-care facilities in, 41
  hazard from smoke in, 35
  history of, 34
  inception of, 34
  in relation to building codes, 35
  protection of exits in, 34
  provisions for alarm systems in, 34
  provisions of, 34
  requirements for exit drills in, 34
  requirements for exits in, 35
  requirements for occupancies in, 37
  requirements for one- and two-family dwellings in, 38
  standards in, 34
  twelve requirements in, 34
Lightning
  as a source of ignition, 59
  safeguard for, 59
Limited water supply systems, 176
Line functions, 247
Lines of authority, 247
Liquefied gas, 87
  Halon 1211 as, 193
  Halon 1301 as, 193
  portable fire extinguishers, 193
Liquid
  as a form of matter, 71
  boiling point of a, 84
  evaporation rate of a, 84–85
  ignition temperature of a, 85
  specific gravity of a, 84
  vapor pressure of a, 84
Local fire marshal, as organization for fire prevention, 261
Local signaling systems
  essentials for, 213
  power supply for, 213
  uses of, 213
Lodging houses, definition of, 39
London Transport Board, studies by, 32, Fig. 2.1

Loudspeaker, as communication equipment, 197
Low hazard occupancies, Table 9.2

## M

Management of fire departments, responsibility of, 248
Manufacturing plants, as ordinary hazard occupancies, 171
Masonry
  as interior finish, 154
  effect of, on fire statistics, 154
Master planning
  for fire defenses, 234
  for fire protection, 235–236
    Phase I, 235
    Phase II, 236
    Phase III, 236
    Phase IV, 237
  goals for, 234
Matter
  definition of, 71
  forms of, 71
  properties of, 71
Means of egress
  elevators as a, 31
  numbers of, 33–34
Mechanical energy, 59
  as a cause of fire, 59
  friction as a form of, 59
  compression of gas as a form of, 59
Medicinal gases, 88
Mercantile occupancies
  considerations for, 42
  definition of, 42
  examples of, 42
Metal-clad doors, 159
Metals, combustible, 191
Met-L-X powder, 192
Minimum hazard occupancies, Table 9.2
Mobile homes
  as hindrance to firesafety, 37
  as residential occupancies, 40
  classification of, 40
  standard for, 40
Model Arson Law, degrees of arson in, 136–137
Moisture content, of wood, 74–74
Monoammonium phosphate, in dry chemical extinguishing systems, 188
Multipurpose dry chemical
  in portable fire extinguishers, 194

potassium chloride as a base for, 194
Municipal fire alarm systems, installation of first, 9
Municipal fire defenses
  evaluation of, 226, 227
  factors in evaluating, 226
  factors in planning, 226
  *ISO Grading Schedule* for, 230
  master plan for, 235–236
  personnel in, 228
  water for, 236
Mutual aid
  as fire department operation, 262–263
  functions of, 258–259
Mutual fire societies
  equipment used by, 7
  establishment of, 7
  membership of, 7

## N

National Board of Fire Underwriters (NBFU), 9, 229
*National Building Code*, 269, 281–282
National Bureau of Standards (NBS), 302
National Commission on Fire Prevention and Control, study by, 234
National Fire Prevention and Control Administration (NFPCA), Fig. 12.2
National Fire Protection Association (NFPA)
  activities of
    firesafety technical standards as, 292
    information exchange as, 293
    public education as, 293
    services to public protection agencies as, 294
    technical advisory services as, 293
  articles of organization of, 292
  Fire Analysis Department of, 22
  Fire Service Section of, 294
  public firesafety materials of, 293
  standards developed by, 292
  Uberto C. Crosby, as president of, 1
*National Fuel Gas Code*, 287
National Park Service (NPS), 304
Natural act, as a factor of fire ignition sequence, 113
Natural fiber textiles, 78
NBFU (*see* National Board of Fire Underwriters)
NBS (*see* National Bureau of Standards)

Subject Index     333

NFPCA (see National Fire Prevention and Control Administration)
Nitric acid, as an inorganic acid, 93
Nitrogen dioxide
  as a fire gas, 64
  properties of, 64
Nocturnes, duties of, 3
Nonflammable gases, 86
Nontactical operations
  prefire planning as, 257
  training as, 257–258
Nozzles, used in water spray fixed systems, 181
NPS (see National Park Service)
Nuclear energy
  as a source of electrical power, 59
  definition of, 59
  formation of, 59

O

Occupancies
  assembly, 41
  business, 42
  classification of, 171
  educational, 41
  extra hazard, 171
  fire loading by, 157–158
  fire severity expected in, Table 6.5
  hazards of, 35
  health care, 40
  high hazard, Table 9.2
  industrial, 43
  Life Safety Code, requirements for, 37
  light hazard, 171
  losses from fires in, 35
  low hazard, Table 9.2
  mercantile, 42
  minimum hazard, Table 9.2
  ordinary hazard, 171
  residential, 38–40
  types of, 37–43
Occupancies, assembly (see assembly occupancies)
Occupancies, business (see business occupancies)
Occupancies, educational (see educational occupancies)
Occupancies, health care (see health care occupancies)
Occupancies, industrial (see industrial occupancies)

Occupancies, mercantile (see mercantile occupancies)
Occupancies, residential (see residential occupancies)
Occupational Health and Safety Administration (OSHA), 302
Odorizing, 91–92
Office buildings, as light hazard occupancies, 171
Officers
  for fire department, 232
  promotion of, 247
  role of, in size-up, 255
Oil refineries, as extra hazard occupancies, 171
Oklahoma State University, 308
One- and two-family dwellings
  as residential occupancies, 38–39
  fires in, 39, Table 2.3
  Life Safety Code, requirements for, 38
Open flames and sparks, as cause of fires, 127–128
Open-plan buildings, 41
Operation EDITH, 293
Operations
  nontactical, 257–258
  tactical, 255–257
Organization, principles of, 246–247
Organizations for fire suppression, 254–255
OSHA (see Occupational Health and Safety Administration)
Our Lady of the Angels School fire
  description, 16
  illustrated, Fig. 1.5
  results of, 126
Outside sprinkler systems, 176
Overhaul
  as tactical operation, 257
  at radiation emergencies, 101
Oxidation, 52, 53, 57
Oxygen
  as a component of cotton, 78
  as a component of wood, 73
  as an element of fire, 52
  removal of, 67
Oxygen deficiency, as a product of combustion, 66

P

Paramedic, 311
Parliamentary Act of 1853, 4

Parmelee, Henry S., automatic sprinkler head by, 10, Fig. 1.2
Passageways, design of, 32
People factors
 characteristics to consider in, 26
 for firesafe building design, 26
 in a fire situation, 25
Perchloric acid, as an inorganic acid, 93
Perforated pipe systems, 10
Personnel
 additional, 228
 administration and management of, 249
 for fire protection, 228
 scope of activities of, 249
 recruitment of, 249
 training of, 228
 utilization of, 253
 work schedules for, 253
Personnel department, in fire departments, 252
Philadelphia Contributionship, 8
Phosgene, as a fire gas, 64
Photoelectric flame detectors, 221
Photoelectric smoke detectors, 220
Physical properties, of water, 164–165
Piers, as ordinary hazard occupancies, 171
Plastics
 as combustible solids, 76–77
 as interior finish, 154
 fire behavior of, 77
 manufacturer of, 77
 storage of, 77–78
 thermoplastics, 78
Police, relationship of, to fire departments, 252
Population distribution, as a factor for fire protection, 226
Portable fire extinguishers, 192–196
 application of, 194–195
 carbon dioxide, 193
 dry chemical, 194
 dry powder, 194
 liquefied gases, 193
 markings for, 195, Fig. 7.4
 multipurpose dry chemical, 194
 role of, in fire protection, 192
 standard for, 193
 tests for establishing the ratings of, 195
 types of, 193–194
 vaporizing liquids, 193

*Portable Fire Extinguishers, Standard for*, 193, 194, 195
Portable pump, 197
Potassium bicarbonate, in dry chemical extinguishing systems, 188
Potassium chloride
 as base for multipurpose dry chemical, 194
 effectiveness of, 189
 in dry chemical systems, 188
*Praefectus Vigilum*, duty of, 3
Pratt, Philip W., first American sprinkler system by, 9
Preaction systems, 174
Prefire planning
 as nontactical operation, 257
 factors for, 257
 steps in, 258
Pressure characteristics, of water systems, 238–239
*Principles of Fire Protection*, by Richard L. Tuve, 52
Products of combustion
 burns as, 65
 fire gases as, 62
 flame as, 64
 heat as, 64
 oxygen deficiency as, 66
 smoke as, 65
Promotion practices, of fire departments, 250
Property loss
 determination of, 227
 from fire, 23
Property protection
 as a factor of firesafety design, 142–143
 design considerations of, 142
 effect of sprinklers on, 169–170
 question concerning, 142
 requirements for, 142
Proprietary signaling systems
 features of, 213
 operation of, 213
 uses of, 213
Protective clothing, 199
Protective equipment, for fire fighters
 breathing apparatus as, 198
 life guns as, 199
 life nets as, 199
 protective clothing as, 199
 resuscitators as, 198
 smoke ejectors as, 198
Protective signaling systems

power supplies for, 215–216
primary, 216
secondary, 216
Public firesafety education
as activity of NFPA, 293
fire as a failure of, 113
fire prevention programs in, 46
goals of, 46
inspection as part of, 47
Junior Fire Marshal Program as part of, 47
Learn Not to Burn as a part of, 47
methods of, 47
role of fire departments in, 262–263
role of publicity campaigns in, 47
to prevent fire losses, 131
Public fire defenses, 227
Pumper
for rural areas, 232
for water supply, 227
pressure-volume ratings of, 12
Pumper companies, effectiveness of, 228
Pumping system, for water distribution, 240
Purchasing department, of fire departments, 252
Pyrolysis reaction, 54

## Q

*Quarstionarius*, duties of, 3

## R

Radiation
as a means of heat transfer, 56
quality and quantity of, 56
Radiation shielding, 190
Radio phone, as communication equipment, 197
Radioactive emergencies
fire fighting at, 98–100
extinguishment of fire at, 100
overhaul at, 101
Radioactive materials
characteristics of, 95
description of, 95
fire protection for, 96
handling and storage of, 95–96
transportation of, 96–98
Radio-type fire alarm box, 203
Rankine degree, definition of, 55
Rate compensation devices, 219

Rate-of-rise detectors
advantages of, 218
pneumatic tube, 211
types of, 218
Reactive gases, 86
Regular dry pipe systems, 174
Remote station signaling systems
location of, 213–214
uses of, 213
Rescue, as tactical operation, 255
Rescue company, as organization for fire suppression, 254
Residential district, alarm response for, 227–228
Residential occupancies
apartment buildings as, 38, 39
definition of, 38
dormitories as, 38, 39
hotels as, 38, 39
inspections of, 131–132
mobile homes as, 40
one- and two-family dwellings as, 38, 39
rooming houses as, 38, 39
Resistance, as source of ignition, 58
Resistance bridge smoke detectors, 221
Resuscitation, 198
Rolling steel doors, 159
Roman Empire
fire protection during, 3
role of *Aquarii* during, 3
role of *Corps of Vigiles* during, 3
role of *Familia Publica* during, 3
role of *Nocturnes* during, 3
role of *Praefectus Vigilum* during, 3
role of *Quarstionarius*, during, 3
role of *Siponarii* during, 3
Roof hatches, for ventilation, 148
Room contents, role in investigation, 114
Rooming houses
as residential occupancies, 38
definition of, 39
Rural fire defenses
minimum standard protection by, 229
water supply for, 228
Rural fire protection, 228–229
Rural operations, Table 1.2

## S

Salvage, as tactical operation, 256
Salvage corps, formation of, 7

## 336    Principles of Fire Protection

Sampling smoke detectors, 221
San Francisco fire, 269
Schools, as light hazard occupancies, 171
SFPE (*see Society of Fire Protection Engineers*)
Sheet-metal doors, 159
Shopping mall, fire in, 29
Signaling systems
　automatic and manual, 210–216
　auxiliary, 210, 211, 213
　central station, 210, 211–212, Fig. 8.4
　classification of, 210, 215
　direct circuit type, 211
　functions performed by, 210
　household fire warning, 211, 214–215
　local, 210, 212–213
　local energy type, 211
　proprietary, 210, 214
　remote station, 210, 213–214
　shunt type, 211
　types of, 210–211
*Siponarii*, duty of, 3
Site planning
　considerations for hydrants in, 149
　features for, 148
　water supply for, 149
Size-up, as tactical operation, 255
Skylights, for ventilation, 148
Smoke
　as cause of death, 143
　as a fire hazard in buildings, 143
　as a major killer in fires, 65
　as a product of combustion, 76
　confinement of, 156
　control of, 159–160
　from wood, 75
　human behavior concerning, 27
　in high-rise buildings, 160
　movement of, 31
　properties of, 66
　ventilation for, 148
　venting of, 160
Smoke control
　confinement as a method of, 160
　dilution as a method of, 160
　exhaust as a method of, 160
　methods of, 160
Smoke detectors
　as a means of life safety, 37
　beam-type, 220
　for prevention of life loss, 132
　ionization, 220
　photoelectric, 220
　resistance bridge, 221

　sampling, 221
　types of, 220
　value of, 219
Smoke ejectors, 198
Smokey the Bear, 303
Smoking, as cause of fires, 126
Society of Fire Protection Engineers (SFPE), 295, Fig. 12.2
Sodium bicarbonate, in dry chemical extinguishing systems, 188
Solid, as a form of matter, 71
Sources of ignition
　chemical, 57
　electrical, 57–58
　forms of, 57–59
　mechanical, 59
　nuclear, 59
Sparking
　as a source of ignition, 58–59
　danger from, 59
Sparky the Fire Dog, 293, Fig. 12.1
Specific gravity, of a liquid, 84
Spontaneous heating, 57
Spray nozzles, adjustable, 19
Sprinkler heads, temperature ratings of, 177
Sprinklers, connections for, 148
Sprinkler systems
　combined dry-pipe and preaction, 175
　deluge, 175
　development of, 168
　effect on life safety, 169
　effect on property protection, 169–170
　factors of, 171
　installation of, 170–171
　limited water supply in, 176
　NFPA *Sprinkler System Standard* for, 172
　outside, 176
　Parmelee, 168
　performance of, 170
　preaction, 174
　reasons for unsatisfactory performance of, 170
　regular dry-pipe, 174
　types of, 172–177
　value of, 169
　wet-pipe, 173
"Stack effect," in high-rise buildings, 31
Staff functions, 248
Staffing practices
　cost of, 250
　laws and regulations regarding, 250
　of fire departments, 250–251

Subject Index 337

procedures for, 250
Stairways
  evaluation of, in large-loss fire prevention, 131
  exit requirements for, 30
*Standard Building Code*, 281, 283
*Standard for Automotive Fire Apparatus*, 287
*Standard for Household Fire Warning Equipment*, 288
*Standard for Mobile Homes*, 40
*Standard for Portable Fire Extinguishers*, 287
*Standard for Recreational Vehicles*, 287
*Standard for the Installation of Sprinkler Systems*, 287
Standards
  adoption into law, 279
  consensus, 279
  developed by NFPA, 292
  enforcement of, 275–279
  formation of, 272–279
  role of, in building codes, 283–284
  scope and content of, 274
  types of, 279–284
Standards-making process, 272–275
Standard time-temperature curve, 157, Table 6.5
Standpipes
  Class 1, 177
  Class 2, 177–178
  Class 3, 178
  connections for, 148
  dry, 178–179
  "primed" systems, 179
  types of, 178–179
  water for, 178
  wet, 178
State fire marshal
  as organization for fire prevention, 260–261
  creation of first, 17
  powers of, 277
  role of, in investigations, 116, 135
Static sparking
  as a source of ignition, 59
  description of, 59
Statute of Limitations, for arson, 137
Steam, conversion of water to, 165
Steam sprinkler systems, 9–10
Steel
  as type of interior finish, 152
  critical temperature of, 152
  encasement of, 152–153

Structural defects
  as cause of large-loss fires, 123, Table 5.2
  elevator shafts as, 123
  open stairways as, 123
Structural integrity, of wood, 152
Structure, compartmentation of, as a factor of fire development, 114
Sulfur dioxide
  as a fire gas, 63
  properties of, 63
Sulfuric acid, 93
Surface water supply, 239
Synthetic textiles, 79

## T

Tactical control unit, as organization for fire suppression, 254
Tactical operations, 255–257
Task force
  as organization for fire suppression, 254–255
  concept of, 228
Telegraph-type fire alarm box, 205
Telephone-type fire alarm box
  parallel arrangement of, 207
  series arrangement of, 207
Temperature, definition of, 55
Temperature ratings, of sprinkler heads, 177
Temperature units
  British thermal units, 55
  calorie, 55
  Celcius degree, 55
  Fahrenheit degree, 55
  Kelvin degree, 55
  Rankine degree, 55
Test procedures
  for fire resistance of buildings, 154
  for interior finish materials, 155
Textiles
  combustibility of, 78
  flame retardant, 80
  natural fiber
    derived from animals, 78
    derived from plants, 78
  storage of, 80
  synthetic, 79
*The Extinguishment of Fire*, by Walter M. Haessler, 2
Thermal conductivity of materials, Table 4.1

Thermostats, 217
"Thick Water," 168
Time factors, role of, in fire development, 114
Tin-clad doors, 159
Toxic chemicals, action at emergencies involving, 106
Toxic gases, 86
Traffic and transportation in firesafety planning, 149
Traffic control
 external, 45
 internal, 45
 preplanning for external, 45
 preplanning for internal, 45
Training
 as nontactical operation, 257–258
 for fire department, 232
 Internal Fire Service Training Association (IFSTA), 296
Transportation
 Department of (DOT), 304–305
 of chemicals, 97
 of combustible metals, 192
 of radioactive materials, 97–98
 shipping containers used in, 98
Transportation emergencies
 actions at, 105–106
 involving hazardous materials, 103–104
 life safety hazards during, 105
Transportation incidents, as fire department operations, 258
Tunnel tests, for flame spread, 155
Tuve, Richard L., *Principles of Fire Protection* by, 52
"Two-sided" god, fire as, 1
Two-way radios, as communication equipment, 197

## U

UFIRS (*Uniform Fire Incident Reporting System*), 117
UL (*see Underwriters Laboratories Inc.*)
Ultraviolet flame detectors, 221
Underwriters Laboratories Inc. (UL)
 labels of, Fig. 12.3
 objective of, 298
 publications of, 298
*Uniform Building Code*, 281, 282
Union Fire Company, 8

United States Forest Service (USFS), 303
University of Maryland, 308
Urban fire defenses, 227–228
 adequate water supply for, 227
 minimum requirements for, 227–228
 water supply for, 227
Urea-potassium bicarbonate, in dry chemical extinguishing systems, 188
USFS (*see United States Forest Service*)

## V

Vapor
 corrosive, 94–95
 explosion characteristics of, 81
 flammability limit of, 53
Vapor pressure, of liquids, 84
Vaporizing liquids, in portable fire extinguishers, 123
Ventilation
 as tactical operation, 256
 functions of, 148
 importance of, 148
 in buildings, 148

## W X Y Z

Wallboards, as interior finish, 152
Warehouses, as ordinary hazard occupancies, 171
Water (*see also water systems*)
 as a means of fire and explosion control, 66–67
 as an extinguishing agent, 164–168
 conversion to steam, 165
 cooling effect of, 166
 distribution systems for, 239
 effect on wood fires, 167
 extinguishing properties of, 165
 for fire fighting, 228
 for fire protection, 236–239
 for standpipes, 178
 heat of vaporization of, 165
 heat transfer in cooling action of, 166
 in spray form, 166
 physical properties of, 164–165
 rates of consumption of, 237
 size of droplets, 166
 smothering capabilities of, 167
 sources of supply of
  ground water supply as, 239
  surface supplies as, 239

use of, in electrical fires, 167
uses of municipal, 237
Water department
relation of, to fire department, 252
responsibility of, 252
Water distribution systems
combination systems, 240
gravity system, 240
pumping system, 240
sources of supply, 239
types of, 240
Water flow, 230
Waterflow alarm, 215
Water fog, 167–168
Water pressure, for fire fighting, 238–239
Water spray fixed systems, 179–181
application of, 180, 181
design of, 180
nozzles used in, 181
use of, 180
Water supply
for rural fire defenses, 228
for site planning, 149
for urban fire defenses, 227
in *ISO Grading Schedule*, 230
insurance grading schedule, 230
minimum requirements of, 230–231
pumper for, 227
to building, 149
Water system
evaluation of, 238
fire protection requirements in, 238–239
pressure characteristics of, 238–239
required capacity of, 238
supply pipes in, 238
types of
combination as, 240
gravity as, 240
pumping as, 240
uses of, 237
Wet-pipe systems, 173
"Wet water," 168
Wharves, as ordinary hazard occupancies, 171
Windows
as means for life safety, 37
condition of, at fire scene, 116
exit requirements for, 30
Wood
as a construction material, 72
as interior finish, 152
components of, 73
decomposition of, 74
effect of water on, 167
fire loading, 76
flame spread characteristics of, 75–76
flammability of, 72
flashover, 75
ignition of, 74
ignition temperature of, 74
low heat conductivity of, 72
moisture content of, 73, 152
physical properties of, 72
size of, 152
smoke produced by, 76
storage of, 76
Wood, Peter G., studies by, 22–23
Wood-shingle roofs, as cause of conflagrations, 13, Table 1.2